MECHANICAL TECHNOLOGY
FOR HIGHER ENGINEERING
TECHNICIANS

MECHANICAL TECHNOLOGY FOR HIGHER ENGINEERING TECHNICIANS

Peter Black

Senior Lecturer in Mechanical Engineering, Mid-Essex Technical College,
Chelmsford, England

PERGAMON PRESS
Oxford · New York · Toronto · Sydney · Braunschweig

Pergamon Press Ltd., Headington Hill Hall, Oxford
Pergamon Press Inc., Maxwell House, Fairview Park, Elmsford, New York 10523
Pergamon of Canada Ltd., 207 Queen's Quay West, Toronto 1
Pergamon Press (Aust.) Pty. Ltd., 19a Boundary Street, Rushcutters Bay, N.S.W. 2011, Australia
Vieweg & Sohn GmbH, Burgplatz 1, Braunschweig

Copyright © 1972 Peter Black

All Rights Reserved. No part of this publication may be reproduced, stored in a retrieval system, or transmitted, in any form or by any means, electronic, mechanical, photocopying, recording or otherwise, without the prior permission of Pergamon Press Ltd.

First edition 1972

Library of Congress Catalog Card No. 73-133885

C J 2. 9 0 6. 7 6. !. A

Printed in Hungary

08 015681 9

CONTENTS

Preface	vii
The Engineering Technician	ix
The International System of Metric Units	x
Some Conversion Factors	xiv
1 Torsion	1
2 Bending	11
3 Complex Stress	56
4 Struts and Cylinders	81
5 Dynamics	101
6 Mechanisms	127
7 Fluid Mechanics	153
8 Combustion and Heat Transfer	189
9 Triboengineering	242
10 Vibration	276
11 Control	312

CONTENTS

Appendices

Notes on Methods of Assessing Power Output for I.C. motors prior to the Introduction of SI · 1 — 331

Dimensions · 2 — 333

Some Books for Further Reading · 3 — 334

Some More Conversion Factors · 4 — 335

A Few Interesting Dates · 5 — 337

The 0·1 per cent Proof Stress for a Few Materials · 6 — 338

International Paper Sizes · 7 — 339

Values of $e^{u\theta}$ · 8 — 340

Some Notes on Proportion · 9 — 341

Index — 343

PREFACE

THE student aiming at the new HNC in Mechanical Engineering, the now recognised technical qualification for top engineering technicians, will find the contents of this book sufficient for the two-year study of the subject Mechanical Technology, it being assumed that he already possesses an ONC, or that he is similarly qualified in Applied Heat, Mathematics and Applied Mechanics.

Selected material, sometimes in condensed or revised form, has been extracted from my two previous books *(Strength of Materials, Mechanics of Machines)* and integrated with the basic theory of Thermodynamics and Mechanics of Fluids as contained in the "guide syllabuses" issued by the Joint Committee[*] set up to administer the new scheme. The more difficult of the worked examples are intended to give the student some idea of the examination standard.

In several important industries the science of lubrication has been something of a cinderella, not least because comprehension of it was, and still is, hindered by its "image" which, in the minds of a significant proportion of people, is that of a man with an oilcan. Since, from experience, it is almost impossible to destroy a long-held concept, it is usually preferable to introduce a new one entirely and this was the reason for the inclusion in the Jost Report on the state of lubrication education and research—*Lubrication (Tribology)*, HMSO, 1966—of a recommendation for the introduction of the new term "tribology", the connotation of which is as follows:

> The science and technology of interacting surfaces in relative motion and of the practices related thereto.

This term includes the subjects of friction, lubrication and wear; the properties and operational behaviour of bearing materials and the engineering of bearings and their environments; the quality control and inspection of lubricants and the management and organisation of lubrication.

[*] The Joint Committee is composed of representatives nominated by the following:
Institution of Mechanical Engineers,
Institution of Production Engineers,
Royal Aeronautical Society,
Association of Teachers in Technical Institutions,
Association of Technical Institutions,
Association of Principals of Technical Institutions.

Its joint secretary is housed in the I.Mech.E. in London.

PREFACE

It will be evident that, although triboengineering design—i.e. the engineering application—is the ultimate target, the field itself is interdisciplinary and embraces both the efforts of chemists and physicists (trioscience) and those of metallurgists, lubrication engineers and friction technologists (tribotechnology).

In view of the foregoing, relevant and related topics have been grouped in this book to form the chapter entitled "Triboengineering".

The name "pascal" has been suggested for the unit of pressure (and of stress) and although not yet agreed officially, I make no apology for using it in this text, not only because Pascal† made early and important contributions to mathematics and to mechanics of fluids, but also because the word is shorter than the combination "newton—per—metre—squared", itself even longer than the obsolete "cycle—per—second".

To avoid confusion elsewhere I have either written the word "litres" in full or used a capital letter for the abbreviation.

PETER BLACK

---

† Blaise Pascal (1623–62) was the son of a mathematician who educated him at home but forbade him to begin mathematics until considered (by this father) able to master it. However, by the age of eleven he had developed in secret the first 23 propositions of Euclid, calling circles "rounds" and straight lines "bars". (1 bar = 10^5 pascals!) In 1639 (at the age of 16) he published a treatise on conic sections; in 1647 he patented a calculating machine; in 1648 he noted experimentally the reduction in height of a mercury column with increase in altitude (which led to the invention of the barometer) and in 1661 he published a paper on the equilibrium of fluids. Together with Pierre Fermat he founded the theory of probability.

THE ENGINEERING TECHNICIAN

THE following definition has been adopted by the Engineering Societies of Western Europe and the USA (known as EUSEC) for the purposes of conference discussions:

An engineering technician is one who can apply in a responsible manner proven techniques which are commonly understood by those who are expert in a branch of engineering, or those techniques specially prescribed by professional engineers.

Under general professional engineering direction, or following established engineering techniques, he is capable of carrying out duties which may be found among the list of examples set out below.

In carrying out many of these duties, competent supervision of the work of skilled craftsmen will be necessary. The techniques employed demand acquired experience and knowledge of a particular branch of engineering, combined with the ability to work out the details of a task in the light of well-established practice.

An engineering technician requires an education and training sufficient to enable him to understand the reasons for and purposes of the operations for which he is responsible.

The following duties are typical of those carried out by engineering technicians:

Working on design and development of engineering plant and structures; erecting and commissioning of engineering equipment and structures; engineering drawing; estimating, inspecting and testing engineering construction and equipment; use of surveying instruments; operating, maintaining and repairing engineering machinery, plant and engineering services and locating defects therein; activities connected with research and development, testing of materials and components and sales engineering, servicing equipment and advising consumers.

THE INTERNATIONAL SYSTEM OF METRIC UNITS

On behalf of the UK, the Ministry of Technology takes part in the work of the international body concerned with the working of the metric system, namely the Conférence Générale des Poids et Mésures or CGPM. In 1954 this body adopted a system of units based on the metre, kilogramme, second, ampère, candela and degree Kelvin and, in 1960, gave it the name "Système Internationale d'Unités". The common abbreviation, SI, is the same in most languages. Units of and symbols for the basic quantities are as shown in Table 1.

TABLE 1

Quantity	Unit	Symbol
Length	metre	m
Mass	kilogramme	kg
Time	second	s
Electric current	ampère	A
Luminous intensity	candela	cd
Temperature	degree Kelvin	°K

In the SI, the names of multiples (multiplying factors in powers of 10) of units are formed by means of fourteen standard prefixes (see BS 350 and BS 3763) the most useful of which are given in Table 2.

TABLE 2

Factor	Prefix	Symbol
10^{12}	tera	T
10^{6}	mega	M
10^{3}	kilo	k
10^{-3}	milli	m
10^{-6}	micro	μ
10^{-12}	pico	p

If the product or quotient of any two quantities in a system is the unit of the resultant quantity, the system is said to be *coherent*. In the SI, the force necessary to impart an acceleration of 1 m/s² to a mass of 1 kg is taken as the unit of force and called the newton (N) so that the newton is a coherent unit and the SI is a coherent system.

Thus $1\ [N] = 1\ [kg] \times 1\ [m/s^2]$, i.e. $N = kg\ m\ s^{-2}$.

The constant in the equation representing Newton's 2nd Law of Notion is therefore unity so that the motion of unit mass when acted upon by unit force may be represented by Figure 1.

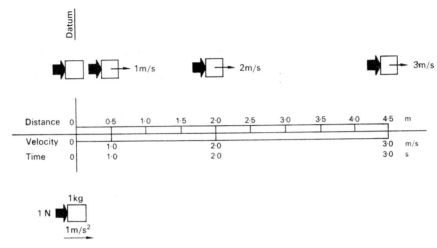

Fig. 1

Since, under the attraction of the earth, a mass of 1 kg will fall freely with an acceleration of, approximately, 9·81 m/s² ($\simeq$ 32·2 ft/s²) it is evident that the approximate gravitational force on 1 kg is 9·81 N. (The actual value depends on position since the earth is not spherical.) This force, known as 1 kgf, is the present unit of force in certain metric-based countries and is the so-called "technical" unit of force. It is to be hoped that, ultimately, it will be superseded by the newton.

In the SI, the work done when the point of application of a force of 1 N is displaced (in the direction of the force) through a distance of 1 m is 1 Nm. This is taken as the unit of energy and called the joule (J).

Thus $1\ [J] = 1\ [N] \times 1\ [m]$, i.e. $J = Nm = kg\ m^2\ s^{-2}$.

In the SI, a rate of energy expenditure of 1 J/s is taken as the unit of power and called the watt (W).

Thus $1\ [W] = \dfrac{1\ [J]}{1\ [s]}$, i.e. $W = Js^{-1} = kg\ m^2\ s^{-3}$.

Units such as the newton, the joule and the watt are known as *derived units* since they are derived in terms of basic units. The units of other quantities can be derived in the same way and the most useful of these from the point of view of mechanical engineering are shown in Table 3.

Where a unit is named after a person it is written in small letters (called "lower case") while a large letter (capital) is used for the abbreviation. Thus the abbreviated form of watt

Table 3

Engineering quantity	Special name of unit (if any)	Symbol for unit	Definition of unit
Force	newton	N	kg m s^{-2}
Energy	joule	J	kg m^2 s^{-2} (Nm)
Power	watt	W	kg m^2 s^{-3} (Js^{-1})
Torque			kg m^2 s^{-2} (Nm)
Area			m^2
Volume			m^3
Specific volume			m^3 kg^{-1}
Density			kg m^{-3}
Linear velocity			m s^{-1}
Angular velocity			rad s^{-1}
Linear acceleration			m s^{-2}
Angular acceleration			rad s^{-2}
Frequency	hertz		cycle s^{-1}
Stress/pressure	pascal	Pa‡	kg m^{-1} s^{-2} (Nm^{-2})
Dynamic viscosity			kg m^{-1} s^{-1} (Nsm^{-2})
Kinematic viscosity			m^2 s^{-1}
Thermal conductivity			kg ms^{-3} °K^{-1} (Js^{-1} m^{-1} °K^{-1})*
Specific heat capacity			m^2 s^{-2} °K^{-1} (J kg^{-1} K^{-1})
Calorific value			kg m^5 s^{-2} (Jm^{-3})
Potential difference	volt	V	kg m^2 s^{-3} A^{-1} (JA^{-1} s^{-1})
Magnetic flux	weber	Wb	kg m^2 s^{-2} A^{-1} (Vs)
Flux density	tesla	T	kg s^2 A^{-1} (Vs m^{-2})
Inductance	henry	H	kg m^2 s^{-2} A^{-2} (Vs A^{-1})
Charge	coulomb	C	As
Capacitance	farad	F	kg^{-1} m^{-2} s^4 A^2 (V^{-1} sA)
Resistance	ohm	Ω	kg m^2 s^{-3} A^{-2} (VA^{-1})

* For °K^{-1} substitute deg K^{-1}.
‡ Not yet agreed internationally.

is W and of microwatt is μW. The radian, or measure of plane angle, is a dimensionless supplementary unit.

In conjunction with the basic and derived SI units of Tables 1 and 3, the "traditional" units shown in Table 4 may be used.

Table 4

Engineering quantity	Name of unit	Symbol for unit	Definition of unit
Mass	tonne	t	10^3 kg
Volume	litre	L	10^{-3} m^3
Area	hectare	ha	10^4 m^2 (about 2·5 acre)
Stress/pressure	bar	bar	10^5 Nm^{-2}
Dynamic viscosity	poise	P	10^{-1} kg m^{-1} s^{-1}
Kinematic viscosity	stokes	St	10^{-4} m^2 s^{-1}
Flux density	gauss	G	10^{-4} T
Energy	electron volt	eV	1·6021 × 10^{-19} J

THE INTERNATIONAL SYSTEM OF METRIC UNITS

Note that:
$$1 \text{ bar} = 10^5 \text{ N/m}^2 = 14 \cdot 5038 \text{ lbf/in}^2$$
$$1 \text{ MPa} = 10^6 \text{ N/m}^2 = 10 \text{ bar} \simeq 145 \text{ lbf/in}^2.$$

The standard atmosphere (atm) is defined as follows:
$$1 \text{ atm} = 101\ 325 \text{ N/m}^2 = 1 \cdot 01325 \text{ bar} \simeq 760 \text{ mm Hg}.$$

The kgf/cm² is often referred to as the "technical atmosphere" (at) so that
$$1 \text{ at} = 1 \text{ kgf/cm}^2 = 14 \cdot 2233 \text{ lbf/in}^2 \simeq 736 \text{ mm Hg}.$$

Hence $\qquad 1 \text{ kgf/mm}^2 = 1422 \cdot 33 \text{ lbf/in}^2 \simeq 9 \cdot 81 \text{ MPa}.$

Note also: $\qquad 1 \text{ MPa} = 1 \text{ MN/m}^2 = 1 \text{ N/mm}^2$

and $\quad 1 \text{ W} = 1 \text{ J/s} = 1 \text{ Nm/s} = 1(\text{kg m}^2 \text{ s}^{-3}) = 1(\text{kg m}^2 \text{ s}^{-3} \text{ A}^{-1})\text{ A} = 1(\text{V}) \times 1(\text{A}).$

SOME CONVERSION FACTORS[†]

1 Btu/lb	= 2325 J/kg	(calorific value)
1 Btu	= 1055 J	
1 Btu/h	= 0·293 W	(power)
1 Btu/ft^2	= 11 350 J/m^2	
1 Btu/ft^2 h	= 3·16 W/m^2	
1 Btu/ft^2 h °F	= 5·68 W/m^2 deg K	(surface coefficient)
1 Btu ft/ft^2 h °F	= 1·73 Wm/m^2 deg K	(thermal conductivity)
1 Btu in/ft^2 h °F	= 0·144 Wm/m^2 deg K	(thermal conductivity)
1 ft^2 h °F/Btu	= 0·1765 m^2 deg K/W	(thermal resistance)
1 ft lbf	= 1·356 J	(energy)
1 ft lbf/lb deg K	= 2·99 J/kg deg K	(characteristic constant)
1 lb/bhp-hour	= 0·169 kg/MJ	(specific fuel consumption)
1 lb/ft s	= 1·488 kg/m s	(dynamic viscosity)
1 ft^2/s	= 0·0929 m^2/s	(kinematic viscosity)
1 slug ft^2	= 1·356 kg m^2	(moment of inertia)
1 in^4	= 0·416×10^{-6} m^4	(2nd Moment of Area)
1 mile/gallon	= 282·5 litres/100 km	
1 gallon	= 4·546×10^{-3} m^3	
1 lbf/in^3	= 157 N/m^3	(specific weight)
1 lb/in^3	= 27·68×10^3 kg/m^3	(specific mass)
1 lb/ft^3	= 16·02 kg/m^3	(specific mass)
1 ft^3/lb	= 0·0624 m^3/kg	(specific volume)
1 lbf/in^2	= 6900 Pa (N/m^2)	(stress, pressure)
1 tonf/in^2	= 15·44 MPa (MN/m^2)	(stress, pressure)
1 tonf in^2	= 6·42 Nm^2	(flexural rigidity)
1 tonf ft	= 3030 Nm	(torque, bending moment)

[†] See also Appendix 4.

1

TORSION

THE application as shown in Figure 2 of parallel forces *FF* to a cube such as *ABCD* causes the face *AB* to turn through a small angle φ to the dotted position AB_1. This angle, when measured in radians, is defined as the *shear strain*.

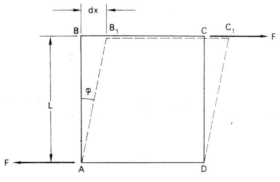

FIG. 2

Thus, Shear strain $\varphi = \dfrac{dx}{L}$ radians.

Within the elastic limit, the ratio Shear stress/Shear strain is constant and is called the Modulus of Rigidity. It is denoted by G. Denoting the shear stress by f_s we have, therefore,

$$G = \frac{f_s}{\varphi} \quad \text{or} \quad \varphi = \frac{f_s}{G}.$$

Consider a solid shaft of radius R to be acted on by equal and opposite couples as shown in Figure 3 so that the radius OA twists through θ radians with respect to the other end and adopts the dotted position OB (Fig. 4). The state is one of pure shear.

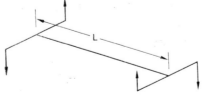

FIG. 3

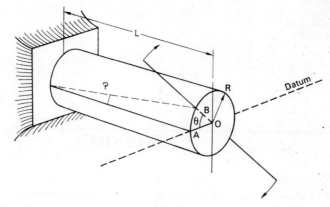

Fig. 4

Since arc $AB = R\theta = L\varphi$

∴ $$\varphi = \frac{R\theta}{L}$$

∴ $$\frac{f_s}{G} = \frac{R\theta}{L}, \quad \text{i.e.} \quad f_s = \frac{G\theta}{L} \times R.$$

Thus for a given twist on a given shaft, the shear stress is proportional to the radius. It is therefore zero at the axis. The equation is normally written

$$\frac{f_s}{r} = \frac{G\theta}{L}$$

where r has any value between zero and R.

It follows that, referring to Figure 5,
if f_s' = stress at any radius r
and f_s = stress at radius R (i.e. the maximum value)

then $$\frac{f_s'}{r} = \frac{f_s}{R}, \quad \text{i.e.} \quad \underline{f_s' = \frac{f_s}{R} r.}$$

Shear force on elemental ring

$$= f_s' 2\pi r\, dr = \frac{f_s}{R} 2\pi r^2\, dr$$

Moment of this force about the polar axis

$$= \frac{f_s}{R} 2\pi r^2\, dr \times r = \frac{f_s}{R} 2\pi r^3\, dr$$

Total moment of resistance to shear

$$T = \frac{f_s}{R} \times 2\pi \int_0^R r^3\, dr = \frac{f_s}{R}\left(\frac{\pi R^4}{2}\right) = \frac{f_s}{R}\left(\frac{\pi D^4}{32}\right).$$

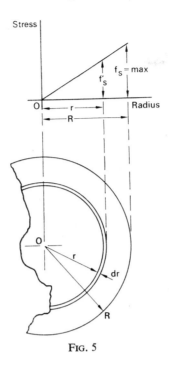

Fig. 5

The quantity in the brackets is the polar 2nd Moment of Area of the shaft section. Denoting this by J we have

$$T = \frac{f_s}{R} J \quad \text{or} \quad \frac{f_s}{R} = \frac{T}{J}$$

but, from above,

$$\frac{f_s}{R} = \frac{G\theta}{L}$$

so that we can write

$$\frac{f_s}{R} = \frac{T}{J} = \frac{G\theta}{L}.$$

This is known as the *Torsion Equation* since, within the elastic limit, the moment of resistance to shear, T, is equal and opposite to the torque applied. In this equation, for a *hollow* shaft having respectively external and internal diameters of d_1 and d_2 the value of J is $(\pi/32)(d_1^4 - d_2^4)$.

STRAIN ENERGY OF TORSION

From the torsion equation $T = (GJ/L) \times \theta$ so that the graph of T against θ has the linear form shown in Figure 6a. The slope of this graph is the *torsional stiffness* of the shaft and is denoted by q.

Thus

$$q = \frac{T}{\theta}.$$

The shaded area represents the work done in twisting one end of a shaft through an angle

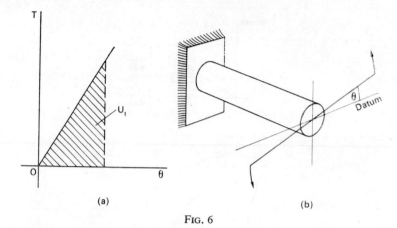

Fig. 6

θ relative to the other end, this work being, within the elastic limit, stored in the material as strain energy.

Torsional strain energy, U = Average torque × angle of twist

$$= \frac{T}{2}\theta \quad \text{and} \quad T = \frac{GJ}{L}\theta$$

$$= \frac{GJ\theta^2}{2L}$$

Alternatively

$$U = \frac{T}{2}\theta \quad \text{where} \quad \theta = \frac{TL}{GJ}$$

$$= \frac{T^2L}{2GJ}$$

Again,

$$U = \frac{T}{2}\theta \quad \text{where} \quad T = \frac{f_s J}{r} \quad \text{and} \quad \theta = \frac{f_s L}{rG}$$

$$= \frac{f_s^2}{G} \times \frac{LJ}{2r^2} \quad \text{and} \quad J = \frac{\pi r^4}{2}$$

$$= \frac{f_s^2}{4G} \times \pi r^2 L$$

i.e.

$$U = \frac{f_s^2}{4G} \times \text{volume of shaft.}$$

COIL SPRINGS

A spring may be compressed or extended or twisted axially so as either to wind or unwind it. In each case, within the elastic limit, the work done is conserved in the form of strain energy. A spring may do one of the following things:

(a) minimise shock by absorbing kinetic energy (e.g. vehicle spring);
(b) return a deflected mechanism to its original position (e.g. valve spring);
(c) release energy under controlled conditions (e.g. clock spring).

EFFECTS OF AXIAL LOAD

Referring to Figure 7, an axial load F can be resolved into two components acting at point P:

1. $F \cos \theta$ normal to the axis of the wire.
2. $F \sin \theta$ parallel to the axis of the wire.

At any point Q on the wire axis the component $F \cos \theta$ tends to produce the effect shown in Figure 8a, i.e. it induces a direct shear stress

$$\frac{F \cos \theta}{A} \quad \text{where} \quad A = \frac{\pi}{4} d^2$$

d being the wire diameter.

Since $F \cos \theta$ also acts at a distance R from point Q, it applies a torque

$$T = F \cos \theta \times R$$

to any section XX (Fig. 7b) and hence, over all such sections there will be an additional shear stress increasing from zero at the wire axis to a maximum given by

$$f_s = \frac{Tr}{J} \quad \text{where} \quad r = \frac{d}{2}.$$

This is relatively much greater than the direct shear stress.

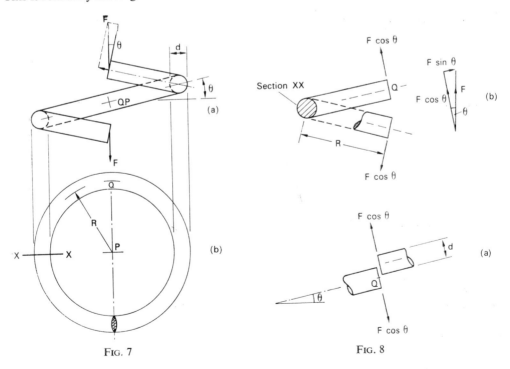

Fig. 7

Fig. 8

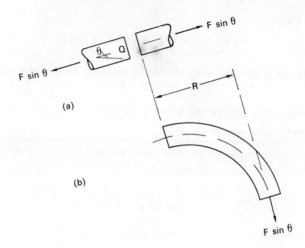

FIG. 9

Considering again point Q on the wire axis, the component $F \sin \theta$ tends to produce the effect shown in Figure 9a, i.e. it induces a direct tensile stress $(F \sin \theta)/A$ where $A = (\pi/4)d^2$. This component is also responsible for a bending moment $M = F \sin \theta \times R$ (Fig. 9b) which tends to reduce the coil radius R. The maximum bending stress (see Chapter 2) is given by

$$f = \frac{My}{I} \quad \text{where} \quad y = \frac{d}{2} \quad \text{and} \quad I = \frac{\pi}{64}d^4.$$

Torsional strain energy $\quad U_t = \dfrac{T^2 L}{2GJ} \quad$ where $\quad T = FR \cos \theta$

$$= \frac{F^2 R^2 L}{2}\left(\frac{\cos^2 \theta}{GJ}\right)$$

Bending strain energy $\quad U_b = \dfrac{M^2 L}{2EI} \quad$ since M is constant (see Chapter 2)

$$= \frac{(FR \sin \theta)^2 L}{2EI} \quad \text{substituting for } M$$

$$= \frac{F^2 R^2 L}{2}\left(\frac{\sin^2 \theta}{EI}\right)$$

Total strain energy $\quad = U_t + U_b$

i.e. $$U = \frac{F^2 R^2 L}{2}\left(\frac{\cos^2 \theta}{GJ} + \frac{\sin^2 \theta}{EI}\right).$$

Now, Work done in deflection = Average force × Deflection

or $$U = \frac{F}{2}\delta.$$

Equating: $$\frac{F}{2}\delta = \frac{F^2R^2L}{2}\left(\frac{\cos^2\theta}{GJ} + \frac{\sin^2\theta}{EI}\right)$$

Hence Axial deflection $$\delta = FR^2L\left(\frac{\cos^2\theta}{GJ} + \frac{\sin^2\theta}{EI}\right).$$

If the number of coils is denoted by n, then, since $\cos\theta = (2\pi Rn)/L$,

Length of wire in spring, $$L = \frac{2\pi Rn}{\cos\theta}.$$

Substituting for L gives $$\delta = \frac{2\pi FR^3n}{\cos\theta}\left(\frac{\cos^2\theta}{GJ} + \frac{\sin^2\theta}{EI}\right).$$

Although this and previous expressions are based on the assumption that R and θ are constant (whereas they both vary with F) they are in practice sufficiently accurate.

When θ is small (<10 deg say) $\sin^2\theta$ is negligible while $\cos\theta \simeq 1\cdot0$. The material may then be considered to be in a state of pure torsion and the spring may be described as "close-coiled".

Putting $\cos\theta = 1\cdot0$ and ignoring $\sin^2\theta$ gives

$$\delta = \frac{2\pi FR^3n}{GJ} \quad \text{where} \quad J = \frac{\pi d^4}{32}$$

i.e. $$\delta = \frac{64FR^3n}{Gd^4}.$$

This expression may be obtained more simply for a close-coiled helical spring by assuming pure torsion to start with. Then:

Work done by axial force = Torsional strain energy only

or $$\frac{F}{2}\delta = \frac{T}{2}\theta \quad \text{(referring to Fig. 10)}$$

whence $$\delta = \frac{T}{F}\theta.$$

Now $\theta = \frac{TL}{GJ}$, $T = FR$, $J = \frac{\pi d^4}{32}$ and $L = 2\pi Rn$ nearly.

Substituting: $$\delta = \frac{FR}{F} \times \frac{FR(2\pi Rn)}{G} \times \frac{32}{\pi d^4}$$

i.e. $$\delta = \frac{64FR^3n}{Gd^4} \quad \text{as before.}$$

EXAMPLE. Wire 2 m long and 5 mm diameter is formed into a helical spring of 50 mm mean coil diameter, a total of 200 mm being allowed for the end fastenings. Assume pure torsion and determine the axial stiffness. Find the axial force which will induce a maximum shear stress of 350 MPa given that for the wire material $G = 80\,000$ MPa (1 MPa = 1 MN/m²). See Fig. 11, p. 9.

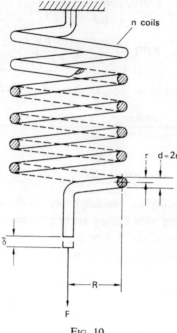

Fig. 10

Solution. Effective length of wire, $L = 2\pi R n = 2000 - 200 = 1800$ mm

∴ Number of turns, $\quad n = \dfrac{1800}{2\pi \times 25} = 11.5$

Axial stiffness, $\quad \lambda = \dfrac{F}{\delta} = \dfrac{Gd^4}{64R^3n}$

$$= \dfrac{(80\,000 \times 10^6)}{64 \times 11.5} \left(\dfrac{5}{1000}\right)^4 \left(\dfrac{1000}{25}\right)^3$$

$$= 4350 \text{ N/m}$$

For wire section, $\quad J = \dfrac{\pi}{32} d^4$

$$= \dfrac{\pi}{32}\left(\dfrac{5}{1000}\right)^4$$

$$= 61.3 \times 10^{-12} \text{ m}^4$$

Torque, $\quad T = FR = \dfrac{f_s J}{r} \quad$ where $\quad F =$ required axial force.

Hence $\quad F = \dfrac{f_s J}{rR}$

$$= \dfrac{(350 \times 10^6)(61.3 \times 10^{-12})}{0.0025 \times 0.025}$$

$$= \underline{344 \text{ N.}}$$

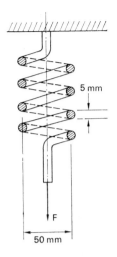

Fig. 11

EXAMPLE. An open-coiled spring has 20 turns made of steel 12 mm diameter. The mean coil radius is 125 mm and the coils make an angle of 30 deg with the horizontal when the unloaded spring is suspended vertically. Assume $E = 206\,000$ MPa and $G = 80\,000$ MPa and calculate the axial deflection under an axial force of 180 N (a) assuming pure torsion, (b) taking bending effect into account. By what percentage does assumption (a) underestimate the deflection?

(a) Deflection,
$$\delta = \frac{64FR^3n}{Gd^4} = \frac{64 \times 180 \times 0.125^3 \times 20}{(80\,000 \times 10^6)\,0.012^4}$$
$$= 0.271 \text{ m}$$

$$\delta = \frac{2\pi FR^3n}{\cos\theta}\left(\frac{\cos^2\theta}{GJ} + \frac{\sin^2\theta}{EI}\right) \quad \text{and} \quad J = 2I \quad \text{for a circular section}$$

$$= \frac{2\pi FR^3n}{\cos\theta}\left(\frac{\cos^2\theta}{2GI} + \frac{\sin^2\theta}{2.58GI}\right) \quad \text{since} \quad E = 2.58G$$

$$= \frac{2\pi FR^3n}{GI\cos\theta}\left(\frac{\cos^2\theta}{2} + \frac{\sin^2\theta}{2.58}\right) \quad \text{and} \quad I = \frac{\pi}{64}(0.012)^4 = 1.02 \times 10^{-9} \text{ m}^4$$

$$\cos\theta = 0.866, \quad \cos^2\theta = 0.75$$
$$\sin\theta = 0.500, \quad \sin^2\theta = 0.25$$

$$\therefore \quad \delta = \frac{2\pi \times 180 \times 0.125^3 \times 20 \times 10^9}{(80\,000 \times 10^6)\,1.02 \times 0.866}\left(\frac{0.75}{2} + \frac{0.25}{2.58}\right)$$

$$= 0.623(0.375 + 0.097)$$
$$= 0.294 \text{ m}$$

Percentage error resulting from assumption (a)
$$= \frac{(0.294 - 0.271)}{0.294} 100 = \underline{7.8} \text{ (low)}.$$

EXAMPLES 1

1. The power input to one end of a uniform shaft is 22·4 kW at 150 rev/min. Three identical machines are belt-driven from identical pulleys at distances of 3·65, 6·10 and 9·14 m respectively from the driven end. If $G = 83\,000$ MPa and the shear stress is not to exceed 63 MPa, find, neglecting bending effects: (a) a suitable shaft diameter (50 mm), (b) the total twist (10·7 deg).
2. Find the strain energy stored in the material of a hollow shaft 1·22 m long, 100 mm external diameter, 25 mm thick when subjected to a torque of 1085 Nm. Assume $G = 81\,500 \times 10^6$ MPa (0·7 J approx.).
3. Allow a maximum shear stress of 103 MPa and design a close-coiled spring having an axial stiffness of 1750 N/m and a safe deflection of 25 mm. Assume a mean coil diameter of ten times the wire diameter and take the value of G as $90\,000 \times 10^6$ MPa (21 turns, 32·5 mm diameter).
4. A hollow steel shaft 63·5 mm external diameter is to be connected via a clutch to a solid alloy shaft of the same diameter. If the torsional stiffness of the steel shaft is to be 0·8 of that of the alloy shaft, determine the value of its internal diameter given that $G_a = 0·4 G_s$ (53·5 mm).
5. The coils of a helical spring make an angle θ with planes normal to the spring axis. If the deflection is calculated using the formula for close-coiled springs, find the value of θ corresponding with an error of 1·0 per cent (8·1 deg).

2

BENDING

FORCE PRODUCING BENDING AND SHEAR

The moment of a force about a point is defined as the product of the force and the perpendicular distance of the point from its line of action. Such a moment when applied to a component in such a way as to result in bending is referred to as a *bending moment* and denoted by M.

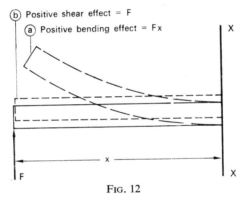

FIG. 12

Bending convention: To the left of any section such as XX (Fig. 12) a clockwise moment is considered positive and vice versa.

Inseparable from the bending is the shear effect which tends to produce a relative motion between the parts separated by the plane XX.

Shear convention: If the resultant force to the left of a section is upwards, it is considered positive.

SIMPLE BENDING

A bending moment applied to a beam produces tensile and compressive strains as shown in Figure 13. Evidently there exists an intermediate unstrained or "neutral" plane.

Where the beam section and this neutral plane intersect is known as the *neutral axis* of the section and it will be shown later that this axis passes through the centroid of the section.

It is reasonable to assume that the strain (and hence the stress) at any point is proportional to the distance, y, from the neutral plane, so that the graph of f against y is a straight line.

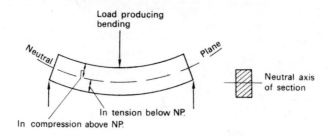

Fig. 13

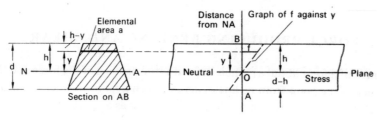

Fig. 14

This is represented by the dotted line drawn from point O, Figure 14. (That this is so will be shown later.)

Let f be the stress on an elemental area a distant y from (above) the neutral axis. The compressive force on this element is fa and this will increase from zero at O up to a positive maximum where $y = h$. For elements below the neutral axis the force will be tensile, increasing from zero at O down to a negative maximum where $y = -(d-h)$.

If the section AB is to be in horizontal equilibrium, the resultant compressive force must be equal and opposite to the resultant tensile force,

i.e.
$$\underline{\Sigma(fa) = 0}.$$

It will be shown later that $f = \dfrac{E}{R} y$ where E = elastic modulus, R = beam radius at AB.

Hence we can write
$$\Sigma \left(\frac{E}{R} y \times a \right) = 0$$

or
$$\frac{E}{R} \Sigma(ay) = 0$$

∴
$$\underline{\Sigma(ay) = 0} \quad \text{since} \quad \frac{E}{R} \neq 0.$$

Now, by definition, $\Sigma(ay)$ is the 1st Moment of Area of the section about the neutral axis, and this can only be zero if this axis passes through the centroid of the section. The depth of the centroid—and hence of the N.A.—is given by

$$h = \Sigma(ar)/\Sigma a$$

where Σa = area of section = A (Fig. 15) and r = distance of element a from the top.

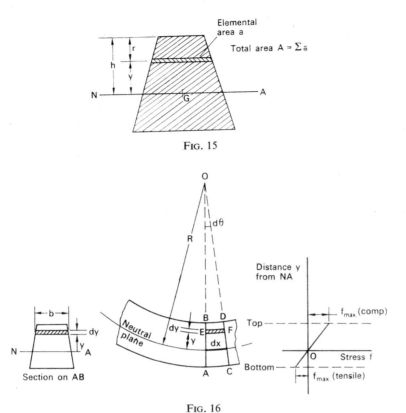

FIG. 15

FIG. 16

An expression for the stress at any distance y from the neutral axis will now be derived, the following being the assumptions made:

1. The material is homogeneous and isotropic.
2. The effects of shear force are neglected so that a plane section normal to the axis remains plane.
3. In neither tension nor compression is the limit of proportionality exceeded, and in both the value of E is the same.

Let the plane sections AB and CD, Figure 16, be a small distance dx apart when the beam is straight. The application of a bending moment M will cause AB and CD produced to intersect at some point O so that BD subtends the small angle $d\theta$.

Elements such as EF above the neutral axis will be reduced in length while similar elements below it will undergo tensile strain. For the element EF:

Original length $= dx = R\, d\theta$ where $R =$ neutral plane radius

Length after bending $= (R-y)\, d\theta$

$\qquad\qquad\qquad\quad = R\, d\theta - y\, d\theta$

Reduction in length $= y\, d\theta$

Compressive strain $= \dfrac{y\, d\theta}{R\, d\theta} = \dfrac{y}{R}$

Compressive stress $= E \times \text{strain}$

i.e.
$$f = \dfrac{E}{R} y.$$

Similarly, for elements below the neutral axis, the tensile stress is
$$f = -\dfrac{E}{R} y.$$

Since R is constant for this small length of beam it follows that, as assumed earlier, f is proportional to y. It is therefore zero at the neutral axis.

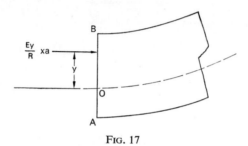

Fig. 17

Referring to Figure 16, the transverse area of the element EF is $b.dy$ and this may, for convenience, be denoted by a.

Compressive force on this element $= \text{Stress} \times \text{section}$

$$= \dfrac{E}{R} y \times a$$

This acts at a distance y from the neutral axis as shown in Figure 17 so that, about the N.A.,

Moment of this force $= \dfrac{E}{R} ya \times y = \dfrac{E}{R}(ay^2)$

If all such similar moments are added (including those of the tensile forces below the N.A. which are also clockwise) the result will be what is called the *Moment of Resistance* of the section.

This can be written $\Sigma\left(\dfrac{E}{R} ay^2\right)$

or $\dfrac{E}{R} \Sigma(ay)^2$ since $\dfrac{E}{R}$ is constant.

Now, by definition, the quantity $\Sigma(ay^2)$ is the 2nd Moment of Area of the section about the neutral axis. This is denoted by I and can be found by one of the usual methods. Hence,

Moment of resistance to bending $= \dfrac{E}{R} I.$

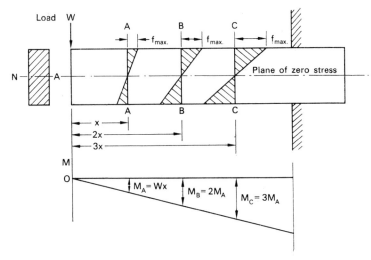

Fig. 18

Since at any section this is equal and opposite to the applied bending moment M (provided the elastic limit is not exceeded) we can write:

$$M = \frac{E}{R}I \quad \text{or} \quad \frac{M}{I} = \frac{E}{R}$$

But

$$f = \frac{E}{R}y \quad \text{or} \quad \frac{f}{y} = \frac{E}{R}$$

so that

$$\underline{\frac{f}{y} = \frac{M}{I} = \frac{E}{R}}.$$

This is known as the *Bending Equation*. Since $f = \frac{M}{I}y$, it follows that, for a given beam section, the stress varies along the beam in proportion to M and across the section in proportion to y. This is illustrated in Figure 18 by graphs of stress distribution drawn at equal intervals along a simply loaded cantilever.

RELATION BETWEEN w, F AND M

Consider an element dx of cantilever carrying a load of w per unit length. Its weight is $w.dx$ and the graphs of F and M are shown in Figure 19.

Since the element is in equilibrium under the forces exerted on it by the remaining parts of the beam,

we have $\qquad F + dF = F + w.dx$

i.e. $\qquad dF = w.dx$

or $\qquad \dfrac{dF}{dx} = w.$

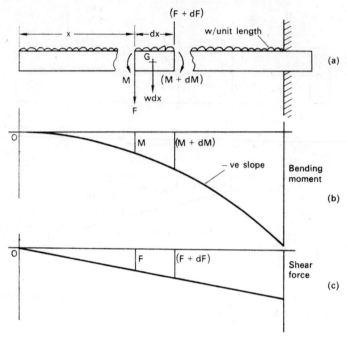

FIG. 19

If w is constant then dF/dx (i.e. the slope of the shear force graph) is constant. Since from Figure 19c this slope is evidently negative, a downward load must be taken as negative, i.e.

$$\frac{dF}{dx} = -w.$$

The element is also in equilibrium under the couples acting on it so that, taking moments about point G,

we have $$M + F\frac{dx}{2} + (F+dF)\frac{dx}{2} = M + dM$$

i.e. $$F.dx + dF\frac{dx}{2} = dM$$

or $$dM = F.dx \quad \text{(ignoring the product of small quantities)}$$

hence $$\frac{dM}{dx} = F.$$

If the shear force, F, is constant, the slope of the bending moment graph is constant. It follows that

$$F = \int w.dx = \text{Area under load intensity diagram}$$

and $$M = \int F.dx = \text{Area of shear force diagram}.$$

If the load per unit length is variable (i.e. w is not constant) then w must be expressed as some function of x. If this is not possible, then the graph of w against x must be plotted and the area found by planimeter, adding squares or mid-ordinate rule.

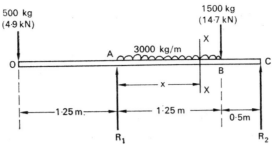

Fig. 20

EXAMPLE. Draw the graphs of F and M for the system shown in Figure 20 and find the position and value of the maximum bending moment.

Solution. Converting masses (kg) to gravitational force (N) by multiplying by 9·81, treating the distributed load as concentrated at its mid-point and taking moments about point A:

$$(R_2 \times 1{\cdot}75) + (500 \times 9{\cdot}81)1{\cdot}25 = 1{\cdot}25(3000 \times 9{\cdot}81)\frac{1{\cdot}25}{2} + (1500 \times 9{\cdot}81)1{\cdot}25$$

or
$$1{\cdot}75 R_2 + (4900 \times 1{\cdot}25) = 29\,400\left(\frac{1{\cdot}25^2}{2}\right) + (14\,700 \times 1{\cdot}25)$$

whence $\qquad R_2 = 20\,100 \text{ N} \quad \text{or} \quad 20{\cdot}1 \text{ kN}$

so that $\qquad R_1 = [500 + (3000 \times 1{\cdot}25) + 1500] - 20\,100$
$\qquad\qquad = 36\,200 \text{ N} \quad \text{or} \quad 36{\cdot}2 \text{ kN}.$

The graph of F is as shown in Figure 21:

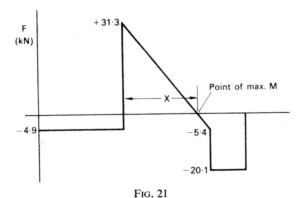

Fig. 21

From the graph, the position of the point of zero shear (i.e. of maximum bending moment) is given by

$$X = \left(\frac{31{\cdot}3}{36{\cdot}2}\right)1{\cdot}25 = 1{\cdot}08 \text{ m} \text{ to right of point } A.$$

At point A, $\qquad M_a = -(4900 \times 1{\cdot}25) = -6120 \text{ Nm}$

At point B, $\quad M_b = -(4900 \times 2\cdot 5) + (36\ 200 \times 1\cdot 25) - (1\cdot 25 \times 29\ 400)\dfrac{1\cdot 25}{2}$

$\qquad\qquad\qquad = +10\ 050$ Nm

$\qquad\qquad$ (*Check*: $M_b = 20\ 100 \times 0\cdot 5 = 10\ 050$ Nm.)

At any section XX between points A and B,

$$M_x = -4900(1\cdot 25 + x) + 36\ 200x - (29\ 400x)\left(\dfrac{x}{2}\right)$$

$$= -14\ 700x^2 + 31\ 300x - 6120$$

This is a maximum when $F = 0$, i.e. when $x = X = 1\cdot 08$ m

Thus $\qquad M_{\max} = -(14\ 700 \times 1\cdot 08^2) + (31\ 300 \times 1\cdot 08) - 6120$

$\qquad\qquad\qquad = 10\ 530$ Nm

Alternatively, since the equation for M over the part AB is a continuous function, the value of x which makes this function a maximum can be found by differentiating and equating to zero.

Thus $\qquad\qquad \dfrac{d(M_x)}{dx} = 0 = -(2 \times 14\ 700)x + 31\ 300$

i.e. $\qquad\qquad x = 1\cdot 07$ m as before (nearly!)

The graph of M is as shown in Figure 22:

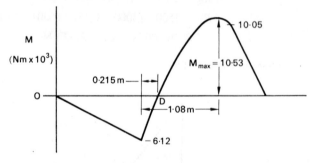

FIG. 22

Putting $M_x = 0$ gives the position of D, the point of contraflexure, or point where M changes sign.

Thus $\qquad\qquad -14\ 700x^2 + 31\ 300x - 6120 = 0$

or $\qquad\qquad x^2 - 2\cdot 13x - 0\cdot 416 = 0$

whence $\qquad\qquad x = 0\cdot 215$ m

The deflected shape of the beam is shown in Figure 23.
Note that at point D the beam is straight since M is there zero.

EXAMPLE. The simply supported beam shown in Figure 24 carries a distributed load which increases uniformly from zero at the left-hand end up to a maximum of 4500 kg/m. Draw the graphs of F and M, indicating the position of $M_{\max}$ and its value.

BENDING

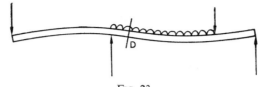

FIG. 23

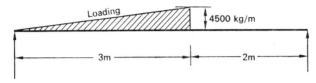

FIG. 24

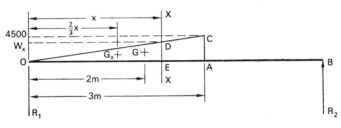

FIG. 25

Solution.

Mean loading $= 2250$ kg/m

Total load $= 2250 \times 3 \times 9 \cdot 81 = 66\,200$ N or $66 \cdot 2$ kN

The total load may be considered to act at point G, Figure 25 (the centroid of the triangle OAC), which is 2 m from point O. Moments about this point give:

$$R_2 \times 5 = 6620 \times 2$$

whence $\qquad R_2 = 26\,480$ N and $R_1 = 39\,720$ N

For any section XX between O and A the load intensity is given by

$$w_x = -\frac{x}{3}(4500 \times 9 \cdot 81) = -\left(\frac{x}{3}\,44\,200\right)$$

Shear force, $\qquad F_x = R_1 -$ weight of piece OED

$$= 39\,720 - \frac{x}{2}\left(\frac{x}{3}\,44\,200\right)$$

i.e. $\qquad F_x = 39\,720 - 7370x^2 \quad (= 32\,350$ when $x = 1$
$\qquad\qquad\qquad\qquad\qquad\qquad\quad = 10\,240$ when $x = 2)$

The graph of F is shown in Figure 26:

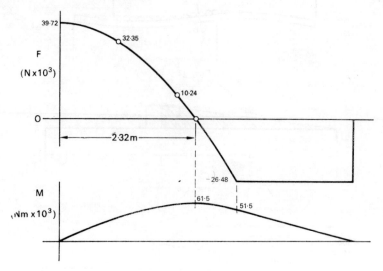

FIG. 26

The bending moment is a maximum when

$$F_x = 0$$

i.e. when $\quad 7370x^2 = 39\,720 \quad$ or when $\quad x = 2\cdot32$ m.

The weight of part OED may be considered to act at point G_x (the centroid of the triangle OED) which is at a distance $x/3$ from XX.

Hence
$$M_x = R_1 x - (\text{weight of piece } OED)\frac{x}{3}$$
$$= 39\,720x - 7370x^2\left(\frac{x}{3}\right)$$
$$= 39\,720x - 2457x^3$$

When $x = 2\cdot32$ m we have
$$M_{\max} = (39\,720 \times 2\cdot32) - (2457 \times 2\cdot32^3)$$
$$= 61\,500 \text{ Nm}$$

When $x = 3\cdot00$ m we have
$$M_x = (39\,720 \times 3) - (2457 \times 3^3)$$
$$= 51\,500 \text{ Nm}$$

Other values may be put in the equation for M_x and used to obtain the graph of M shown in Figure 26. Note that the value of x which makes M a maximum may be obtained (without drawing the graph of F) from

$$\frac{d(M_x)}{dx} = 39\,720 - 7370x^2$$

whence $x = 2\cdot32$ m as before.

BENDING COMBINED WITH DIRECT STRESS

The direct compressive stress induced in a short strut of section A by a compressive load F acting at the section centroid G, Figure 27a, is given by

$$f_d = \frac{F}{A}.$$

This may be represented by a horizontal straight line mn drawn normal to the strut axis and distant an amount f_d (to some convenient scale) from a datum MN.

In practice, no load acts truly at the centroid of a section, so that there will be, inevitably, some eccentricity, x, Figure 27b, and this will introduce a bending moment given by

$$M = Fx.$$

This causes the strut to deflect as shown in Figure 27c. The bending stress on the load side of the neutral axis is evidently of the same sign as the direct stress, i.e. is compressive, while on the opposite side it is tensile. It may be represented by the straight line pq intersecting mn at the neutral axis.

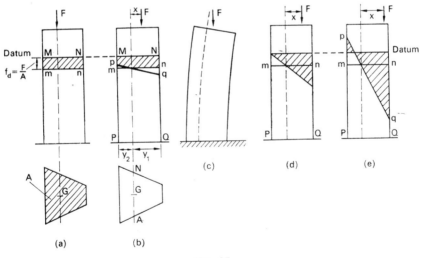

Fig. 27

The two maximum values are, respectively,

$$f_b = \frac{My_1}{I} \quad \text{at } Q \text{ (compressive)},$$

and

$$f_b = \frac{My_2}{I} \quad \text{at } P \text{ (tensile)}.$$

The resultant stress is obtained by superimposing these values on the direct stress. Thus

at Q:
$$f_{max} = \frac{F}{A} + \frac{My_1}{I} \quad \text{compressive,}$$

at P:
$$f_{min} = \frac{F}{A} - \frac{My_2}{I} \quad \text{compressive, since} \quad \frac{F}{A} > \frac{My_2}{I}.$$

If the eccentricity, x, is increased for a given value of F until the tensile bending stress (My_2/I) is equal to the direct stress, then the resultant stress at P will be zero as in Figure 27d. Any further increase in x will result in a net tensile stress at P (as in Figure 27e) although the applied load is compressive. The stress distribution is represented by the shaded area in Figure 27 (a to e), showing the effect of increasing the bending moment. In general, the resultant stress is the algebraic sum of direct and bending stresses and is given by

$$f = \frac{F}{A} \pm \frac{My}{I}.$$

If the bending moment is increased by increasing F while keeping x constant, the direct stress (F/A) is increased in the same proportion as the bending stress (Fxy/I) so that the shape of the stress distribution diagram is unaltered, although all the values are increased as shown in Figure 28.

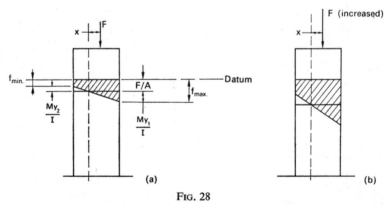

Fig. 28

SECTION MODULUS

In the theory just given, the maximum compressive bending stress is My_1/I. This may also be written $M/(I/y_1)$. The quantity I/y_1 is referred to as the *Compressive Section Modulus* and denoted by Z_1. Similarly, I/y_2 is denoted by Z_2 and called the *Tensile Section Modulus*. The equation for the stress resulting from combined bending and direct effects may therefore be written

$$f = \frac{F}{A} \pm \frac{M}{Z}$$

where Z is the relevant **modulus**.

BENDING

The positive sign will give the maximum stress and this will be of the same kind as the direct stress. The negative sign will give the least stress, the sign of which will depend on the relative magnitudes of the direct and bending effects.

Evidently, for a section which is symmetrical about the neutral axis, the two values of Z are equal.

EXAMPLE. The axis of the load on a test piece is 0·2 mm from the geometric axis. If the diameter of the test piece is 20 mm and the load at the first sign of yielding is 100 000 N, find the maximum corresponding stress. Sketch the stress distribution.

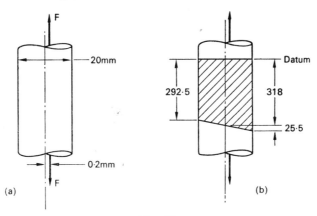

FIG. 29

Solution.

Section of test piece, $A = \dfrac{\pi}{4}(0\cdot020)^2 = 0\cdot000314 \text{ m}^2$

Direct stress, $\dfrac{F}{A} = \dfrac{100\,000}{0\cdot000314}\left(\dfrac{1}{10^6}\right) = 318 \text{ MPa}$

Maximum distance from N.A., $y = \dfrac{20}{2} = 10 \text{ mm} = 0\cdot010 \text{ m}$

2nd Moment of Area, $I = \dfrac{\pi}{64}\left(\dfrac{20}{1000}\right)^4 = 78\cdot5\times10^{-10} \text{ m}^4$

Bending moment, $M = 100\,000\times0\cdot0002 = 20 \text{ Nm}$

Bending stress, $\dfrac{My}{I} = \dfrac{20\times0\cdot010}{78\cdot5}\left(\dfrac{10^{10}}{10^6}\right) = 25\cdot5 \text{ MPa}$

Maximum stress, $f_{max} = \dfrac{F}{A} + \dfrac{My}{I} = 318+25\cdot5 = 343\cdot5 \text{ MPa}$

Minimum stress, $f_{min} = 318-25\cdot5 = 292\cdot5 \text{ MPa}.$

The stress is distributed across the section as shown in Figure 29b.

FLEXURE AND RADIUS OF CURVATURE

A clockwise bending moment on the left of a beam section, which according to convention is positive, induces a curvature which is concave when viewed from above as shown in Figure 30. This is known as a "sagging" curvature.

Part of a sagging beam of varying radius of curvature, i.e. deflected by a varying positive bending moment, is shown in Figure 31.

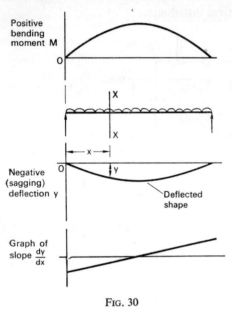

Fig. 30

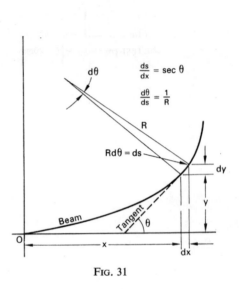

Fig. 31

Slope of beam over element ds, $dy/dx = \tan \theta$

$$\therefore \quad \frac{d^2y}{dx^2} = \sec^2 \theta \frac{d\theta}{dx} \quad \text{differentiating,}$$

$$= \sec^2 \theta \frac{d\theta}{ds} \cdot \frac{ds}{dx}$$

$$= \sec^2 \theta \left(\frac{1}{R}\right) \sec \theta \quad \text{and} \quad \sec^2 \theta = 1 + \tan^2 \theta$$

$$\text{or} \quad \sec^3 \theta = (1 + \tan^2 \theta)^{3/2}$$

so that
$$\frac{d^2y}{dx^2} = \frac{(1+\tan^2 \theta)^{3/2}}{R} \quad \text{where} \quad \tan \theta = \frac{dy}{dx}$$

Transposing and substituting for $\tan \theta$ we obtain:

$$\frac{1}{R} = \frac{d^2y}{dx^2} \bigg/ \left[1 + \left(\frac{dy}{dx}\right)^2\right]^{3/2}$$

Since deflection is small within the elastic limit, powers of dy/dx may be neglected, so that we may write

$$\frac{1}{R} = \frac{d^2y}{dx^2} \quad \left(\text{or} \quad \frac{1}{R} = -\frac{d^2y}{dx^2} \quad \text{if } M \text{ is negative}\right).$$

From the bending equation, $1/R = M/EI$ so that in general we have

$$\frac{1}{R} = \pm \frac{d^2y}{dx^2} = \frac{M}{EI}$$

or
$$M = \pm EI \frac{d^2y}{dx^2}.$$

This is known as the *Differential Equation of Flexure*. (Note that, if I is not constant, it must be moved to the left-hand side of the equation.)

Integrating both sides of this equation with respect to x we have

$$\pm EI \frac{dy}{dx} = \int M \, dx, \quad M \text{ being expressed in terms of } x,$$

i.e. Slope,
$$\frac{dy}{dx} = \pm \frac{1}{EI} \int M \, dx$$

and Deflection,
$$y = \pm \frac{1}{EI} \int\int M \, dx.$$

The foregoing are general expressions and may be applied to specific cases, care being taken to obtain the correct sign for d^2y/dx^2. Figure 30 shows that dy/dx increases from negative through zero to positive, so that (in this case) d^2y/dx^2 is positive.

HORIZONTAL CANTILEVER WITH CONCENTRATED LOAD AT FREE END

For any section XX distant x from the origin, O:

$$M_x = EI \frac{d^2y}{dx^2} = -W(L-x)$$

$$\therefore \quad EI \frac{dy}{dx} = -W\left(Lx - \frac{x^2}{2}\right) + A,$$

but when $x = 0$, $dy/dx = 0$ so that $A = 0$.

$$\therefore \quad EI \frac{dy}{dx} = -W\left(Lx - \frac{x^2}{2}\right)$$

from which the slope at any point can be found.

Again,
$$EIy = -W\left(L\frac{x^2}{2} - \frac{1}{2}\frac{x^3}{3}\right) + B,$$

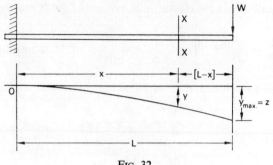

Fig. 32

but, when $x = 0$, $y = 0$ so that $B = 0$.

$$\therefore \quad EIy = -W\left(\frac{L}{2}x^2 - \frac{x^3}{6}\right)$$

from which the deflection at any point can be found.

At the load point y is a maximum and $x = L$, so that

$$EIy_{\max} = -W\left(\frac{L}{2}L^2 - \frac{L^3}{6}\right).$$

Denoting the maximum deflection by z and transposing:

$$z = -\frac{1}{3}\frac{WL^3}{EI}.$$

(Negative because it is downward from the origin considered.)

EXAMPLE. For the cantilever shown in Figure 33, $E = 206\,000$ MPa and $I = 85 \times 10^{-6}$ m^4. Assume only the differential equation of flexure and determine (a) the slope at the load point, (b) the deflection at the load point, (c) the deflection at the free end.

Solution. At any section XX: $M_x = EI\dfrac{d^2y}{dx^2} = -50\,000(3-x)$

$$\therefore \quad EI\frac{dy}{dx} = -50\,000\left(3x - \frac{x^2}{2}\right) + A$$

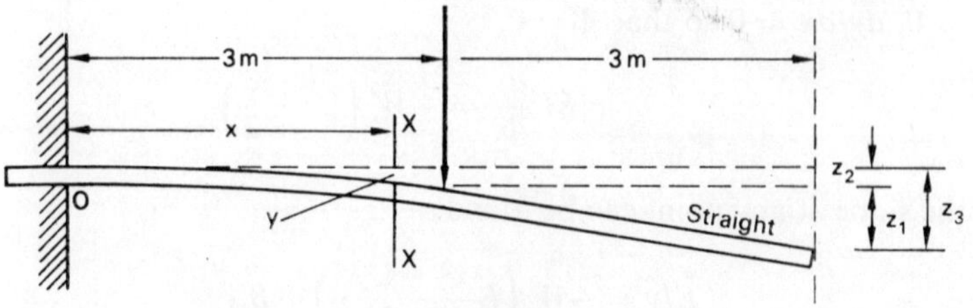

Fig. 33

Since dy/dx is zero when x is zero, it follows that $A = 0$.

Hence, when $x = 3$,
$$EI\frac{dy}{dx} = -50\,000\left(3^2 - \frac{3^2}{2}\right) = -22\,500$$

so that
$$\frac{dy}{dx} = -\frac{225\,000 \times 10^6}{(206\,000 \times 10^6)\,85}$$
$$= -0\cdot01286$$

Integrating again:
$$EIy = -50\,000\left(3\frac{x^2}{2} - \frac{1}{2}\frac{x^3}{3}\right) + B$$

Since y is zero when x is zero, it follows that $B = 0$.

Hence
$$EIy = -75\,000x^2 + 8333x^3$$

Thus, when $x = 3$,
$$EIy = -(75\,000 \times 3^2) + (8333 \times 3^3) = -451\,000$$

so that
$$y = -\frac{451\,000 \times 10^6}{(206\,000 \times 10^6)\,85}$$
$$= -0\cdot0258 \text{ m } (25\cdot8 \text{ mm})$$

This is denoted by z_2 in Figure 33. In the same figure,
$$z_1 = \text{slope} \times 3$$
$$= -0\cdot01286 \times 3$$
$$= -0\cdot03858 \text{ m } (38\cdot6 \text{ mm})$$

The deflection at the free end is therefore given by
$$z_3 = z_1 + z_2$$
$$= \underline{64\cdot4 \text{ mm.}}$$

HORIZONTAL CANTILEVER WITH UNIFORMLY DISTRIBUTED LOAD

For any section XX distant x from the origin, O, Figure 34:
$$M_x = EI\frac{d^2y}{dx^2} = -w(L-x)\left(\frac{L-x}{2}\right)$$
$$= -\frac{w}{2}(L^2 - 2Lx + x^2)$$

$\therefore$
$$EI\frac{dy}{dx} = -\frac{w}{2}\left(L^2x - 2L\frac{x^2}{2} + \frac{x^3}{3}\right) + A$$

Since dy/dx is zero when x is zero, it follows that $A = 0$.

Hence
$$EI\frac{dy}{dx} = -\frac{w}{2}\left(L^2x - Lx^2 + \frac{x^3}{3}\right)$$

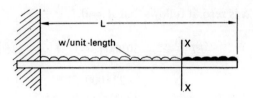

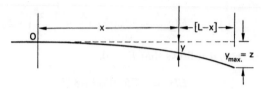

Fig. 34

from which the slope at any point can be found.

Again, $$EIy = -\frac{w}{2}\left(L^2\frac{x^2}{2} - L\frac{x^3}{3} + \frac{1}{12}x^4\right) + B$$

Since y is zero when x is zero, it follows that $B = 0$.

Hence $$EIy = -\frac{w}{2}\left(\frac{L^2}{2}x^2 - \frac{L}{3}x^3 + \frac{1}{12}x^4\right)$$

from which the deflection at any point can be found.
At the free end y is a maximum and $x = L$, so that

$$EIy_{\max} = -\frac{w}{2}\left(\frac{L^2}{2}L^2 - \frac{L}{3}L^3 + \frac{1}{12}L^4\right)$$
$$= -\frac{wL^4}{2}\left(\frac{1}{2} - \frac{1}{3} + \frac{1}{12}\right)$$
$$= -\frac{wL^4}{8} = -\frac{WL^3}{8}$$

where $W = wL$.
Denoting the maximum deflection by z and transposing, we obtain

$$z = -\frac{1}{8}\cdot\frac{WL^3}{EI}.$$

(Negative because downwards from 0.)

SIMPLY SUPPORTED BEAM WITH CONCENTRATED LOAD AT CENTRE

Only by measuring the deflection upwards from a central origin can the constants of integration be made zero. Referring to Figure 35, the actual downward deflection (relative to a support) at any point is obtained by subtracting y from z.

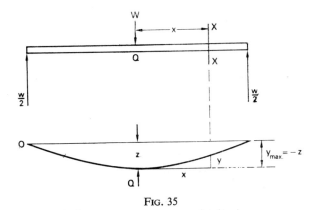

FIG. 35

At any section XX distant x from Q:

$$M_x = EI\frac{d^2y}{dx^2} = \frac{W}{2}\left(\frac{L}{2}-x\right)$$

$$\therefore \quad EI\frac{dy}{dx} = \frac{W}{2}\left(\frac{L}{2}x - \frac{x^2}{2}\right) + A$$

Since dy/dx is zero when x is zero, it follows that $A = 0$.

Hence
$$EI\frac{dy}{dx} = \frac{W}{2}\left(\frac{L}{2}x - \frac{x^2}{2}\right)$$

from which the slope at any point can be found.

Again,
$$EIy = \frac{W}{2}\left(\frac{L}{2}\cdot\frac{x^2}{2} - \frac{1}{2}\cdot\frac{x^3}{3}\right) + B$$

Since y is zero when x is zero, it follows that $B = 0$.

Hence
$$EIy = \frac{W}{2}\left(\frac{L}{4}x^2 - \frac{1}{6}x^3\right)$$

from which the value of y can be found.

At a support, y is a maximum and $x = L/2$, so that

$$EIy_{max} = \frac{W}{2}\left(\frac{L}{4}\cdot\frac{L^2}{4} - \frac{1}{6}\cdot\frac{L^2}{8}\right)$$

$$= WL^3\left(\frac{1}{32} - \frac{1}{96}\right)$$

$$= \frac{WL^3}{48}$$

y being measured *upwards* from Q.

Putting $y_{max} = -z$ and transposing, we obtain an expression for the deflection at the centre relative to a support, viz.

$$z = -\frac{1}{48}\cdot\frac{WL^3}{EI}.$$

The fractions which precede the quantity WL^3/EI are known as *Deflection Coefficients*.

EXAMPLE. Two self-aligning bearings 3 m apart support a shaft 65 mm diameter. If a transverse load of 2700 N were to be applied at the centre of the span, find from first principles (in degrees) the inclination of the shaft at the bearings. Assume $E = 206\,000$ MPa.

Solution. For the section of the shaft, $I = \dfrac{\pi}{64}\left(\dfrac{65}{1000}\right)^4 = 0.8 \times 10^{-6}$ m^4.

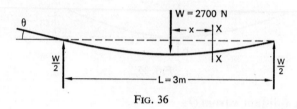

FIG. 36

At any section XX, Figure 36,

$$M_x = EI\frac{d^2y}{dx^2} = \frac{W}{2}\left(\frac{L}{2}-x\right)$$

∴

$$EI\frac{dy}{dx} = \frac{W}{2}\left(\frac{L}{2}x - \frac{x^2}{2}\right),$$

the constant of integration being zero. Putting $x = L/2$ and transposing gives

$$\frac{dy}{dx} = \frac{W}{2EI}\left(\frac{L^2}{4} - \frac{L^2}{8}\right)$$

$$= \frac{WL^2}{16EI}$$

$$= \frac{2700 \times 3^2 \times 10^6}{16(206\,000 \times 10^6)0.8}$$

$$= 0.0092$$

Inclination, $\theta = \tan^{-1} 0.0092 = 0° 32'$.

SIMPLY SUPPORTED BEAM WITH UNIFORMLY DISTRIBUTED LOAD

Referring to Figure 37 and taking the origin at the centre in order to make the constants of integration zero, we have, on any section XX distant x from O:

$$M_x = EI\frac{d^2y}{dx^2} = \frac{wL}{2}\left(\frac{L}{2}-x\right) - w\left(\frac{L}{2}-x\right)\frac{1}{2}\left(\frac{L}{2}-x\right)$$

whence

$$EI\frac{d^2y}{dx^2} = \frac{w}{2}\left(\frac{L^2}{4} - x^2\right)$$

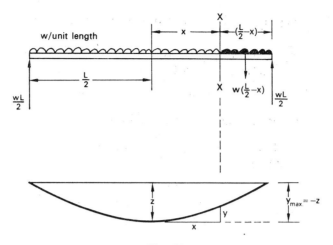

FIG. 37

$$EI \frac{dy}{dx} = \frac{w}{2}\left(\frac{L^2}{4}x - \frac{x^3}{3}\right) \quad \text{(giving the slope at any point)}$$

and

$$EIy = \frac{w}{2}\left(\frac{L^2}{8}x^2 - \frac{x^4}{12}\right) \quad \text{(giving } y \text{ at any point)}$$

At a support, y is a maximum and $x = L/2$, so that

$$EIy_{\max} = \frac{w}{2}\left(\frac{L^2}{8} \cdot \frac{L^2}{4} - \frac{1}{12} \cdot \frac{L^4}{16}\right) = \frac{5}{384}wL^4 \quad \text{(and } wL = W\text{)}$$

Putting $y_{\max} = -z$ as before and transposing, we obtain

$$z = -\frac{5}{384}\frac{WL^3}{EI} \quad \text{downwards from a support.}$$

EXAMPLE. An I-section beam is 350 mm deep, has $I = 180 \times 10^{-6}$ m⁴ and carries a load of 15 000 kg distributed uniformly over a simple span of 10 m. Estimate the deflection at midspan assuming $E = 206\,000$ MPa. If the beam is propped at the centre so that the deflection there is zero, find the load carried by the prop and the positions of the points of contraflexure.

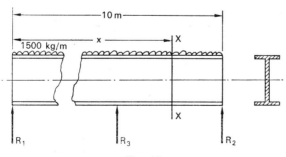

FIG. 38

Solution. Before the prop, R_3, is introduced, the deflection at the centre of the span is given by

$$z = \frac{5}{384} \frac{WL^3}{EI} = \frac{5}{384} \frac{(15\,000 \times 9 \cdot 81)10^3 \times 10^6}{(206\,000 \times 10^6)180}$$

$$= 0 \cdot 0516 \text{ m } (= 51 \cdot 6 \text{ mm})$$

The prop must produce therefore an upward deflection equal to this, so that

$$0 \cdot 0516 = \frac{1}{48} \frac{R_3 L^3}{EI} \quad \text{or} \quad R_3 = \frac{0 \cdot 0516 \times 48(206\,000 \times 10^6)180}{10^3 \times 10^6}$$

$$= 91\,800 \text{ N}$$

Hence $\quad R_1 = R_2 = \dfrac{(15\,000 \times 9 \cdot 81) - 91\,800}{2} = 27\,600 \text{ N}$

For any section XX, working from the left-hand end,

$$M_x = 27\,600x - x(1500 \times 9 \cdot 81)\frac{x}{2} + 91\,800(x-5) = 0 \quad \text{at points of contraflexure}$$

or
$$x^2 - 16 \cdot 3x + 62 \cdot 5 = 0,$$
whence
$$x = 6 \cdot 2 \text{ m}.$$

The points of contraflexure are therefore at $6 \cdot 2 - 5$ or $1 \cdot 2$ m from the centre of the span.

BEAM SUBJECTED TO A UNIFORM BENDING MOMENT

Such a system is shown in Figure 39, the part AB sustaining a constant bending moment of $-Wl$. Since $M = EI/R$ it follows that R is constant between the supports, i.e. AB bends into a circular arc.

For any section XX between A and B:

$$M_x = -EI\frac{d^2y}{dx^2} = -Wl$$

$$-EI\frac{dy}{dx} = -Wlx + A.$$

Since dy/dx is zero when $x = L/2$, substitution in the above equation gives

$$A = Wl\frac{L}{2}.$$

Hence,
$$-EI\frac{dy}{dx} = -Wlx + Wl\frac{L}{2}$$

and
$$-EIy = -Wl\frac{x^2}{2} + Wl\frac{L}{2}x + B.$$

BENDING

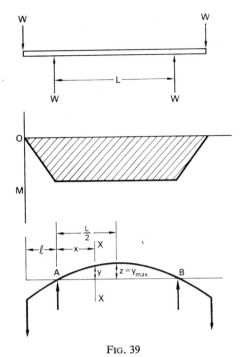

Fig. 39

Since y is zero when x is zero, it follows that $B = 0$.

Hence,
$$-EIy = \frac{Wl}{2}(Lx - x^2)$$

i.e.
$$-y = \frac{Wl}{2EI}(Lx - x^2).$$

The maximum deflection occurs at the centre so that putting $x = L/2$ and writing z for y_{max} we obtain

$$-z = \frac{Wl}{2EI}\left(L\frac{L}{2} - \frac{L^2}{4}\right) = \frac{WlL^2}{8EI} = -\frac{ML^2}{8EI}$$

since $Wl = -M$.

Hence
$$z = \frac{ML^2}{8EI}$$
upwards relative to a support.

SIMPLY SUPPORTED BEAM WITH CONCENTRATED NON-CENTRAL LOAD

Referring to Figure 40 and taking moments about point O, we obtain

$$R_2 L = Wa, \quad \text{i.e.} \quad R_2 = \frac{Wa}{L}.$$

Similarly,
$$R_1 = \frac{Wb}{L}.$$

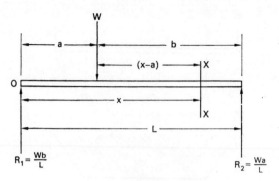

Fig. 40

For any section XX to the left of W, $\quad M_x = \dfrac{Wb}{L} x.$

For any section XX to the right of W, $\quad M_x = \dfrac{Wb}{L} x - W(x-a).$

The drawback of having two equations for M can be overcome by writing

$$M_x = EI \frac{d^2y}{dx^2} = \frac{Wb}{L} x - kW(x-a), \qquad (1)$$

where $k = 0$ for $x < a$ (i.e. where the bracket in the 2nd term is negative) and $k = 1$ for $x > a$ (i.e. where the bracket is positive).

This device is due to Macaulay. If there are two loads there will be three equations, the last of which can be employed in the same manner.

Integrating: $\quad EI \dfrac{dy}{dx} = \dfrac{Wb}{L} \int x \, dx - kW \int_0^x (x-a) \, dx.$

But, $\quad kW \int_0^x (x-a) \, dx = kW \int_0^a (x-a) \, dx + kW \int_a^x (x-a) \, dx.$

The first term on the right-hand side is zero because k is zero when $x < a$, while the second term can be written

$$kW \int_0^{x-a} (x-a) \, d(x-a).$$

Hence, $\quad EI \dfrac{dy}{dx} = \dfrac{Wb}{L} \int x \, dx - kW \int_0^{x-a} (x-a) \, d(x-a).$

Thus the first term of eq. (1) must be integrated with respect to x, while the second term must be integrated with respect to the bracket, $(x-a)$,

i.e. $\quad EI \dfrac{dy}{dx} = \dfrac{Wb}{L} \cdot \dfrac{x^2}{2} + A - kW \dfrac{(x-a)^2}{2}. \qquad (2)$

Integrating again in the same way we obtain

$$EIy = \frac{Wb}{2L} \cdot \frac{x^3}{3} + Ax + B - k\frac{W}{2}\frac{(x-a)^3}{3}. \quad (3)$$

At the left-hand end where $x = 0$, $y = 0$ and $k = 0$ so that the term in $(x-a)$ is omitted. Putting these values in eq. (3) shows that the constant B is zero.

At the right-hand end where $x = L$, $y = 0$ and $k = 1$ so that the term in $(x-a)$ is included. Putting these values in eq. (3) gives

$$0 = \frac{Wb}{2L} \cdot \frac{L^3}{3} + AL - \frac{W}{6}(L-a)^3$$

$$\therefore \quad A = \frac{W}{6L}(L-a)^3 - \frac{Wb}{6L}L^3.$$

But $\quad L - a = b, \quad$ and $\quad L^2 = a^2 + 2ab + b^2,$

so that $\quad A = \frac{Wb^3}{6L} - \frac{Wba^2}{6L} - \frac{2Wab^2}{6L} - \frac{Wb^3}{6L} = \underline{-\frac{Wba^2}{6L} - \frac{2Wab^2}{6L}}.$

Substitution of this value in eq. (3) will give, after some algebra:

$$EIy = -\frac{W}{6}\left[\frac{bx}{L}(L^2 - b^2 - x^2) + (x-a)^3\right].$$

This equation will give the downward deflection, y, at any distance, x, from the left-hand end, the term $(x-a)^3$ being omitted when $x < a$, i.e. to the left of the load.

Putting $L = a + b$ and $x = a$ and transposing EI, we have, for the load point,

$$y = -\frac{W}{6EI}\left[\frac{ab}{L}(a^2 + 2ab + b^2) - b^2 - a^2\right]$$

or $\quad y = -\frac{Wa^2b^2}{3EIL} \quad$ downwards from point O.

It can be shown (see author's *Strength of Materials*, p. 315) that the maximum deflection occurs when

$$x = \sqrt{\frac{L^2 - b^2}{3}} \quad \text{and that this is greater than } \frac{L}{2}.$$

EXAMPLE. Two self-aligning bearings 3 m apart support a shaft 65 mm diameter. Find the inclination of the shaft at the left-hand bearing when a transverse load of 2700 N is applied at a point 1 m from the left-hand end. $E = 206\,000$ MPa.

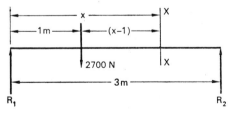

FIG. 41

Solution. Moments about the right-hand end give

$$R_1 \times 3 = 2700 \times 2 \quad \text{whence} \quad R_1 = 1800 \text{ N} \quad \text{and} \quad R_2 = 900 \text{ N}.$$

At any section distant x from the left-hand end,

$$M_x = EI\frac{d^2y}{dx^2} = 1800x - 2700(x-1) \tag{1}$$

$$\therefore \quad EI\frac{dy}{dx} = 1800\frac{x^2}{2} - 2700\frac{(x-1)^2}{2} + A \tag{2}$$

and

$$EIy = 900\frac{x^2}{3} - 1350\frac{(x-1)^3}{3} + Ax + B. \tag{3}$$

To the left to the load, $x < 1$, so that the second term is omitted. Also, when x is zero y is also zero so that in eq. (3) $B = 0$.

To the right of the load, $x > 1$, so that the second term must be included. Also, when $x = 3$, $y = 0$. Putting these values in eq. (3) gives

$$0 = 900\frac{3^3}{3} - 1350\frac{(3-1)^3}{3} + 3A$$

whence $\quad A = -1500.$

Putting this value in eq. (2) gives

$$EI\frac{dy}{dx} = 900x^2 - 1350(x-1)^2 - 1500.$$

At the left-hand end, $x = 0$ and the 2nd term is omitted.

$$EI\frac{dy}{dx} = -1500$$

where

$$I = \frac{\pi}{64}\left(\frac{65}{1000}\right)^4 = 0.8 \times 10^{-6} \text{ m}^4$$

$$\therefore \quad \frac{dy}{dx} = -\frac{1500 \times 10^6}{(206\,000 \times 10^6)0.8} = -0.0091.$$

Hence, inclination of shaft axis $= \tan^{-1} 0.0091 = 0° 31'$.

LEAF SPRINGS

The simplest type has a single "leaf" and is an inverted, simply supported, centrally loaded beam as shown in Figure 42.

Since

$$M = \frac{WL}{4}, \quad y = \frac{t}{2} \quad \text{and} \quad I = \frac{bt^3}{12},$$

the maximum stress is given by $\quad f_{\max} = \frac{WL}{4} \times \frac{t}{2} \times \frac{12}{bt^3} = \underline{\frac{3}{2} \cdot \frac{WL}{bt^2}}.$

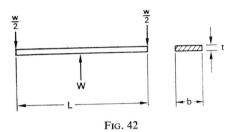

Fig. 42

The deflection at the centre is given by

$$z = \frac{1}{48} \cdot \frac{WL^3}{EI} = \frac{WL^3}{48E} \times \frac{12}{bt^3} = \frac{1}{4} \cdot \frac{WL^3}{Ebt^3}.$$

The transverse stiffness (load per unit deflection) is therefore

$$\lambda = \frac{W}{z} = \frac{4Ebt^2}{L^3}.$$

CURVATURE

For a spring to be flat under a static load, it must be given an initial curvature. If L is known, the radius for a given deflection may be found as follows:

Referring to Figure 43,
$$R^2 = (R-z)^2 + \left(\frac{L}{2}\right)^2$$
$$= R^2 - 2Rz + z^2 + \frac{L^2}{4}$$

i.e. $$2Rz - z^2 = \frac{L^2}{4}$$

whence $$R = \frac{L^2}{8z} + \frac{z}{2}.$$

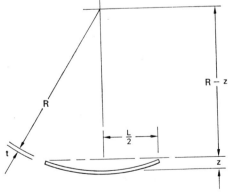

Fig. 43

MULTI-LEAF SPRING

Let a rhombus of thickness t be supported along its diagonal AB and loaded with $W/2$ at the ends of the other diagonal of length L as shown in Figure 44. Let AB be divided into n equal parts of length b.

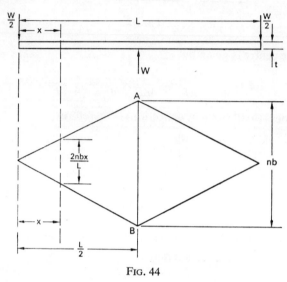

Fig. 44

At any distance, x, from the left-hand end:

Width of section
$$= \frac{x}{L/2} \times nb = \frac{2nbx}{L}$$

2nd moment
$$I = \left(\frac{2nbx}{L}\right)\frac{t^3}{12} = \frac{nbt^3 x}{6L}$$

Bending moment
$$M = \frac{W}{2}x$$

Depth of section,
$$y = \frac{t}{2}$$

∴ Stress
$$f = \frac{My}{I} = \frac{W}{2}x \cdot \frac{t}{2} \cdot \frac{6L}{nbt^3 x} = \underline{\frac{3}{2} \cdot \frac{WL}{nbt^2}}.$$

Since x does not appear in this expression, it follows that the stress is uniform throughout the material. (Note that the stress maximum in a single rectangular leaf is n times as great.)

If the rhombus were cut along the dotted lines as in Figure 45a and the strips of width $b/2$ assembled as shown in Figure 45b, it would become a multi-leaf spring having n leaves of width b, except for the pointed ends.

Since the beam is inverted, $EI(d^2y/dx^2) = -M$.

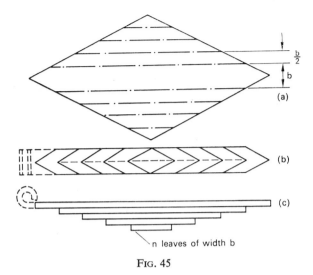

Fig. 45

Transposing and putting in the value for I we obtain

$$\frac{d^2y}{dx^2} = -\frac{W}{2}x \cdot \frac{1}{E} \cdot \frac{6L}{nbt^3x} = -\frac{3WL}{nEbt^3}.$$

Integrating:
$$\frac{dy}{dx} = -\frac{3WL}{nEbt^3} \cdot x + A.$$

Since dy/dx is zero when $x = L/2$ we have,

$$A = \frac{3}{2}\frac{WL^2}{nEbt^3}$$

so that
$$\frac{dy}{dx} = -\frac{3WLx}{nEbt^3} + \frac{3}{2}\frac{WL^2}{nEbt^3}$$

and
$$y = -\frac{3WL}{nEbt^3} \cdot \frac{x^2}{2} + \frac{3}{2}\frac{WL^2}{nEbt^3} \cdot x + B.$$

Since y is zero when x is zero, it follows that $B = 0$.
When $x = L/2$ the deflection is a maximum so that denoting this by z we obtain,

$$z = +\frac{3}{8}\frac{WL^3}{nEbt^3}$$

(positive because the centre of the spring moves upwards relative to the spring eyes).

QUARTER ELLIPTIC SPRING

Such a spring is shown in Figure 46.

Evidently, in the expressions already obtained, W must be written for $W/2$ (i.e. $2W$ for W) and L for $L/2$ (i.e. $2L$ for L). This gives

$$f = \frac{3}{2} \cdot \frac{(2W)(2L)}{nbt^2} = \frac{6WL}{nbt^2}$$

and

$$z = \frac{3}{8} \cdot \frac{(2W)(2L)^3}{nEbt^3} = \frac{6WL^3}{nEbt^3}.$$

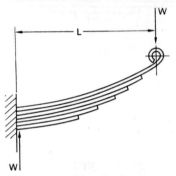

Fig. 46

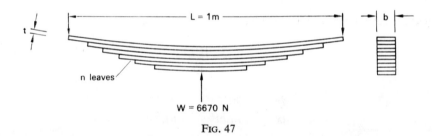

Fig. 47

EXAMPLE. When the spring shown in Figure 47 is deflected at the load point through 125 mm, the stress is not to exceed 525 MPa. Find a suitable number of leaves assuming a value for E of 206 000 MPa.

Solution.

Stress,
$$f = \frac{3}{2} \frac{WL}{nbt^2}$$

∴
$$525 \times 10^6 = \frac{3 \times 6670 \times 1}{2(nbt^2)}$$

whence
$$nbt^2 = 19.05 \times 10^{-6} \text{ m}^3.$$

Deflection,
$$z = \frac{3}{8}\frac{WL^3}{nEbt^3}$$

∴
$$0.125 = \frac{3 \times 6670 \times 1^3}{8(206\,000 \times 10^6)\,nbt^3}$$

whence
$$nbt^3 = 0.097 \times 10^{-6} \text{ m}^4$$

so that
$$t = \frac{0.097}{19.05}$$
$$= 0.0051 \text{ m } (5.1 \text{ mm}).$$

But
$$nbt^2 = \frac{19.05^6}{10^6} \quad \text{where} \quad t = \frac{5.1}{10^3}, \quad \text{i.e.} \quad t^2 = \frac{26}{10^6},$$

∴
$$nb = \frac{19.05}{10^6}\left(\frac{10^6}{26}\right) = 0.73.$$

Assuming $n = 12$ gives $b = 0.733/12 = 0.061$ m (about 60 mm).
This is reasonable and makes the maximum depth of the spring about equal to the leaf width.
If 15 leaves were assumed, the leaves would be about 50 mm in width.

STRAIN ENERGY OF BENDING

In simple tension and compression the energy absorbed (work done) in straining is given by

$$U = \frac{f^2}{2E} \times \text{volume of material},$$

where f is less than the elastic limit value and uniform across the section.

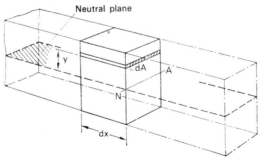

FIG. 48

Consider the element of beam (length dx, section dA) shown in Figure 48 and assume simple bending.

Volume of element $= dx\,.\,dA$

Strain energy of element $= \dfrac{f^2}{2E}(dx\,.\,dA) \quad \left(\text{where} \quad f = \dfrac{My}{I}\right)$

MECHANICAL TECHNOLOGY FOR HIGHER ENGINEERING TECHNICIANS

$$= \frac{M^2 y^2}{2EI^2} dx \cdot dA$$

Strain energy between sections
$$= \sum \frac{M^2 y^2}{2EI^2} dx \cdot dA$$

$$= \frac{M^2}{2EI^2} dx \cdot \sum (dA \cdot y^2).$$

But $\sum (dA \cdot y^2)$ is the 2nd Moment of Area of the section, I.

$\therefore$ Strain energy between sections $= \dfrac{1}{2EI} M^2 dx$.

Hence, for a length of beam L,

Total strain energy, $U = \dfrac{1}{2EI} \displaystyle\int_0^L M^2 \, dx$ (M being expressed in in terms of x).

Note that, if I is not constant, it also must be expressed in terms of x (if possible) and included in the integral.

The general expression just derived may be applied to various specific cases of loading.

SIMPLY SUPPORTED BEAM WITH CONCENTRATED LOAD AT CENTRE

Referring to Figure 49, from a support to mid-span, $M_x = (W/2) x$.

For half the beam,
$$U = \frac{1}{2EI} \int_0^{L/2} \left(\frac{W}{2} x\right)^2 dx$$

For the whole beam,
$$U = \left(\frac{1}{2EI} \int_0^{L/2} \frac{W^2}{4} \cdot x^2 \, dx \right) 2$$

$$= \frac{W^2}{4EI} \left| \frac{x^3}{3} \right|_0^{L/2}$$

$$= \frac{W^2}{12EI} \cdot \frac{L^3}{8}$$

i.e.
$$U = \frac{1}{96} \frac{W^2 L^3}{EI}.$$

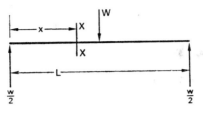

Fig. 49

CANTILEVER WITH UNIFORM LOAD

Referring to Figure 50 the bending moment at any section XX is given by

$$M_x = -\frac{w}{2} x^2$$

which is a continuous function.

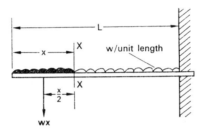

Fig. 50

$$\therefore \quad U = \frac{1}{2EI} \int_0^L \left(\frac{w}{2} x^2\right)^2 dx$$

$$= \frac{w^2}{8EI} \left|\frac{x^5}{5}\right|_0^L$$

$$= \frac{w^2 L^5}{40 EI}$$

where $wL = W$

so that

$$U = \frac{1}{40} \frac{W^2 L^3}{EI}.$$

Similarly it may be shown that

For a simply supported beam, uniformly loaded: $U = \dfrac{1}{240} \dfrac{W^2 L^3}{EI}$,

For a beam subjected to uniform bending moment: $U = \dfrac{1}{2} \dfrac{M^2 L}{EI}$.

SIMPLY SUPPORTED BEAM WITH CONCENTRATED NON-CENTRAL LOAD

Referring to Figure 51, moments about the left-hand end give

$$R_2 L = Wa$$

so that

$$R_2 = \frac{Wa}{L}.$$

Similarly,

$$R_1 = \frac{Wb}{L}.$$

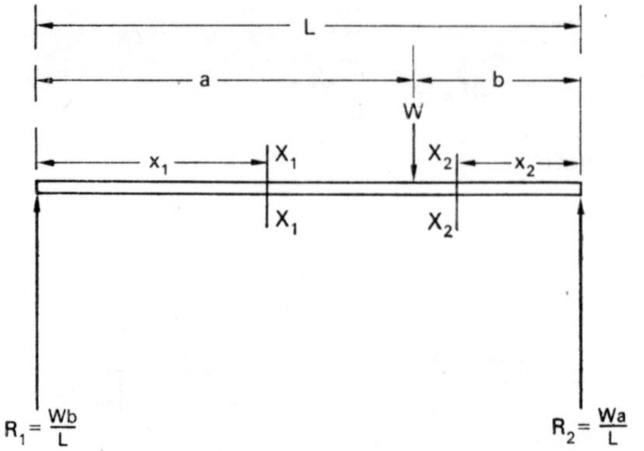

FIG. 51

For any section $X_1 X_1$ to the left of the load:

$$M_{x_1} = \frac{Wb}{L} x_1$$

∴ For this part of the beam,

$$U_a = \frac{1}{2EI} \int_0^a \frac{W^2 b^2}{L^2} \cdot x_1^2 \, dx$$

whence

$$U_a = \frac{W^2 b^2 a^3}{6EIL^2}.$$

Similarly, for the other part of the beam,

$$U_b = \frac{W^2 a^2 b^3}{6EIL^2}.$$

∴ Total strain energy,

$$U = U_a + U_b$$

$$= \frac{W^2 a^2 b^2}{6EIL^2}(a+b) \quad \text{and} \quad a+b = L$$

i.e.

$$U = \frac{W^2 a^2 b^2}{6EIL}.$$

BENDING

Since the work done in straining $= (W/2)y$, where y is the deflection at the load point, we can write:

$$\frac{W}{2}y = \frac{W^2 a^2 b^2}{6EIL} \quad \text{from which} \quad y = \frac{Wa^2b^2}{3EIL} \quad \text{(a result already obtained)}.$$

EXAMPLE. Deduce, from a consideration of strain energy, an expression for the deflection at the load point of the triangular cantilever shown in Figure 52.

Find this deflection when the load is 270 N assuming the value of E to be 206 000 MPa.

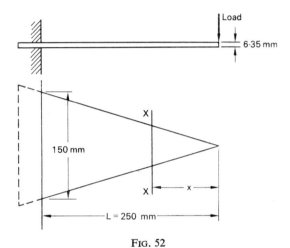

FIG. 52

Solution. If I is the value of the 2nd Moment of Area of the section at the support, then, at any section XX,

2nd Moment, $\qquad I_x = \frac{x}{L}I$

and Bending moment, $\qquad M_x = -Wx$

Strain energy, $\qquad U = \frac{1}{2E}\int \left[\frac{M_x^2}{I_x}\right] dx$

$$= \frac{1}{2E}\int_0^L (-Wx)^2 \frac{L}{xI} dx$$

$$= \frac{W^2 L}{2EI}\int_0^L x \, dx$$

$$\therefore \quad U = \frac{W^2 L^3}{4EI}.$$

MECHANICAL TECHNOLOGY FOR HIGHER ENGINEERING TECHNICIANS

If z is the deflection at the load point, then $U = \dfrac{W}{2} z$

so that $\dfrac{W}{2} z = \dfrac{W^2 L^3}{4EI}$

i.e. $z = \dfrac{1}{2} \dfrac{WL^3}{EI}$, where I = value at support.

From Figure 52: $L = 0.250$ m

$$I = \dfrac{0.150}{12}\left(\dfrac{6.35}{1000}\right)^3 \text{ m}^4$$

$\therefore \quad z = \dfrac{1}{2} \dfrac{270 \times 0.25^3 \times 120 \times 10^9}{(206\,000 \times 10^6) 0.15 \times 6.35^3}$

$= 0.0032$ m (3.2 mm).

THE FIRST THEOREM OF CASTIGLIANO

Suppose a beam to be loaded as shown in Figure 53.

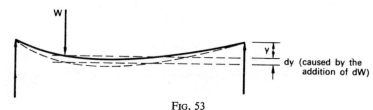

FIG. 53

If y is the deflection at the load point, then

Strain energy = Average force × distance

or $\qquad U = \dfrac{W}{2} y$

If y is increased by an amount dy when W is increased by dW, then,

Increase in strain energy, $\quad dU = \dfrac{dW}{2} dy + W\, dy$

($= W\, dy$ nearly, i.e. neglecting the product of small quantities)

and, Total strain energy,

$$U + dU = \dfrac{W}{2} y + \dfrac{dW}{2} dy + W\, dy. \tag{1}$$

But, if $(W + dW)$ had been applied in the first place, we would have had

Total strain energy $= \dfrac{W + dW}{2}(y + dy)$

$= \dfrac{W}{2} y + \dfrac{W}{2} dy + \dfrac{dW}{2} y + \dfrac{dW}{2} dy. \tag{2}$

Equating (1) and (2):

$$\frac{W}{2}y + \frac{dW}{2}dy + W\,dy = \frac{W}{2}y + \frac{W}{2}dy + \frac{dW}{2}y + \frac{dW}{2}dy$$

i.e. $$W\,dy - \frac{W}{2}dy = \frac{dW}{2}y$$

or $$W.dy = dW.y.$$

But, from above, $$W.dy = dU$$

so that $$dU = dW.y$$

i.e. $$\frac{dU}{dW} = y.$$

Thus the derivative of the strain energy with respect to the load gives the deflection at the load point. If other loads are present (i.e the beam is subjected to a system of forces) then the partial differential of the total strain energy with respect to one of the loads is equal to the movement of that load in its own direction,

i.e. $$y = \frac{\partial U}{\partial W}.$$

This statement is the First Theorem of Castigliano.

In the case of straight beams, the total strain energy is given by

$$U = \frac{1}{2EI}\int_0^L M^2\,dx.$$

Hence, differentiating with respect to some particular load, W, will give the deflection at that load, viz.

$$y = \frac{\partial U}{\partial W} = \frac{1}{2EI}\int_0^L 2M\frac{\partial M}{\partial W}\,dx,$$

or $$y = \frac{1}{EI}\int_0^L M\frac{\partial M}{\partial W}\,dx.$$

This general expression may be applied to specific cases.

HORIZONTAL CANTILEVER WITH CONCENTRATED LOAD AT FREE END

At any section XX, Figure 54,

$$M_x = -Wx$$

$\therefore$ $$\frac{\partial M_x}{\partial W} = -x$$

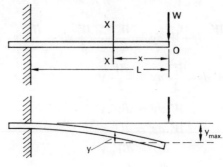

Fig. 54

At the load point,
$$y_{max} = \frac{1}{EI}\int_0^L (-Wx)(-x)\,dx$$
$$= \frac{1}{EI}\int_0^L Wx^2\,dx$$
$$= \frac{W}{EI}\left|\frac{x^3}{3}\right|_0^L$$

i.e. $\quad y_{max} = \frac{1}{3}\cdot\frac{WL^3}{EI}\quad$ (a result already obtained).

The positive sign shows that this is upward from the origin considered.

SIMPLY SUPPORTED BEAM LOADED AS IN FIGURE 55

At any section XX to the left of the load, Figure 55:
$$M_x = \frac{1}{2}(W+wL)x - \frac{w}{2}x^2$$

∴
$$\frac{\partial M_x}{\partial W} = \frac{x}{2}.$$

Putting these values in the general expression and obtaining the total strain energy by doubling that for half the beam (since M is not a continuous function) we can write, for the resul-

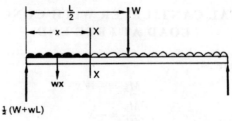

Fig. 55

tant deflection at the load point:

$$y = \frac{1}{EI} 2 \int_0^{L/2} \left[\frac{1}{2}(W+wL)x - \frac{w}{2}x^2\right]\frac{x}{2} dx$$

$$= \frac{1}{EI} 2 \int_0^{L/2} \left[\frac{1}{4}(W+wL)x^2 - \frac{w}{4}x^3\right] dx$$

$$= \frac{2}{EI}\left[\frac{1}{12}(W+wL)x^3 - \frac{w}{16}x^4\right]_0^{L/2}$$

$$= \frac{2}{EI}\left[\frac{1}{12}(W+wL)\frac{L^3}{8} - \frac{w}{16}\frac{L^4}{16}\right].$$

Putting $w = 0$ gives $\quad y = \dfrac{1}{EI}\left[\dfrac{WL^3}{48}\right] \quad$ as obtained on p. 29.

Putting $W = 0$ gives $\quad y = \dfrac{2}{EI}\left[\dfrac{wL^4}{96} - \dfrac{wL^4}{256}\right]$

$$= \frac{wL^4}{EI}\left(\frac{1}{48} - \frac{1}{128}\right)$$

i.e. $\quad y = \dfrac{5}{384}\dfrac{wL^4}{EI} \quad$ as obtained on p. 31.

Other cases may be treated in a similar manner.

SHEAR STRESS DUE TO BENDING

It will be shown in Chapter 3 that the shear stress induced by a transverse shear force is accompanied by an equal "complementary" shear stress in a plane at right angles to it. In a leaf spring this stress is resisted only by friction between the leaves, which, usually, is insufficient to prevent relative motion between them. In the web of an I-section beam, the effects of shear may cause failure by buckling.

Referring to Figure 56 we have, for the element of length dx, width b and thickness dy:

Resultant force parallel to the neutral axis

$$= \frac{(M+dM)y}{I}.b\,dy - \frac{My}{I}.b\,dy$$

$$= \frac{dM}{I}.by\,dy$$

$\therefore$ Total tension on piece of section A and length dx

$$= \int_y^{y_{max}} \frac{dM}{I} by\,dy.$$

MECHANICAL TECHNOLOGY FOR HIGHER ENGINEERING TECHNICIANS

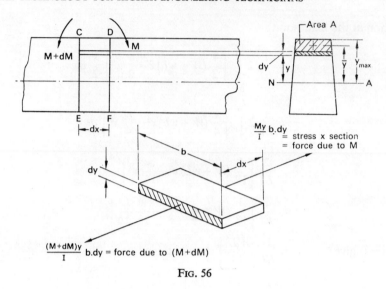

FIG. 56

This is resisted by the complementary shear stress, q, acting on the area $b.dx$ and inducing a resisting force $qb.dx$. Equating total tension to shear resistance we obtain

$$qb\, dx = \int_{y}^{y_{max}} \frac{dM}{I} by\, dy$$

or

$$q = \frac{dM}{dx} \cdot \frac{1}{Ib} \int_{y}^{y_{max}} by\, dy.$$

But, $\int_{y}^{y_{max}} by\, dy$ is the 1st Moment of Area, about the neutral axis, of the shaded area, A. Thus, if $\bar{y}$ is the distance from the neutral axis of the centroid of this area, then

$$\int_{y}^{y_{max}} by\, dy = A\bar{y}.$$

Also, $\qquad \dfrac{dM}{dx} = F,\quad$ the shear force.

Hence, Horizontal shear stress, $\qquad q = \dfrac{FA\bar{y}}{Ib}.$

Evidently this expression also gives the value of shear stress on the vertical section, at y from the neutral axis.

RECTANGULAR SECTION

Referring to Figure 57 we have:

$$I = \frac{bd^3}{12}$$

$$A = b\left(\frac{d}{2} - y\right) \quad \text{(shaded area)}$$

$$\bar{y} = y + \left(\frac{d/2 - y}{2}\right).$$

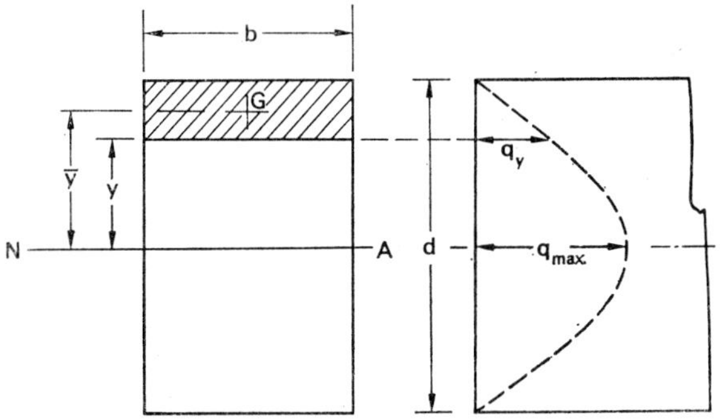

FIG. 57

Substitution in the general expression gives, after some manipulation,

$$q = \frac{6F}{bd^3}\left(\frac{d^2}{4} - y^2\right).$$

If q is plotted against y a parabola will be obtained, as shown dotted in Figure 57. Evidently q is a maximum when $y = 0$ and a minimum when $y = d/2$. Putting $y = 0$ we obtain

$$q_{max} = \frac{3}{2} \cdot \frac{F}{bd}.$$

The quantity F/bd is the average shear stress across the section.

I-SECTION

As shown in Figure 58, the complementary shear in the web is at 90 deg to the applied shear, on longitudinal planes parallel to the neutral axis.

Referring to Figure 59, on any web section XX distant y from the neutral axis, the shear stress is given by

$$q_y = \frac{FA\bar{y}}{Ib}$$

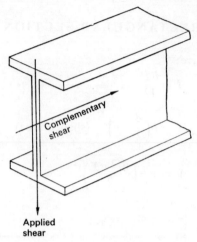

Fig. 58

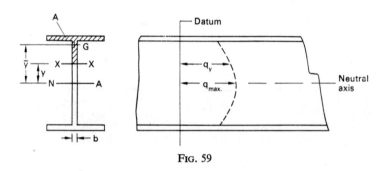

Fig. 59

where A = shaded area,
$\bar{y}$ = distance of centroid of this area from the neutral axis,
b = web thickness.

The graph of q is again parabolic, though q is not zero at the junction of web and flange, since A is not zero at this point.

The shear stress in the flange must be zero at the points farthest from the web, i.e. at the free vertical surface of the flange. It therefore acts over vertical planes such as YY, Figure 60, increasing towards the junction with the web.

For any section YY distant x from the web, the shear stress is given by

$$q_x = \frac{Fa\bar{y}}{Ib}$$

where A = shaded area to left of YY,
$\bar{y}$ = distance of centroid of this area from the neutral axis,
b = flange thickness.

Since A is the only variable, the graph of q against x is a straight line as shown dotted, Figure 60.

Compared to q_y, q_x is negligible.

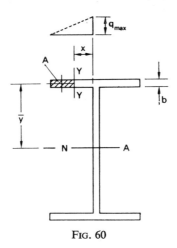

Fig. 60

SOLID CIRCULAR SECTION

Referring to Figure 61, it is evident that the width of the shaded area is not constant, so that $A\bar{y}$ must be expressed in terms of constants. It is shown in *Strength of Materials* (p. 389) that this expression is

$$A\bar{y} = \tfrac{2}{3}r^3 \cos^3 \varphi$$

so that
$$q = \frac{F A \bar{y}}{Ib}$$
(where $I = \frac{\pi}{4} r^4$ and $b = 2r \cos \varphi$)

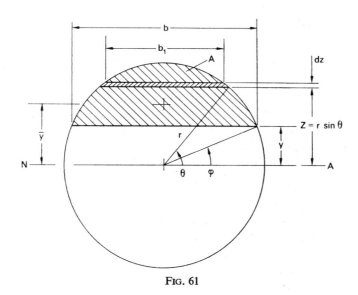

Fig. 61

$$\therefore \quad q = F\left(\frac{2}{3}r^3 \cos^3 \varphi\right) \frac{4}{4r^4} \times \frac{1}{2r \cos \varphi}$$

$$= \frac{4F \cos^2 \varphi}{3\pi r^2} \quad \text{and} \quad \cos^2 \varphi = 1 - \sin^2 \varphi = 1 - \frac{y^2}{r^2}$$

i.e.
$$q = \frac{4F}{3\pi r^2}\left(1 - \frac{y^2}{r^2}\right).$$

This is the equation of a parabola. At the neutral axis, i.e. when $y = 0$, we have

$$q_{max} = \frac{4}{3} \cdot \frac{F}{\pi r^2}.$$

The quantity $F/\pi r^2$ is the average shear stress.

Other cases may be treated in a similar manner.

EXAMPLES 2

1. A uniform beam 6 m long weighs 5000 kg, is hinged at the left-hand end and is maintained in a horizontal position by a vertical prop at a point 3·65 m from the hinge. Between hinge and prop there is a load of 2460 kg/m while at 1·22 m to the right of the prop is concentrated a load of 4000 kg. Draw to scale the graphs of F and M and find:

   (a) the value and position of the maximum bending moment between hinge and prop (24 200 Nm);
   (b) the position of the point of contraflexure (2·44 m from LH end).

2. A beam is simply supported over a span of 6 m and carries a uniform wall of mass 6550 kg/m from the left-hand support to the centre of the span. Draw the graph of F and find the point of zero shear. Use this figure to find the value of M_{max} (2·28 m from LH end, 170 000 Nm).

3. Steel strip 50 mm wide is to be used in the manufacture of a spring. When deflected, the strip is to form a circular arc of radius 1·37 m. Find the strip thickness at which the maximum tensile bending stress attains a value of 124 MPa and calculate the bending moment exerted by the spring at this stress value (1·63 mm, 27·7 Nm).

4. 5000 kg is to be distributed over a simple span of 6 m, the intensity increasing uniformly from zero at the left-hand end. Find the position and size of the maximum bending moment (39 800 Nm at 3·5 m from LH end).

5. Two circles 250 mm and 175 mm diameter with centres 25 mm apart represent a beam section. Calculate the value of I about the neutral axis normal to the line joining the centres (120×10^{-6} m⁴).

6. That part of a G-clamp parallel to the screw axis is of T-section 3·175 mm thick throughout. The flange, which is nearer the screw, is 15·9 mm wide and the total depth is 19 mm. Calculate the maximum values of tensile and compressive stress induced by a clamping load of 335 N given that the screw axis is 70 mm from the tip of the web (39 MPa and 68 MPa).

7. An I-section strut has a section of $66 \cdot 5 \times 10^{-4}$ m², the depth of which is 200 mm. Find the greatest and least stress when a vertical load of 20 000 kg is applied on the web at a distance of 62·5 mm from the neutral axis. Take $I = 46 \times 10^{-6}$ m⁴ (58 MPa and 1·85 MPa).

8. The maximum tensile stress in a tube 75 mm external diameter 12·5 mm thick is not to exceed the mean by more than 30 per cent. Estimate the permissible eccentricity of the load (2·75 mm).

9. No tension is to be caused in the base section of a pillar by wind action. The diameter is to be 1·83 m and the stone to be used has a density of 2240 kg/m³. Assume a wind pressure of 1920 Pa on the projected area and estimate the height to which it may be built with safety (7·5 m).

10. A horizontal cable is to be attached to the upper end of a wooden pole 150 mm diameter 3·65 m high, the lower end of which is embedded in the ground. If the expected tension in the cable is 1345 N, estimate:

    (a) the maximum bending stress (14 MPa);
    (b) the deflection at the top (120 mm).

    Assume a value of E for the timber of 6900 MPa.

11. A horizontal cantilever 0·3 m long, 76 mm wide, and 127 mm deep, of material for which $E = 208\,000$ MPa, carries a concentrated load at the free end. If a spirit level placed half-way along registers an incline of 1 in 2000, estimate the value of the load (38·9 kN).
12. A uniform beam of length L and depth d is to carry a total load of W uniformly distributed. If the maximum deflection is not to exceed $L/400$ when the maximum bending stress is 124 MPa, show that L must not exceed $20d$. The beam is simply supported and $E = 208\,000$ MPa.
13. The flexural rigidity, EI, of a uniform section beam is 770 000 Nm². If the beam is simply supported over a span of 2·44 m and loaded with 5000 kg at a point 0·9 m from the left-hand end, estimate the deflection at the centre of the span (18 mm approx.).
14. Equal end couples of 340 000 Nm are applied to an aluminium shaft 1·83 m long, 50 mm diameter. Assume $E = 69\,000$ PMa and find:

    (a) the slope at the ends (7·9 deg);
    (b) the deflection at the mid-point (63 mm).
15. A single leaf return spring is 2·5 mm thick and has an initial radius of 1·5 m. The span is 0·75 m while the width is 100 mm at the centre and decreases uniformly to zero at the ends. Determine the stress in the material when the spring is just flat (172 MPa).
16. A cantilever spring 610 mm long has 6 leaves 75 mm wide and 9·5 mm thick. Assume $E = 207\,000$ MPa and find what value of load will induce a maximum stress of 275 MPa (3100 N).
17. A 50 mm diameter shaft is simply supported over a span of 760 mm and loaded at the centre until the bending stress is 103 MPa. Calculate the value of the load and estimate the amount of strain energy stored. Assume $E = 207\,000$ MPa (7000 N, 3·32 J).
18. Steel ribbon 3·175 mm wide, 0·5 mm thick, is wound on a drum 500 mm diameter. If $E = 207\,000$ MPa, find the maximum stress in the steel and the strain energy stored per metre (207 MPa, 0·055 J).
19. Derive, in terms of stress and elastic modulus, an expression for the strain energy of a centrally loaded leaf spring having n leaves of width b and thickness t, the length of the main leaf being L ($bntLf^2/12E$).
20. A joist of timber 75 mm wide, 150 mm deep, is simply supported over a span of 6·1 m and loaded centrally with 890 N. Assume $E = 6900$ MPa and calculate:

    (a) the maximum bending stress (4·6 MPa);
    (b) the slope at the supports in degrees (0·76 deg);
    (c) the deflection at the load point (27 mm);
    (d) the maximum shear stress (57 kPa);
    (e) the bending and shear stresses at a point 50 mm above the neutral axis and 1·52 m from a support (1·53 MPa and 32 kPa).
21. An I-section joist is 12·5 mm thick throughout and has overall dimensions of 200 mm by 100 mm. Estimate the maximum shear force which may be applied, given that the longitudinal shear stress is not to exceed 77 MPa (176 kN).

3

COMPLEX STRESS

SHEAR STRESS RESULTING FROM A TENSILE LOAD

Referring to Figure 62, a tensile load induces a direct tensile stress on planes normal to the load axis, such as CB. This stress is given by

$$f = \frac{F}{A},$$

where A = section and F = force.

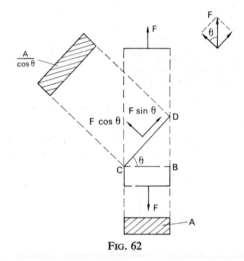

FIG. 62

When considering its effect on an inclined plane, such as CD, the load (force) may be resolved into two components, namely:

1. A normal component, $F \cos \theta$, acting over the section $A/\cos \theta$ and inducing a tensile stress given by

$$f_n = \frac{F \cos \theta}{A/\cos \theta} = \frac{F}{A} \cos^2 \theta.$$

This is a maximum when $\cos \theta = 1$, i.e. when $\theta = 0$. Then

$$f_{n_{\max}} = \frac{F}{A} = f.$$

2. A parallel (or tangential) component, $F \sin \theta$, acting over the same area $A/\cos \theta$ and inducing a shear stress given by

$$f_s = \frac{F \sin \theta}{A/\cos \theta} = \frac{F}{A} \sin \theta \cos \theta = \frac{F}{2A} \sin 2\theta.$$

This is a maximum when $\sin 2\theta = 1$, i.e. when $2\theta = 90$ deg or $\theta = 45$ deg. Then

$$f_{s_{max}} = \frac{F}{2A} = \frac{f}{2} \quad \text{numerically.}$$

Thus, on planes inclined at 45 deg to the axis of the applied load, the shear stress is half the direct tensile stress. Hence, under a tensile load, a material will fail in shear if the elastic limit in shear is less than half the elastic limit in tension. This also holds for a state of compression, cast iron failing in this way as shown in Figure 63.

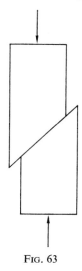

FIG. 63

COMPLEMENTARY SHEAR

Suppose a shear stress, f_{s_1}, to be induced on opposite faces of a rectangular element by forces F_1 applied as shown in Figure 64.

Then   Shear force $F_1 = f_{s_1}(yz)$
and   Clockwise shear couple $= F_1 x$
$$= f_{s_1}(yz)\,x.$$

If the equal anticlockwise couple required for equilibrium is obtained by introducing forces F_2 acting (as shown dotted) on the other two faces, and if f_{s_2} is the resulting shear stress, then

Anticlockwise shear couple $= F_2 y$
$$= f_{s_2}(xz)\,y.$$

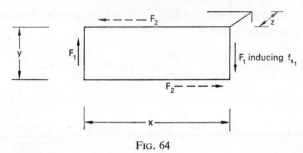

Fig. 64

Equating couples we obtain:
$$f_{s_2}(xz)\,y = f_{s_1}(yz)\,x$$
i.e.
$$f_{s_2} = f_{s_1}.$$

It follows that, if a component remains in equilibrium under a system of forces, two of which induce a shear stress, an equal shear stress (producing an equal and opposite shear couple) is automatically in existence on planes at 90 deg. This is given the name *Complementary Shear*.

It is evident from Figure 65 (which represents a cube) that, due to the two shear couples, the diagonal face AB is under compression. If the normal compressive stress on this face required for equilibrium of part ABD is f_N we have, resolving *forces* horizontally:
$$f_N(\sqrt{(2)}\,S\times S)\cos 45 = f_s \times S^2 \quad (S = \text{side})$$
i.e.
$$f_N\sqrt{(2)}S^2 \times \frac{1}{\sqrt{2}} = f_s S^2$$
whence
$$f_N = f_s \quad \text{numerically.}$$

On the other diagonal face, CD (Fig. 65c), there will be a tensile stress also of equal magnitude to the shear stress.

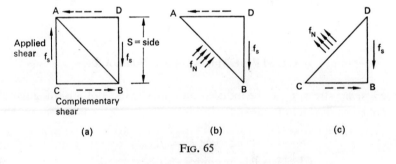

(a)   (b)   (c)

Fig. 65

VOLUMETRIC STRAIN

As a result of an elastic tensile strain on a linear dimension, L,

$$\begin{aligned}\text{New length} &= L + dL \quad \text{where} \quad dL = \text{increase}\\ &= L + eL \quad \text{where} \quad e = \text{strain}\\ &= L(1+e).\end{aligned}$$

Similarly, after linear compression,

$$\text{New length} = L(1-e).$$

Let Figure 66 represent a rectangular element having edges of length x, y and z, and suppose it to be immersed in liquid, i.e. to be subjected to a uniform compressive stress on all three faces. The forces acting on each face will cause a reduction in the length of each edge and hence in the volume, i.e. there will be a *volumetric strain*.

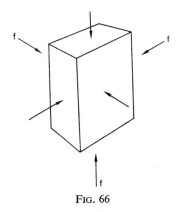

FIG. 66

Representing the strains by e_x, e_y and e_z we can write

$$\begin{aligned}\text{New volume} &= x(1-e_x)\,y(1-e_y)\,z(1-e_z) \\ &= xyz(1+e_xe_y+e_ye_z+e_ze_x-e_x-e_y-e_z-e_xe_ye_z)\end{aligned}$$

If the elastic limit is not exceeded, the strains are very small and their products may be neglected without serious error, so that we may write

$$\begin{aligned}\text{New volume} &= xyz+xyz(-e_x-e_y-e_z) \\ &= xyz-xyz(e_x+e_y+e_z)\end{aligned}$$

Since the original volume was xyz, we have

$$\text{Volume change} = -xyz(e_x+e_y+e_z)$$

$$\therefore \text{Volumetric strain} = \frac{\text{Change in volume}}{\text{Original volume}} = \underline{-(e_x+e_y+e_z)}.$$

Thus the volumetric strain, e_v, is the sum, for practical purposes, of the linear strains. The negative sign in the case considered indicates that the strain is compressive. It follows that, for an elemental cube,

$$x = y = z, \quad \text{so that} \quad e_x = e_y = e_z = e, \quad \text{say.}$$

Then
$$e_v = \underline{-3e = 3 \text{ (linear strain)}}.$$

If the fluid (compressive) stress acting on the element shown in Figure 66 and producing the volumetric strain is denoted by f, we have, within the elastic limit,

$$\frac{\text{Stress}}{\text{Strain}} = \frac{f}{e_v} = \text{a constant.}$$

This constant is called the *Bulk Modulus* and denoted by K.

POISSON'S RATIO

In addition to the change in length in the direction of a direct load, there is a simultaneous dimensional change in the two directions normal to such a load. Thus a tensile load produces a tensile (longitudinal) strain along its own axis and a compressive (lateral) strain along the other two axes. This lateral strain causes a reduction in each lateral dimension.

Within the elastic limit, the ratio of lateral to longitudinal strain is constant for a given material and known as *Poisson's Ratio*. It is denoted by σ, and its value for most metals lies between 0·25 and 0·33.

Thus, $\qquad$ Lateral strain $= \sigma \times$ Longitudinal strain $= \sigma(f/E)$.

Since lateral strains are of opposite sign to the longitudinal strain caused by the same load, it follows that, in the case of three mutually perpendicular loads, the resultant strain in the direction of any one of them is the algebraic sum of the relevant longitudinal and lateral strains.

EXAMPLE. An element is 50 mm wide, 12·5 mm thick and 100 mm long. Assume $E = 208\,000$ MPa and $\sigma = 0.25$ and estimate, for a longitudinal load of 67 250 N:

(a) longitudinal strain,
(b) lateral strain,
(c) volumetric strain,
(d) change in volume.

Solution.

Longitudinal stress, $\qquad f_y = \dfrac{67\,250}{0\cdot 050 \times 0\cdot 0125}\left(\dfrac{1}{10^6}\right) = 107\cdot 5$ MPa

$\therefore \qquad e_y = \dfrac{f_y}{E} = \dfrac{107\cdot 5}{208\,000} = 0\cdot 000517.$

Since $f_x = f_z = 0$,
$\therefore$
$\qquad e_x = e_z = -\sigma e_y$
$\qquad\qquad = -0\cdot 25 \times 0\cdot 000517$
$\qquad\qquad = -0\cdot 000129.$

Also $\qquad e_v = e_y + e_x + e_z$
$\qquad\qquad = e_y + (-2e_x)$
$\qquad\qquad = 0\cdot 000517 - (2 \times 0\cdot 000129)$
$\qquad\qquad = 0\cdot 00026.$

Original volume, $V = 50 \times 12\cdot5 \times 100 = 62\,500$ mm³

Volume change, $e_v V = 0\cdot00026 \times 62\,500$
$= 16\cdot25$ mm³ (increase).

EXAMPLE. If the element in the previous problem is subjected to an additional lateral stress of 62 MPa, the stress in the third direction (f_z) remaining zero, calculate the net strain in each direction when the second stress is (a) compressive (Fig. 67a), (b) tensile (Fig. 67b).

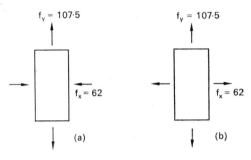

FIG. 67

Solution. Referring to Figure 67a, the horizontal compressive stress produces a reduction in width and hence an increase in length, i.e. the lateral strain due to it is added to the longitudinal strain due to f_y. Taking tensile strain as positive, we have:

Resultant vertical strain, $e_y = \dfrac{f_y}{E} + \dfrac{\sigma f_x}{E}$

$= \dfrac{1}{208\,000}[107\cdot5 + (0\cdot25 \times 62)]$

$= 0\cdot000591$ (tensile).

Referring to Figure 67b, the effect of f_x is now to reduce the strain due to f_y in a vertical direction.

Hence, in this case, $e_y = \dfrac{1}{208\,000}[107\cdot5 - (0\cdot25 \times 62)]$

$= 0\cdot000442$ (tensile).

The horizontal strain in the direction of f_x is given by

Case (a): $e_x = -\dfrac{f_x}{E} - \dfrac{\sigma f_y}{E}$ (since the effects of f_x and f_y are cumulative)

$= -\dfrac{1}{208\,000}[62 + (0\cdot25 \times 107\cdot5)]$

$= -0\cdot000427$ (compressive).

MECHANICAL TECHNOLOGY FOR HIGHER ENGINEERING TECHNICIANS

Case (b):
$$e_x = \frac{f_x}{E} - \frac{\sigma f_y}{E}$$

$$= \frac{1}{208\,000}[62 - (0{\cdot}25 \times 107{\cdot}5)]$$

$$= \underline{0{\cdot}000169 \text{ (tensile)}}.$$

The horizontal strain in the direction of f_z is given by

Case (a):
$$e_z = \frac{f_z}{E} - \frac{\sigma f_y}{E} + \frac{\sigma f_x}{E} \quad \text{(where } f_z = 0\text{)}$$

$$= \frac{1}{208\,000}[-(0{\cdot}25 \times 107{\cdot}5) + (0{\cdot}25 \times 62)]$$

$$= \underline{-0{\cdot}000055 \text{ (compressive)}}.$$

Case (b):
$$e_z = \frac{f_z}{E} - \frac{\sigma f_y}{E} - \frac{\sigma f_x}{E}$$

$$= \frac{1}{208\,000}[-(0{\cdot}25 \times 107{\cdot}5) - (0{\cdot}25 \times 62)]$$

$$= \underline{-0{\cdot}000204 \text{ (compressive)}}.$$

THE RELATION BETWEEN THE ELASTIC CONSTANTS

As already shown, a shear stress f_s applied to opposite faces of a cube will, together with its complementary shear, produce tensile and compressive stresses f_N on the diagonal faces such that $f_N = f_s$ numerically.
Referring to Figure 68, in which the complementary shear stress is shown dotted, we have

$$\text{Net strain in diagonal} = \frac{f_N}{E} + \frac{\sigma f_N}{E}$$

$$= \frac{f}{E}(1 + \sigma).$$

Since the deformation is very small within the elastic range, the increase in the length of the diagonal is approximately y, and y itself is $x/\sqrt{2}$ nearly. Hence we may also write, without serious error:

$$\text{Strain in diagonal} = \frac{\text{Increase}}{\text{Original length}} = \frac{x}{\sqrt{2}} \div L\sqrt{2}$$

$$= \frac{1}{2} \cdot \frac{x}{L} \quad \text{and} \quad \frac{x}{L} = \frac{f_s}{G} \quad \text{(where } G = \text{rigidity modulus)}$$

$$= \frac{1}{2} \cdot \frac{f_s}{G}$$

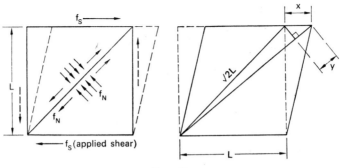

Fig. 68

$$= \frac{1}{2} \cdot \frac{f_N}{G} \quad \text{since} \quad f_s = f_N.$$

Equating the expressions for strain we have:

$$\frac{1}{2} \cdot \frac{f_N}{G} = \frac{f_N}{E}(1+\sigma)$$

i.e. $$E = 2G(1+\sigma).$$

Thus, for an elastic, homogeneous and isotropic material, the value of G depends upon that of E. In the above expression if E is taken as 206 000 MPa and σ as 0·287 (which are the actual values for steel) the value of G will emerge as 80 000 MPa.

If the cube shown in Figure 68 is subjected to a fluid stress f, then, taking compressive stress as positive:

Linear strain in each edge $= \dfrac{f}{E} - \dfrac{\sigma f}{E} - \dfrac{\sigma f}{E}$

$$= \frac{f}{E}(1-2\sigma).$$

But, Volumetric strain $= 3 \times$ linear strain

$$= 3\frac{f}{E}(1-2\sigma).$$

But, Volumetric strain $= \dfrac{f}{K}$

where $K =$ bulk modulus,

so that, equating: $\dfrac{f}{K} = 3\dfrac{f}{E}(1-2\sigma)$

i.e. $$E = 3K(1-2\sigma).$$

Using the same figures for E and σ gives, for steel, a value for K of 161 000 MPa.

By equating the expressions for E, it may be shown that

$$\sigma = \frac{3K-2G}{2G+6K} \quad \text{(see p. 228, Strength of Materials).}$$

If this is now substituted for σ in the first equation for E, it will emerge that

$$E = \frac{9GK}{G+3K}.$$

EXAMPLE. The gauge length and diameter of a tensile test piece were respectively 50 mm and 15 mm. A test load of 40 000 N resulted in measured changes of 0·0705 mm and 0·00465 mm respectively. Determine the values of σ and E and hence estimate the value of G.

Solution.

Lateral strain, $\quad \sigma e = \dfrac{0\cdot00465}{15} = 0\cdot00031$

Longitudinal strain, $\quad e = \dfrac{0\cdot0705}{50} = 0\cdot00141$

$\therefore \quad \sigma = \dfrac{0\cdot00031}{0\cdot00141} = 0\cdot22$

Area of section, $\quad A = \dfrac{\pi}{4}\left(\dfrac{15}{1000}\right)^2 = 177 \times 10^{-6} \text{ m}^2$

$\therefore$ Axial stress, $\quad f = \dfrac{40\,000}{177 \times 10^{-6}}\left(\dfrac{1}{10^6}\right) = 226 \text{ MPa}$

$\therefore$ Elastic modulus, $\quad E = \dfrac{f}{e} = \dfrac{226 \times 10^6}{0\cdot00141} = 160\,000 \text{ MPa}$

$\therefore$ Rigidity modulus, $\quad G = \dfrac{E}{2(1+\sigma)} = \dfrac{160\,000}{2 \times 1\cdot22} = 65\,600 \text{ MPa}.$

PRINCIPAL PLANES AND STRESSES

Referring to Figure 69, let two opposite faces of a block $ABCD$ of unit thickness be subjected to a shear stress, f_s, and let direct tensile stresses, f_x and f_y, be applied as shown, the direct stress in the third direction (f_z) being zero.

For equilibrium there must exist a complementary shear couple of opposite sense to the couple resulting from the applied shear. Also, on any inclined section such as EF there must act a normal stress f_N together with a tangential (shear) stress f_T so that the piece EFD, Figure 70, may itself be in equilibrium. The direction of f_T will depend on the relative magnitudes of f_x and f_y.

The force, shear or direct, acting on each surface will be the product of stress and surface area. Referring to Figure 70 and remembering that the element is of unit thickness we obtain:

COMPLEX STRESS

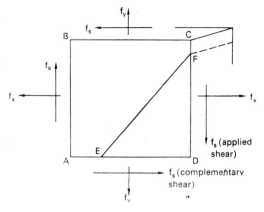

FIG. 69

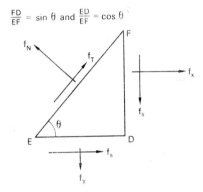

FIG. 70

(a) Resolving forces parallel to f_N:

$$f_N EF = f_x FD \sin\theta + f_s FD \cos\theta + f_s ED \sin\theta + f_y ED \cos\theta.$$

Dividing through by EF, putting $\dfrac{FD}{EF} = \sin\theta$, and $\dfrac{ED}{EF} = \cos\theta$ we obtain

$$f_N = f_x \sin^2\theta + f_y \cos^2\theta + 2f_s \sin\theta \cos\theta$$

whence, putting $\sin^2\theta = \dfrac{1-\cos 2\theta}{2}$, and $\cos^2\theta = \dfrac{1+\cos 2\theta}{2}$ we obtain, after simplifying,

$$f_N = \frac{(f_x+f_y)}{2} + \frac{(f_y-f_x)}{2}\cos 2\theta + f_s \sin 2\theta. \tag{1}$$

(b) Resolving again, perpendicular to f_N:

$$f_T EF = f_s FD \sin\theta - f_s ED \cos\theta - f_x FD \cos\theta + f_y ED \sin\theta.$$

Dividing through by EF and simplifying gives

$$f_T = \frac{(f_y-f_x)}{2}\sin 2\theta - f_s \cos 2\theta. \tag{2}$$

By differentiating eq. (1) and equating to zero we can obtain the values of θ which make f_N either a maximum or a minimum. Thus

$$\frac{(f_y-f_x)}{2}(-2\sin 2\theta)+f_s(2\cos 2\theta) = 0$$

$\therefore$ $\qquad 2f_s \cos 2\theta = (f_y-f_x)\sin 2\theta$

or $\qquad\qquad \tan 2\theta = \dfrac{2f_s}{f_y-f_x}.$ \qquad (3)

This gives two values of 2θ differing by 180 deg, and hence there are two values of θ differing by 90 deg.

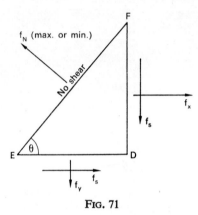

Fig. 71

Now, when the stress on the plane EF is wholly direct, i.e. when $f_T = 0$, we have, from eq. (2):

$$\frac{(f_y-f_x)}{2}\sin 2\theta = f_s \cos 2\theta$$

i.e. $\qquad\qquad \tan 2\theta = \dfrac{2f_s}{f_y-f_x}.$

This is eq. (3), i.e. it is the condition already obtained for making f_N a maximum or a minimum. Representing this condition by Figure 71 and resolving parallel to f_x we obtain

$$f_x FD + f_s ED = f_N \sin \theta EF$$

whence $\qquad\qquad f_s = (f_N - f_x)\tan \theta.$ \qquad (4)

Resolving again, normal to f_x we obtain

$$f_s FD + f_y ED = f_N EF \cos \theta$$

whence $\qquad\qquad f_s = \dfrac{f_N - f_y}{\tan \theta}.$ \qquad (5)

Multiplying together these two equations for f_s we obtain:

$$f_s^2 = (f_N - f_x)(f_N - f_y)$$
$$= f_N^2 - f_N f_y - f_N f_x + f_x f_y$$
$$= f_N^2 - f_N(f_x + f_y) + f_x f_y$$

or $\qquad f_N^2 - (f_x + f_y)f_N + (f_x f_y - f_s^2) = 0,$

COMPLEX STRESS

which is a quadratic. Hence, solving in the usual manner, we obtain,

$$f_N = \tfrac{1}{2}\{(f_x+f_y) \pm \sqrt{[(f_x-f_y)^2 + 4f_s^2]}\}$$

Taking the positive sign, the maximum normal stress will be given by

$$f_{N\,\text{max}} = \tfrac{1}{2}\{(f_x+f_y) + \sqrt{[(f_x-f_y)^2 + 4f_s^2]}\}$$

This will act over a plane making an angle, θ_1 say, with the applied stress f_x, the angle being the first of the two values given by eq. (3). Since the stress is of the same sign as f_x and f_y, it is tensile.

Taking the negative sign, the minimum normal stress will be given by

$$f_{N\,\text{min}} = \tfrac{1}{2}\{(f_x+f_y) - \sqrt{[(f_x-f_y)^2 + 4f_s^2]}\}$$

This will act over a plane making an angle $\theta_1 + 90$, say θ_2, with the applied stress f_x. This is positive (tensile) only if $f_x f_y > f_s^2$.

These two values of f_N are called the *Maximum and Minimum Principal Stresses*, and the planes over which they act (and on which there is no shear) are called the *Principal Planes*. This condition is illustrated in Figure 72.

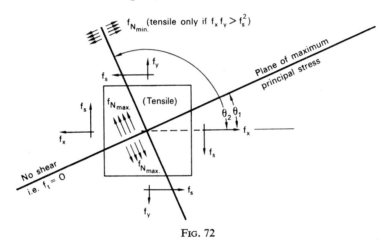

FIG. 72

If, say, f_y is compressive, we have, changing its sign:

$$f_{N\,\text{max}} = \tfrac{1}{2}\{(f_x-f_y) + \sqrt{[(f_x+f_y)^2 + 4f_s^2]}\}.$$

This is still positive (i.e. tensile) even though f_y may be greater numerically than f_x. The corresponding minimum principal stress will be

$$f_{N\,\text{min}} = \tfrac{1}{2}\{(f_x-f_y) - \sqrt{[(f_x+f_y)^2 + 4f_s^2]}\}.$$

This is evidently negative (compressive) always. Thus, when f_x and f_y are of opposite sign, the principal stresses are also. Note that $f_{N\,\text{max}}$ is not necessarily greater numerically than $f_{N\,\text{min}}$, i.e. it is possible for the greatest stress in the material to be compressive.

MAXIMUM SHEAR STRESS

The shear stress on any plane inclined at θ to f_x, Fig. 70, is given by

$$f_T = \frac{(f_y - f_x)}{2} \sin 2\theta - f_s \cos 2\theta \quad \text{(eq. 2)}$$

The values of θ which make this stress a maximum or minimum can be obtained by differentiating and equating to zero. Thus

$$\frac{(f_y - f_x)}{2}(2 \cos 2\theta) - f_s(-2 \sin 2\theta) = 0$$

$$2f_s \sin 2\theta = -(f_y - f_x) \cos 2\theta$$

i.e. $\quad \tan 2\theta = -\dfrac{f_y - f_x}{2f_s}$.

This is the negative reciprocal of eq. (3) so that the two values of θ (90 deg apart) which it gives are 45 deg in advance of θ_1 and θ_2 respectively. The planes of maximum shear are therefore midway between the principal planes as shown in Figure 73.

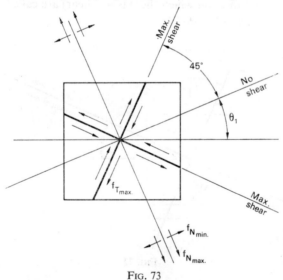

FIG. 73

Writing $\tan 2\theta \cos 2\theta$ instead of $\sin 2\theta$ in eq. (2) and putting $\tan 2\theta = -(f_y - f_x)/2f_s$ we obtain

$$f_{T\,\text{max}} = \frac{(f_y - f_x)}{2}\left[-\frac{(f_y - f_x)}{2f_s}\right] \cos 2\theta - f_s \cos 2\theta$$

from which, after some manipulation, it emerges that

$$f_{T\,\text{max}} = \pm \tfrac{1}{2}\sqrt{[(f_x - f_y)^2 + 4f_s^2]}.$$

Thus the maximum shear stress in the material is half the difference of the principal stresses, i.e.

$$f_{T\,\text{max}} = \tfrac{1}{2}[f_{N\,\text{max}} - f_{N\,\text{min}}].$$

Evidently the worst shear condition obtains when the principal stresses are of opposite sign since the maximum shear stress is then half the numerical sum. Note that the two values of $f_{T\,max}$ are equal and that each is the complementary shear of the other.

Since all components are three-dimensional, there are in fact three mutually perpendicular principal planes although the third principal stress is usually zero. If it is, and if the other two principal stresses are of the same sign, then since the maximum shear stress in the material is half the difference of the greatest and least principal stresses, we have:

either, $$f_{T\,max} = \frac{1}{2}[f_{N\,max} - 0] = \frac{f_{N\,max}}{2}$$

(if $f_{N\,max}$ is the greatest numerical value)

or, $$f_{T\,max} = \frac{1}{2}[f_{N\,min} - 0] = \frac{f_{N\,min}}{2}$$

(if $f_{N\,min}$ is the greatest numerical value).

EXAMPLE. Determine the values of the principal stresses in the element shown in Figure 74, and the angles made with the horizontal by the planes over which they act.

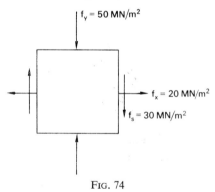

FIG. 74

Solution. The maximum principal stress is given by

$$f_{N\,max} = \tfrac{1}{2}[f_x + f_y + \sqrt{[(f_x - f_y)^2 + 4f_s^2]}]$$
$$= \tfrac{1}{2}\{20 + (-50) + \sqrt{[(20 - (-50))^2 + (4 \times 30^2)]}\}$$
$$= -15 + 46 \cdot 1 = +31 \cdot 1 \text{ MPa} \quad \text{(tensile)}.$$

Hence $f_{N\,min} = -15 - 46 \cdot 1 = -61 \cdot 1$ MPa (compressive).

Now $$\tan 2\theta = \frac{2f_s}{f_y - f_x} = \frac{60}{-50 - (+20)} = \frac{+6}{-7} = -0.857 \quad \text{(see Fig. 75b)}$$

so that $2\theta_1 = 180 - 40 \cdot 6 = 139 \cdot 4$ deg

i.e. $\theta_1 = 69 \cdot 7$ deg

and $\theta_2 = 69 \cdot 7 + 90 = 159 \cdot 7$ deg.

The relative positions of the principal planes are shown in Figure 75a.

MECHANICAL TECHNOLOGY FOR HIGHER ENGINEERING TECHNICIANS

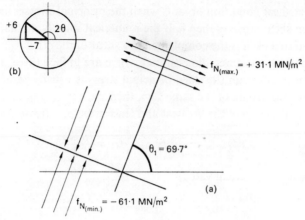

FIG. 75

Alternatively, from eq. (5):

$$\tan \theta_1 = \frac{f_{N\,max} - f_y}{f_s}$$

$$= \frac{31 \cdot 1 - (-50)}{30}$$

$$= 2 \cdot 7$$

so that $\quad 90 - \theta_1 = 20 \cdot 3 \text{ deg}$

i.e. $\quad \theta_1 = 69 \cdot 7 \text{ deg} \quad$ as before.

THE MÖHR CIRCLE

The construction of this (which enables graphical determination of principal stress, maximum shear stress and the stresses on any given section) is as follows, reference being made to Figure 76.

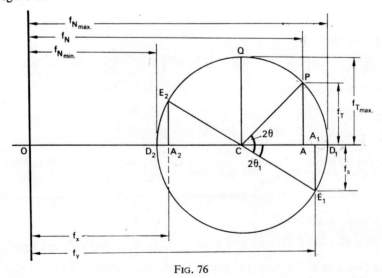

FIG. 76

COMPLEX STRESS

Assuming the applied stresses f_x and f_y to be positive, and assuming the stress in the third direction, f_z, to be zero,

Set off $\quad OA_1 = f_y$ to the right of O to some convenient scale

$OA_2 = f_x$ similarly

$A_1E_1 = f_s$ downwards from A_1 and normal to OA_1

$A_2E_2 = f_s$ upwards from A_2 and normal to OA_2

Bisect A_2A_1 at C and with C as centre draw a circle radius CE_1. Join E_2CE_1 and mark D_2 and D_1.

If θ is the inclination to f_x of the plane over which stress information is required,

Set off radius CP at 2θ to radius CE_1.
Project from P to A, and insert point Q. Then,

$OD_1 = f_{N\,max}$ (acting on a plane at θ_1 to f_x)

$OA = f_N$ (normal stress on plane at θ to f_x)

$OD_2 = f_{N\,min}$ (acting on a plane at (θ_1+90) to f_x)

$AP = f_T$ (shear stress on plane at θ to f_x)

$CQ = f_{T\,max}$ (acting on a plane at 45 deg to $f_{N\,max}$)

Angle $D_1CE_1 = 2\theta_1$.

If one of the applied stresses is negative (compressive) it must be set off to the left of point O, otherwise the construction is the same.

EXAMPLE. Determine graphically the values of the principal stresses in the element shown in Figure 77, and the angles made with the horizontal by the planes over which they act.

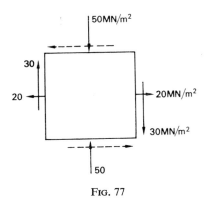

FIG. 77

Solution.

The principal planes and stresses are related as shown in Figure 79, which should be compared to Figure 75.

71

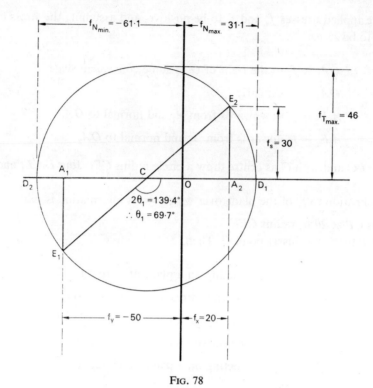

Fig. 78

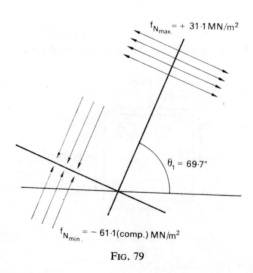

Fig. 79

EXAMPLE. Find, using the figures of the previous example, the effect of (a) reversing the direction of f_y, (b) reversing the directions of f_y and f_x.

Solution (a).

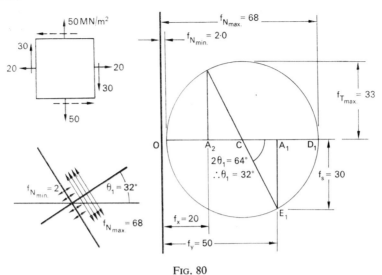

Fig. 80

Solution (b).

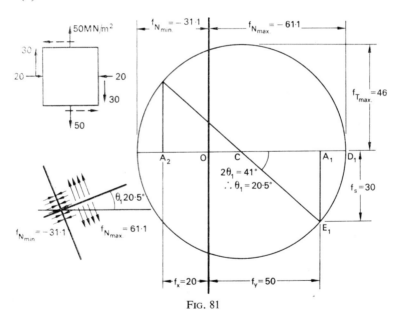

Fig. 81

SIMPLE THEORY OF ELASTIC FAILURE

In simple tension a material is said to have failed when a permanent strain occurs, i.e. when the elastic limit has been exceeded. In simple tension the elastic limit is usually thought of as a definite value of tensile stress, the working (operating) stress being kept to a lower value leaving a margin of safety suitable to the conditions.

In a complex stress system there exist other quantities, one of which, according to the theory adopted, may be the criterion of failure, i.e. may lead to permanent strain. The value of this quantity corresponding with the simple tensile elastic limit is then taken as the limiting value, with which the calculated actual value in a given case is then compared.

The two main hypotheses of failure will now be considered in relation to two-dimensional stress systems.

MAXIMUM PRINCIPAL STRESS THEORY (RANKINE)

This theory appears to hold good for brittle materials generally and, according to it, failure occurs when, irrespective of the values of the other principal stresses, the greatest principal stress (numerically) reaches the simple elastic limit stress,

i.e. when $f_{N\,\max} = f_e$ (where f_e = simple elastic limit), so that, for failure not to occur:

$$\tfrac{1}{2}\{(f_x-f_y)\pm\sqrt{[(f_x-f_y)^2+4f_s^2]}\} \not> f_e.$$

MAXIMUM SHEAR STRESS THEORY (COULOMB, TRESCA AND GUEST)

This theory gives a good approximation where ductile materials are concerned and, according to it, failure occurs when the maximum shear stress equals the shear stress value corresponding with the simple tensile elastic limit, i.e. when

$$f_{T\,\max} = \frac{f_e}{2}.$$

Since $f_{T\,\max}$ is half the difference between greatest and least principal stresses, it follows that, when these are of the same sign (the third principal stress being zero),

either $\qquad f_{T\,\max} = \pm\tfrac{1}{2}(f_{N\,\max}-0) \quad$ if $\quad f_{N\,\max} > f_{N\,\min} \quad$ numerically,

i.e. $\qquad \dfrac{f_{N\,\max}}{2} = \pm f_{T\,\max} = \pm\dfrac{f_e}{2} \quad$ at the elastic limit,

i.e. $\qquad \underline{f_{N\,\max} \not> \pm f_e} \quad$ for safety;

or $\qquad \underline{f_{N\,\min} \not> \pm f_e} \quad$ if $\quad f_{N\,\min} \quad$ is numerically the larger.

When the two principal stresses are unlike, i.e. are of opposite sign, then

$$f_{T\,\max} = \pm\tfrac{1}{2}(f_{N\,\max}-f_{N\,\min}).$$

Since the limiting value of $f_{T\,\max}$ is $f_e/2$, we have, for safety:

$$\underline{f_{N\,\max}-f_{N\,\min} \not> \pm f_e}$$

and, since the minor principal stress is negative,

$$\underline{f_{N\,\max}+f_{N\,\min} \not> \pm f_e}.$$

COMPLEX STRESS

COMBINED BENDING AND TORSION

Cases of pure torsion seldom occur in practice. In the case of a shaft transmitting torque, the weight of the shaft itself produces lateral bending (if horizontal) or axial compression (if vertical), these effects being increased by any component mounted on the shaft. Where the component is a pulley, bending effects are present due to belt tension, while in the case of a gear there is a component of the force between mating teeth which is normal to the shaft. As mentioned in Chapter 1, a helical spring is subjected to a combination of bending and torsion, however the load is applied.

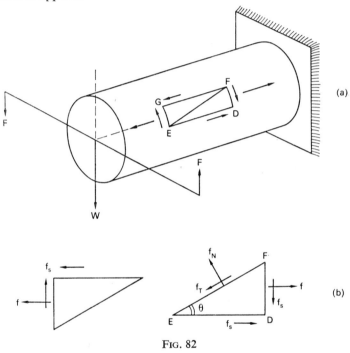

FIG. 82

Referring to Figure 82 and considering the greatly magnified element *EGFD* of a shaft subjected to simultaneous bending and twisting moments, it is evident that there is a tensile bending stress, f, acting normally to *EG* and *FD* and that this is accompanied by a shear stress, f_s, parallel to *EG* and *FD*. A complementary shear of opposite sense acts on *ED* and *FG*. It follows that, for part *EFD* to be in equilibrium, normal and tangential stresses, f_N and f_T respectively, must be present on face *EF*.

Comparing the system with that from which the principal stress formulae were derived, it is clear that those formulae can be used by putting $f_x = f$, and $f_y = 0$.

Thus, $\qquad f_{N\,max} = \tfrac{1}{2}[(f+\sqrt{(f^2+4f_s^2)})].$

and $\qquad f_{N\,min} = \tfrac{1}{2}[(f-\sqrt{(f^2+4f_s^2)})].$

If the bending and twisting moments present in a shaft are denoted by M and T respectively, then, if $d = 2r$ = diameter,

$$f = \frac{Mr}{I} \quad \text{where} \quad I = \frac{\pi}{64} \cdot d^4$$

and

$$f_s = \frac{Tr}{J} \quad \text{where} \quad J = \frac{\pi}{32} \cdot d^4.$$

Note that, since $\sqrt{(f^2+4f_s^2)}$ is always greater than f, it follows that $f_{N\,max}$ is tensile and $f_{N\,min}$ is compressive, i.e. the principal stresses are of opposite sign. Note also that the bending stress is compressive on the other side of the neutral axis. This alters the values of the principal stresses but not the signs.

The angles, θ_1 and θ_2, made with the bending stress by the principal planes can be obtained by substituting in eq. (3),

i.e.

$$\tan 2\theta = \frac{2f_s}{-f}.$$

The maximum shear stress is given by

$$f_{T\,max} = \pm \tfrac{1}{2}\sqrt{(f^2+4f_s^2)}.$$

EQUIVALENT TORQUE

Now, Bending stress, $\quad f = \dfrac{Mr}{I} = \dfrac{4Mr}{\pi r^4} \quad$ putting $\quad I = \dfrac{\pi r^4}{4}$

and $\quad$ Shear stress, $\quad f_s = \dfrac{Tr}{J} = \dfrac{2Tr}{\pi r^4} \quad$ putting $\quad J = \dfrac{\pi r^4}{2}.$

Putting these values in the equation for maximum principal stress and simplifying, we obtain

$$f_{N\,max} = \frac{2}{\pi r^3}[M+\sqrt{(M^2+T^2)}]$$

i.e.

$$\frac{\pi r^3}{2} f_{N\,max} = M+\sqrt{(M^2+T^2)}.$$

Now $\dfrac{\pi r^3}{2} \cdot f_s$ is the torque which can be transmitted by a solid shaft assuming pure torsion. Since the left-hand side of the above expression has the same form, it is denoted by T_e and referred to as the *Equivalent Twisting Moment*. We can therefore write down what is known as the *Rankine Formula* for brittle shafts, viz.

$$T_e = M+\sqrt{(M^2+T^2)}.$$

This expression applies to solid shafts only. For a hollow shaft, $J = \pi(R^4-r^4)/2$, and must be calculated separately. It can then be substituted in the equation

$$\frac{T_e}{J} = \frac{f_{N\,max}}{R}.$$

COMPLEX STRESS

From the foregoing an equivalent torque may be described as that combined bending and twisting moment which will induce a given maximum principal stress at radius r when used in the torsion equation, assuming the criterion of failure to be the maximum principal stress. As already pointed out, this assumption appears to hold good for materials which fail in tension—i.e. which are brittle rather than ductile.

In the case of ductile materials, which fail in shear, the criterion of failure is evidently the maximum shear stress. Putting in the same values as before for f and f_s in the equation for maximum shear stress and simplifying, we obtain

$$f_{T\,max} = \frac{2}{\pi r^3}\sqrt{(M^2+T^2)}$$

i.e.
$$\frac{\pi r^3}{2} f_{T\,max} = \sqrt{(M^2+T^2)} = \text{Equivalent torque,}$$

or
$$\underline{T_e = \sqrt{(M^2+T^2)}.}$$

This is known as the *Coulomb Formula* for ductile shafts, the equivalent torque in this case being that combined bending and twisting moment which will induce a given maximum shear stress at the shaft surface.

EXAMPLE. Find the principal stresses in a propeller shaft and the inclination to its axis of the planes over which they act, given that the thrust induces an axial compressive stress of 10 MPa, and that the motor torque induces a shear stress at the shaft surface of 40 MPa.

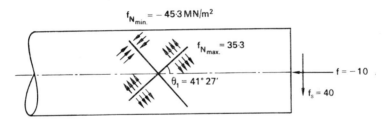

FIG. 83

Solution.

$$f_{N\,max} = \tfrac{1}{2}[f+\sqrt{(f^2+4f_s^2)}]$$
$$= \tfrac{1}{2}[-10+\sqrt{\{10^2+(4\times 40^2)\}}]$$
$$= \tfrac{1}{2}(-10+80\cdot 6)$$
$$= 35\cdot 3 \text{ MPa.}$$

Hence
$$f_{N\,min} = \tfrac{1}{2}(-10-80\cdot 6)$$
$$= -45\cdot 3 \text{ MPa}$$

$$\tan 2\theta = \frac{2f_s}{-f} = \frac{2\times 40}{-(-10)} = 8$$

∴
$$2\theta_1 = 82°\,54'$$
$$\theta_1 = 41°\,27'$$
$$\theta_2 = 41°\,27'+90° = 131°\,27'.$$

The stress system is as illustrated in Figure 83.

77

MECHANICAL TECHNOLOGY FOR HIGHER ENGINEERING TECHNICIANS

EXAMPLE. A shaft is to transmit a torque of 12 000 Nm while resisting a bending moment of 10 000 Nm. Find the shaft diameter which will limit the maximum shear stress to 50 MPa. Calculate the maximum tensile stress in use.

Solution. Since the criterion is maximum shear stress, the Coulomb Formula must be used, so that

$$T_e = \frac{\pi r^3}{2} f_{T\,max} = \sqrt{(M^2+T^2)}$$

$$= \sqrt{[(100\times10^6)+(144\times10^6)]}$$

$$= 15\,650 \text{ Nm} \quad \text{and} \quad f_{T\,max} = 50 \text{ MPa}$$

∴ $$r^3 = \frac{15\,650\times 2}{(50\times 10^6)\pi} = \frac{200}{10^6}$$

i.e. $$r = \frac{5\cdot 85}{10^2} = 0\cdot 0585 \text{ m}$$

so that $$d = 0\cdot 117 \text{ m} = \underline{117 \text{ mm}}.$$

Using the Rankine Formula to find the maximum principal stress:

$$T_e = \frac{\pi r^3}{2} f_{N\,max} = M+\sqrt{(M^2+T^2)}$$

$$= 10\,000+15\,650$$

$$= 25\,620 \text{ Nm}$$

and $$r = 0\cdot 0585 \text{ m}$$

∴ $$f_{N\,max} = \frac{25\,620\times 2}{\pi\times 0\cdot 0585^3}$$

$$= \underline{81\cdot 5 \text{ MPa}}.$$

EXAMPLE. 35 kW is to be transmitted at 2500 rev/min by a hollow shaft which must also withstand simultaneously a bending moment of 30 Nm. If the maximum principal stress is not to exceed 75 MPa, find suitable diameters given that the internal diameter is to be 0·65 of the external.

Solution. Power = Torque×Angular velocity

$$W = T\omega = \frac{2\pi NT}{60}$$

where N is the speed in rev/min.

Hence $$T = \frac{60W}{2\pi N} = \frac{60(35\times 1000)}{2\pi\times 2500} = 134 \text{ Nm}$$

$$T_e = M+\sqrt{(M^2+T^2)}$$
$$= 30+\sqrt{(30^2+134^2)}$$
$$= 167 \text{ Nm}$$

$$J = \frac{\pi}{2}(R^4-r^4)$$

where $\quad r = 0.65\, R.$

$\therefore \quad \tau = \dfrac{\pi}{2} R^4 (1 - 0.65^4)$

$\quad = 1.29\, R^4.$

But $\quad \dfrac{T_e}{J} = \dfrac{f_{N\,\text{max}}}{R}$

where $\quad f_{N\,\text{max}} \triangleright 75 \times 10^6 \text{ Pa}$

$\therefore \quad \dfrac{167}{1.29 R^4} = \dfrac{75 \times 10^6}{R}$

whence $\quad R^3 = \dfrac{167}{1.29(75 \times 10^6)} = \dfrac{1.73}{10^6}$

i.e. $\quad R = \dfrac{1.2}{10^2} = 0.012 \text{ m}$

and $\quad D = 0.024 \text{ m (24 mm)}$

$\therefore \quad d = 0.0156 \text{ m (15.6 mm)}.$

EXAMPLES 3

1. Calculate the reduction in volume of a solid metal cylinder 64 mm diameter 255 mm long when submerged in the sea to a depth of 11 000 m. Assume the density of sea water to remain constant at 998 kg/m³ and assume a figure of 160 000 MPa for the bulk modulus (0.54×10^{-6} m³).
2. The three mutually perpendicular stresses acting on the faces of a rectangular element are:

$\quad f_y = 77.25$ MPa, vertical and tensile;
$\quad f_x = 30.9$ MPa, horizontal and compressive;
$\quad f_z = 61.8$ MPa, horizontal and tensile.

Calculate the strain in each of the three directions assuming $\sigma = 0.286$ and $E = 201\,000$ MPa ($e_y = 0.00034$, $e_x = -0.000352$, $e_z = 0.000242$).
3. The shank of a punch is 25 mm diameter and, during the punching operation, sustains an axial compressive stress of 155 MPa. If the socket restricts the lateral (i.e. radial) strain to one-third of the value when unconstrained, find:

(a) the radial stress imposed by the socket (34·3 MPa);
(b) the actual radial strain (−0·00011);
(c) the increase in the shank diameter (0·0028 mm).

4. On a 50 mm gauge length of a test piece 12·5 mm diameter, a load of 20 000 N produced an extension of 0·13 mm. The twist on a 200 mm length of similar material produced by a torque of 54 Nm, was 3·2 deg. Estimate from these figures the approximate value of Poisson's Ratio (0·3).
5. The strains registered by two gauges mounted at 90 deg on the surface of a brass pressure vessel 3·175 mm thick were respectively 0·00071 and 0·000167. Estimate the reduction in thickness assuming $E = 83\,000$ MPa and taking $\sigma = 0.3$ (0·0012 mm).
6. Calculate the increase in area of a piece of steel plate 0·3 m square when subjected to tensile stresses of 124 MPa and 77 MPa at 90 deg. Assume $\sigma = 0.3$, $E = 204\,000$ MPa (42×10^{-6} m²).
7. Find a suitable diameter for a solid shaft which is required to transmit 6·75 kW at 1300 rev/min, the permitted maximum shear stress being 41·5 MPa. Assume the presence of a bending moment equal in magnitude to the torque (20 mm).
8. The crankshaft of a marine diesel motor is to transmit a torque of 121 000 Nm while sustaining a bending moment of 85 000 Nm. If the permitted tensile and shear stresses are respectively 93 MPa and 54 MPa, find the minimum diameter necessary:

(a) by Rankine Formula (234 mm);
(b) by Coulomb Formula (241) mm.

9. Find the values of the principal and maximum shear stresses and the positions of the planes over which they act, given that

   $f_x = 80$ MPa tensile and horizontal;
   $f_z = 30$ MPa tensile and vertical;
   $f = 0$;
   $f_s = 30$ MPa clockwise in the plane of f_x and f_y.

   Check the values obtained by using the Möhr Circle, and insert them on a suitable sketch (94, 16, 39 MPa, $\theta_1 = 25\cdot 1$ deg).

10. The bending moment in a 50 mm diameter solid shaft rotating at 300 rev/min is numerically equal to 40 per cent of the torque. If the maximum shear stress in the material is 77·4 MPa, calculate the power being transmitted (560 kW).

11. The flywheel of a press has a mass of 2000 kg and radius of gyration of 0·762 m and is mounted on a shaft 152 mm diameter which overhangs the bearing by 0·38 m. If, as a result of fluctuation in torque, the flywheel has a maximum instantaneous acceleration of 4·8 rad/s², find the maximum shear stress induced in the shaft (13·75 MPa).

12. The maximum shear stress induced in a 50 mm diameter shaft by a pure torque is 77·5 MPa. Find the new value of maximum shear stress when an axial tension of 200 000 N is superimposed (92 MPa).

4

STRUTS AND CYLINDERS

CRITICAL VALUE OF AXIAL LOAD

When the length of a strut is great in relation to its sectional dimensions, failure will not occur in compression but in bending. The reason for this is that, in practice, no material is truly homogeneous, no strut is truly straight and no load is truly axial. It follows that a bending moment is always present and that this increases with the load.

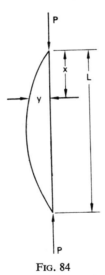

FIG. 84

Such bending under axial load is called *buckling* and the load which produces it is referred to as the *Buckling, Crippling* or *Critical Load*. When such failure occurs, the strut remains in equilibrium in the bent position as shown in Figure 84. Referring to this figure, assuming that the ends are pin-jointed (i.e. free to change their slope) and that the direct compressive stress is negligible, we have:

$$M = EI\frac{d^2y}{dx^2} = -Py$$

where P = load producing deflection y,

or

$$EI\frac{d^2y}{dx^2} + Py = 0$$

or
$$\frac{d^2y}{dx^2} + \frac{P}{EI}y = 0$$

i.e.
$$\frac{d^2y}{dx^2} + k^2y = 0 \text{ (putting } P = k^2EI\text{).}$$

Solving this equation gives

$$\sin kL = 0, \quad \text{i.e.} \quad kL = \pi \text{ (taking least value).}$$

Hence $\quad k = \dfrac{\pi}{L}, \quad$ i.e. $\quad P = \dfrac{\pi^2}{L^2}EI.$

This is known as the *Euler Critical Load*,* the safe load being this value divided by a suitable safety factor. Note that, if the section is not symmetrical, the value of I is the least possible.

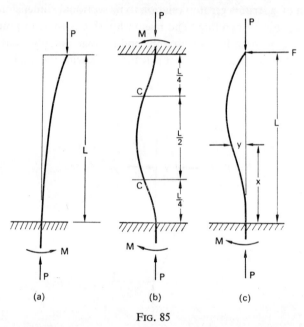

FIG. 85

If a change in slope at one end is prevented, i.e. the end is "direction-fixed" as shown in Figure 85a, then, if there is no lateral restraint at the other end, a bending moment, M, must be introduced to maintain equilibrium. The strut is now equivalent to half a strut of length $2L$ loaded as in Figure 84. Hence

$$\text{Critical load} = \frac{\pi^2}{(2L)^2}EI, \quad \text{i.e.} \quad P = \frac{1}{4} \cdot \frac{\pi^2}{L^2}EI.$$

If both ends are direction-fixed as shown in Figure 85b, there will be two points of contraflexure, CC, distant $L/4$ from each end. The piece of strut between them, of length $L/2$, is identical in shape to the pin-jointed strut of Figure 84 since the bending moment at points

* Leonhard Euler, Swiss mathematician, 1707–83.

C is zero, Hence

$$\text{Critical load} = \frac{\pi^2}{(L/2)^2} EI, \quad \text{i.e.} \quad P = 4\frac{\pi^2}{L^2} EI.$$

If the free end of the strut in Figure 85a is prevented from moving laterally by a horizontal force, F, as shown in Figure 85c, then, for any point distant x from the fixed end:

$$M_x = EI\frac{d^2y}{dx^2} = -Py + F(L-x)$$

or

$$\frac{d^2y}{dx^2} + k^2y = \frac{F}{EI}(L-x) \text{ (putting } P = k^2EI \text{ as before)}.$$

Solving this equation gives

$$kL = \tan^{-1} kL \quad \text{(see \emph{Strength of Materials}, p. 398)}$$

whence $kL = 4\cdot49$ radians

$$\therefore \quad k = \frac{4\cdot49}{L}, \quad \text{i.e.} \quad P = \frac{20\cdot2}{L^2} EI.$$

Since $2\pi^2 = 19\cdot75$, it may be assumed that the critical load is given approximately by

$$P = 2\frac{\pi^2}{L^2} EI.$$

Thus the axial crippling load (Euler value) is $(\pi^2/L^2)EI$ multiplied by a constant, the value of which depends on the end conditions of the strut.

The results are summarised in Figure 86, the crippling loads being expressed as a multiple of that for the first (simplest) case.

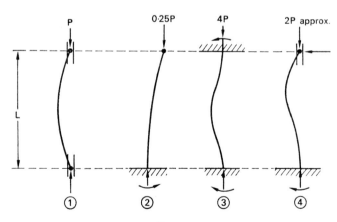

FIG. 86

In practice it is impossible to prevent *all* change in the slope at "direction-fixed" ends, so that the coefficients must be used with some reserve.

Now, if the area of section of the strut is denoted by A, and if the least radius of gyration of this area is denoted by k (not to be confused with the k in the differential equation), we can substitute Ak^2 for I in the basic Euler Formula, so obtaining

$$P = \frac{\pi^2}{L^2} EAk^2, \quad \text{i.e.} \quad \frac{P}{A} = \frac{\pi^2 E}{(L/K)^2} = \text{compressive stress at failure.}$$

Alternatively, $\quad \dfrac{L}{k} = \sqrt{\dfrac{\pi^2 E}{P/A}}.$

The quantity L/K is known as the *Slenderness Ratio* and, if the compressive stress at failure, P/A, is plotted against it for a given material and end conditions, a curve having the form of Figure 87 will be obtained. From this can be deduced the value of L/k corresponding to any desired compressive stress at buckling up to, say, the stress at the elastic limit (in the case of steel) or, say, the 0·2 per cent proof stress in other cases.

Assuming mild steel and pin-jointed ends,

$$\pi^2 E = 9{\cdot}87 \times 206\,000 \times 10^6 = 203 \times 10^{10}.$$

The following table has been compiled using this figure and the curve of Figure 87 plotted from it.

$\dfrac{P}{A}$ (N/m²)	$\dfrac{203 \times 10^{10}}{P/A}$	$\sqrt{\dfrac{203 \times 10^{10}}{P/A}} = \dfrac{L}{k}$
30×10⁶	6·77×10⁴	260
50×10⁶	4·06×10⁴	201
100×10⁶	2·03×10⁴	141
200×10⁶	1·01×10⁴	100
310×10⁶	0·65×10⁴	81
400×10⁶	0·51×10⁴	71
500×10⁶	0·41×10⁴	64

From this curve it is evident that, according to the Euler Theory, the limiting stress of 310 MPa (in the case of Mild Steel) corresponds to a slenderness ratio of about 81. In other words, the strut material will fail in compression before buckling occurs if $L < 81k$.

For a solid circular section of diameter d,

$$k^2 = \frac{d^2}{16} \quad \text{or} \quad k = \frac{d}{4}$$

so that, for critical conditions, $L = 20d$ approx.

From this curve it is also evident that the ratio L/k must not be less than about 120 if the direct stress (which the Euler Theory neglects) is not to exceed about 40 per cent of the elastic limit. This gives a minimum value (for circular sections of mild steel) of about 30 to the ratio L/d.

Limiting values for other sections and elastic materials may be determined in a similar way.

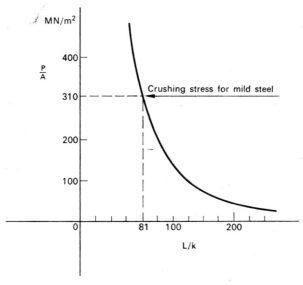

Fig. 87

EXAMPLE. Estimate the value of the Euler Critical Load for a steel strip $1800 \times 25 \times 5$ mm when loaded axially. Assume guided and pin-jointed ends and take $E = 206\,000$ MPa. Find also the direct stress at this load. Determine the greatest deflection possible at the centre assuming an elastic limit of 310 MPa.

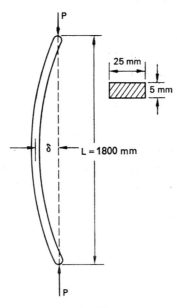

Fig. 88

Least value of $I = \dfrac{25}{1000} \left(\dfrac{5}{1000}\right)^3 \dfrac{1}{12} = 260 \times 10^{-12}$ m^4

Area of section, $A = 0.025 \times 0.005 = 125 \times 10^{-6}$ m^2

MECHANICAL TECHNOLOGY FOR HIGHER ENGINEERING TECHNICIANS

Euler critical load, $P = \dfrac{\pi^2}{L^2} EI = \dfrac{\pi^2}{1\cdot 8^2}(206\,000 \times 10^6)\dfrac{260}{10^{-12}} = 164$ N

Corresponding direct stress, $\dfrac{P}{A} = \dfrac{164}{125 \times 10^{-6}}\left(\dfrac{1}{10^6}\right) = 1\cdot 32$ MPa

$\Bigg($This low value is explained by the high value of the slenderness ratio. Thus

$$k^2 = \dfrac{I}{A} = \left(\dfrac{bd^3}{12}\right)\dfrac{1}{bd} = \dfrac{d^2}{12} = \dfrac{5^2}{12} = 2\cdot 082 \ \text{giving} \ \dfrac{L}{k} = 1240.\Bigg)$$

Section modulus, $Z = \dfrac{I}{y} = \dfrac{260}{10^{12}}\left(\dfrac{1000}{2\cdot 5}\right) = 104 \times 10^{-9}$ m³

Bending moment, $M = P\delta = 164\delta$, where δ = deflection

Permissible stress $= \dfrac{P}{A} + \dfrac{M}{Z}$

∴ $\quad 310 \times 10^6 = (1\cdot 32 \times 10^6) + 164\delta \left(\dfrac{10^9}{104}\right)$

so that $\quad\quad\quad\quad \delta = \dfrac{308\cdot 7 \times 10^6}{164}\left(\dfrac{104}{10^9}\right) = \underline{0\cdot 196 \text{ m (196 mm)}}.$

RANKINE–GORDON FORMULA

This is an empirical formula, i.e. based on experimental results, and is used where both direct and bending stresses are significant, i.e. where $L < 30d$ for a solid circular section axially loaded. The crippling load is given by

$$P = \dfrac{f_c A}{1 + a(L/K)^2}$$

where A, L and k have their previous significance, f_c is the stress in the material at failure (yield stress for mild steel) and a is a constant dependent on material and end conditions.

For mild steel, the values of the constant a for the four cases considered and represented in Figure 86 are:

(1) $\quad \dfrac{1}{7500}$

(2) $\quad 4\left(\dfrac{1}{7500}\right)$

(3) $\quad \dfrac{1}{4}\left(\dfrac{1}{7500}\right)$

(4) $\quad \dfrac{4}{9}\left(\dfrac{1}{7500}\right)$

(Note that case (3), which is the stiffest, has the least value of *a* and so gives the greatest value of *P*.)

The "safe" load is the crippling load divided by a factor of "ignorance" suitable to the circumstances.

EXAMPLE. Thrust is transmitted to an aircraft control via a tube 1 m long, 25 mm external diameter and 2 mm thick. Find the slenderness ratio. Assume the ends to be pin-jointed and guided and use the Rankine Formula to determine the failing load taking $a = 1/3000$ and $f_c = 300$ MPa.

Solution.

$$I = \frac{\pi}{64}(0.025^4 - 0.021^4) = 9.66 \times 10^{-9} \text{ m}^4$$

$$A = \frac{\pi}{4}(0.025^2 - 0.021^2) = 144 \times 10^{-6} \text{ m}^2$$

$$k^2 = \frac{I}{A} = \frac{9.66 \times 10^6}{144 \times 10^9} = 67.2 \times 10^{-6}$$

$L = 1.0$ so that $\dfrac{L^2}{k^2} = \dfrac{1^2 \times 10^6}{67.2} = 15\,600$ and $\dfrac{L}{k} = 122$

$$P = \frac{f_c A}{1 + a(L/k)^2} = \frac{(300 \times 10^6)(144 \times 10^{-6})}{1 + (15\,600/3000)}$$

$$= 6970 \text{ N}.$$

THICK CYLINDER UNDER PRESSURE

Let Figure 89a represent the section of a cylinder having a thickness which is of the same order as its internal diameter and suppose both surfaces to be subjected to pressure, i.e. to be under compression. An element at any radius r will be subject to a radial stress, f_y, together with a so-called "hoop" stress, f_x, while, due to the internal pressure on the ends of the cylinder, there will also be present an axial (tensile) stress, f_z.

Referring to Figure 89b, the cylindrical element is in equilibrium so that

Downward force due to stress on section *AB*

= Upward force tending to fracture cylinder across *AB*,

i.e. $\quad 2f_x \, dr = f_y(2r) - (f_y + df_y) \, 2(r + dr)$

Multiplying out and neglecting the product of small quantities we obtain

$$2f_x \, dr = -2r \cdot df_y - 2f_y \cdot dr$$

i.e. $\quad f_x = -r \cdot \dfrac{df_y}{dr} - f_y \quad$ (dividing through by dr)

or $\quad r \cdot \dfrac{df_y}{dr} = -f_x - f_y.$ \hfill (1)

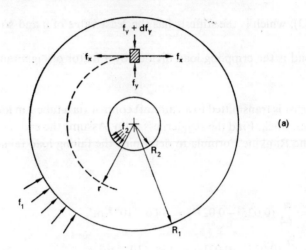

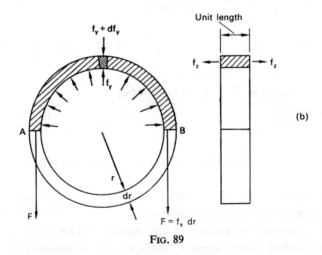

Fig. 89

Assuming transverse sections to remain plane, we also have:

Axial strain,
$$e_z = \frac{f_z}{E} - \frac{\sigma f_x}{E} + \frac{\sigma f_y}{E}$$

i.e.
$$e_z = \frac{1}{E}[f_z - \sigma(f_x - f_y)]. \qquad (2)$$

Since E, σ and f_z are constant, the quantity $(f_x - f_y)$ in eq. (2) must be constant.

Let $\qquad f_x - f_y = 2a \quad$ so that $\quad f_x = f_y + 2a.$

Substitution of this value for f_x in eq.(1) gives

$$r\frac{df_y}{dr} = -(f_y + 2a) - f_y$$
$$= -2(f_y + a)$$

so that,
$$\frac{1}{f_y+a}\cdot df_y = -2\cdot\frac{1}{r}\cdot dr.$$

Integrating both sides of this equation we obtain
$$\int \frac{1}{f_y+a}\, df_y = -2 \int \frac{1}{r}\, dr,$$

i.e.
$$\log_e (f_y+a) = -2 \log_e r + (\text{a constant}).$$

Since $2\log_e r = \log_e r^2$ we can write

$$\log_e (f_y+a) + \log_e r^2 = \text{a constant}$$

or
$$\log_e r^2(f_y+a) = \text{a constant}$$

$\therefore$
$$r^2(f_y+a) = \text{a constant}.$$

Let
$$r^2(f_y+a) = b.$$

Then
$$f_y = \frac{b}{r^2} - a. \tag{3}$$

But
$$f_x = f_y + 2a,$$

so that
$$f_x = \frac{b}{r^2} + a. \tag{4}$$

Equations (3) and (4) are known as *Lamé's Formulae* and give the radial and hoop stresses respectively at any radius r between R_2 and R_1.

At the inner surface where $r = R_2$,

$$f_y = f_2 = \text{applied internal pressure}$$

so that
$$f_2 = \frac{b}{R_2^2} - a.$$

At the outer surface where $r = R_1$,

$$f_y = f_1 = \text{applied external pressure}$$

so that
$$f_1 = \frac{b}{R_1^2} - a.$$

Thus if the pressures and radii are known, the values of a and b can be found and used to determine the radial and hoop stresses at any radius r.

In practice, the external pressure (f_1) is usually zero in which case the distribution of stress is as shown in Figure 90. Note that the difference in height of the two curves is $2a$.

EXAMPLE The delivery pipe to a diesel motor injector is 6·0 mm external diameter and has a bore of 1·4 mm. Determine the maximum hoop stress in the material if the orifice opens at a pressure of 14 MPa. Neglect atmospheric pressure.

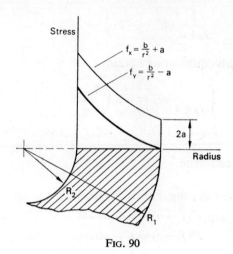

Fig. 90

Solution.

$$R_1 = 0·0030 \text{ m}$$
$$R_2 = 0·0007 \text{ m}$$

Radial stress, $$f_y = \frac{b}{r^2} - a$$

∴ at R_2, $$14 \times 10^6 = \frac{b}{0·0007^2} - a \qquad (1)$$

and at R_1, $$0 = \frac{b}{0·0030^2} - a$$

i.e. $$a = \frac{b}{0·003^2}. \qquad (2)$$

Substitution of this value of a in eq.(1) gives

$$14 \times 10^6 = \frac{b}{0·0007^2} - \frac{b}{0·003^2}$$

$$= b \left(\frac{10^8}{7^2} - \frac{10^8}{30^2} \right)$$

$$= 10^6 b (2·04 - 0·111)$$

so that $$b = \frac{14}{1·929} = 7·25.$$

Substitution of this value in eq. (2) gives:

$$a = \frac{7·25 \times 10^6}{3^2}$$

$$= 806\,000.$$

The hoop stress is a maximum at R_2 and is given by

$$f_{x\ max} = \frac{b}{R_2^2} + a$$

$$= \frac{7\cdot 25}{0\cdot 0072} + 806\,000$$

$$= \underline{15\cdot 7 \text{ MPa}}.$$

EXAMPLE. A steel tube 100 mm inside diameter, 150 mm outside diameter is to have a second tube of the same material and 180 mm outside diameter shrunk on, the shrinkage allowance being such as to induce between the tubes a radial pressure of 30 MPa. Assume $E = 206\,000$ MPa and $\sigma = 0\cdot 287$ and calculate:

(a) hoop stress at inner surface of outer tube,
(b) increase in internal diameter of outer tube,
(c) hoop stress at outer surface of inner tube,
(d) reduction in external diameter of inner tube,
(e) required shrinkage allowance on diameter.

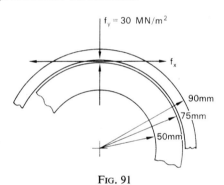

FIG. 91

Solution.

$f_y = 0$ at $r = 0\cdot 090$

so that: $0 = \dfrac{b}{0\cdot 092} - a$

whence $a = 123\cdot 5b$

and $f_y = 30 \times 10^6$ at $r = 0\cdot 075$

so that: $30 \times 10^6 = \dfrac{b}{0\cdot 075^2} - 123\cdot 5b$

whence $b = 55 \times 10^4$

so that $a = 68 \times 10^6$.

Also, at $r = 0\cdot 075$, $f_x = \dfrac{55 \times 10^4}{0\cdot 075^2} + (68 \times 10^6)$

$= \underline{166 \text{ MPa}}.$

Circumferential (hoop) strain, $e_x = \dfrac{f_x}{E} + \dfrac{\sigma f_y}{E} = \dfrac{1}{E}(f_x + \sigma f_y)$

Increase in circumference $= e_x(\pi d)$

where $d = $ diameter $= 0{\cdot}15$ m

∴ Increase in diameter $= e_x d$

$$= \dfrac{1}{206\,000 \times 10^4}[(166 \times 10^6) + 0{\cdot}0287(30 \times 10^6)]0{\cdot}15$$

$= 0{\cdot}000127$ m $\;(0{\cdot}127$ mm$)$.

For the inner tube:

$f_y = 0$ at $r = 0{\cdot}050$

so that: $0 = \dfrac{b}{0{\cdot}05^2} - a$

whence $a = 400b$

and $f_y = 30 \times 10^6$ at $r = 0{\cdot}075$

so that $30 \times 10^6 = \dfrac{b}{0{\cdot}075^2} - 400b$

whence $b = -13{\cdot}5 \times 10^4$

so that $a = -54 \times 10^6$.

Also, at $r = 0{\cdot}075$, $f_x = -\dfrac{13{\cdot}5 \times 10^4}{0{\cdot}075^2} + (-54 \times 10^6)$

$= -78$ MPa $\;($compressive$)$.

∴ Reduction in diameter $= \dfrac{1}{206\,000 \times 10^6}[(78 \times 10^6) - 0{\cdot}287(30 \times 10^6)]0{\cdot}15$

$= 0{\cdot}0000505$ m $\;(0{\cdot}05$ mm$)$

Required shrinkage allowance $= 0{\cdot}000127 + 0{\cdot}00005$

$= 0{\cdot}00018$ m

$= \underline{0{\cdot}18\text{ mm}}.$

SOLID SHAFT UNDER EXTERNAL PRESSURE

Referring to Figure 91a, external pressure (which is equivalent to compressive stress) may result from the shrinking on of a component, or from a force fit.

If the shaft is regarded as a tube of zero internal diameter, the equilibrium equation will be identical to that obtained for a thick cylinder, viz.

$$f_x + f_y + r\dfrac{d(f_y)}{dr} \qquad\qquad \text{(eq. 1)}$$

Assuming the sections to remain plane and hence the strain in a longitudinal direction to

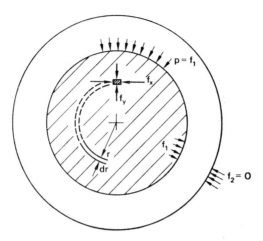

Fig. 91a.

remain constant, then because

$$e_z = \frac{f_z}{E} - \frac{\sigma f_x}{E} + \frac{\sigma f_y}{E} \qquad (\text{eq.2})$$

therefore $f_x - f_y =$ constant, and, as before

$$f_x = \frac{b}{r^2} + a, \quad f_y = \frac{b}{r^2} - a.$$

Since neither stress can be infinite when $r = 0$ (on the axis), it follows that $b = 0$.

Hence $\qquad f_x = a = -f_y \quad$ at all radii.

Thus, in the shaft, the radial and hoop stresses are equal numerically, are equal to the pressure on the shaft, and are uniform, i.e. are the same at all radii. The shrunk-on tube is merely a thick cylinder under internal pressure.

THIN CYLINDERS SUBJECT TO INTERNAL PRESSURE

When the thickness is small relative to the diameter, the hoop stress may be considered uniform and the radial stress negligible. Referring to Figure 92 and considering an element of length L which subtends an angle $d\theta$ at the axis:

Area of element $\qquad = r.d\theta \times L$

Radial force on it $\qquad = Lr.d\theta \times p$

Horizontal component $\qquad = pLr.d\theta \times \sin\theta$

$\qquad\qquad\qquad\qquad\qquad = pLr \sin\theta.d\theta$

∴ Total force on half cylinder to the right of section YY,

$$F = pLr \int_0^\pi \sin\theta \, d\theta$$
$$= 2pLr.$$

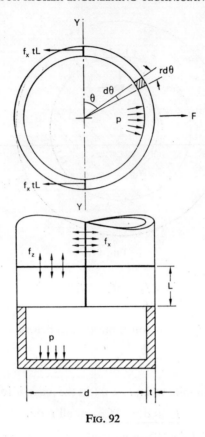

Fig. 92

This is resisted by the hoop stress acting over the section tL on each side, so that

$$2(f_x tL) = 2pLr$$

i.e.
$$f_x = \frac{pr}{t} = \frac{pd}{2t}. \tag{1}$$

Total force on an end $= p\left(\dfrac{\pi}{4} d^2\right)$ and this is resisted by the longitudinal stress, f_z, acting over the section $(\pi d)t$, so that

$$f_z(\pi d . t) = p\left(\frac{\pi}{4} d^2\right), \quad \text{i.e.} \quad f_z = \frac{pd}{4t}. \tag{2}$$

Thus $f_z = \dfrac{f_x}{2}$, which indicates that the cylinder will fail under the hoop stress, i.e. will split along its length.

THIN CYLINDER SUBJECT TO ROTATION

In this case the hoop stress is the result of centrifugal action and, if the thickness is small relative to the diameter, may be considered uniform. Referring to Figure 93 and considering an element of length b which subtends an angle $d\theta$ at the axis:

STRUTS AND CYLINDERS

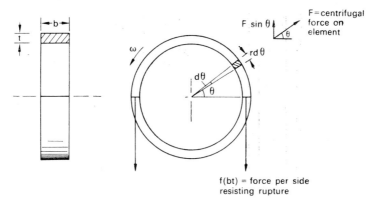

FIG. 93

Section of element	$= r.d\theta \times t$
Volume of element	$= tr.d\theta \times b$
Mass of element	$= btr.d\theta \times \varrho$ (where ϱ = density).

When the peripheral speed is u, ($= \omega r$), then

Centrifugal force $= \varrho btr.d\theta \left(\dfrac{u^2}{r}\right) = \varrho btu^2.d\theta$

Vertical component of this $= \varrho btu^2.d\theta.\sin\theta$.

The total vertical force tending to rupture the rim across a horizontal diameter is therefore

$$\varrho btu^2 \int_0^\pi \sin\theta\, d\theta, \quad \text{i.e.} \quad 2\varrho btu^2$$

But the total force resisting rupture is the product (stress in rim)×(section of rim) so that, denoting the hoop stress by f_x, we can write

$$2[f_x(bt)] = 2\varrho btu^2$$

whence
$$f_x = \varrho u^2.$$

EXAMPLE. The flywheel of a petrol motor is to be in the form of a hollow cylinder and is to be made of metal having $\varrho = 7750$ kg/m³. Determine the permissible external diameter if the hoop stress is not to exceed 35 MPa at 4170 rev/min. If the axial width of the flywheel is to be 50 mm and the polar moment of inertia is to be 0·225 kg m², find from first principles a suitable radial thickness.

Solution.

Angular velocity, $\omega = \dfrac{2\pi \times 4170}{60} = 437$ rad/s

Rim speed, $u = \omega R_1 = 437 R_1$

where R_1 = outer radius.

Hoop stress, $f_x = \varrho u^2$

∴ $35 \times 10^6 = 7750(437 R_1)^2$

95

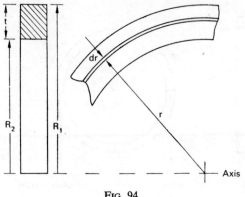

Fig. 94

whence
$$R_1^2 = 0.0237$$
$$R_1 = 0.154 \text{ m}$$
i.e.
$$D_1 = 308 \text{ mm.}$$

Referring to Figure 94 and considering the element of radius r, radial thickness dr and axial thickness b,

Elemental mass $= 2\pi r \cdot dr \cdot b\varrho$

2nd Moment $= 2\pi r \cdot dr \cdot b\varrho \times r^2 = 2\pi b\varrho r^3 \, dr$.

Hence, for the whole flywheel,

$$I = 2\pi b\varrho \int_{R_2}^{R_1} r^3 \, dr = 2\pi b\varrho \left(\frac{R_1^4 - R_2^4}{4}\right).$$

In the given case, therefore,

$$0.225 = 2\pi \times 0.050 \times 7750 \left(\frac{0.0154^4 - R_2^4}{4}\right)$$

whence
$$R_2^4 = 1.93 \times 10^{-4}$$
$$R_2^2 = 1.39 \times 10^{-2}$$
$$R_2 = 0.118 \text{ m} = 118 \text{ mm}$$
i.e.
$$D_2 = 236 \text{ mm.}$$

Hence, Radial thickness $= R_1 - R_2 = 36 \text{ mm.}$

CHANGE IN VOLUME OF A THIN CYLINDER UNDER INTERNAL PRESSURE

As already shown, the hoop stress, f_x, in a thin cylinder under internal pressure is given by $pd/2t$ and is twice the longitudinal stress, f_z.

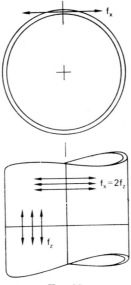

Fig. 95

Referring to Figure 95,

Axial strain,
$$e_z = \frac{f_z}{E} - \frac{\sigma f_x}{E}$$

$$= \frac{f_z}{E} - \frac{2\sigma f_z}{E}$$

$$= \frac{f_z}{E}(1-2\sigma).$$

Putting $f_z = \dfrac{pd}{4t}$ we obtain

$$e_z = \frac{pd}{4tE}(1-2\sigma).$$

Hoop strain,
$$e_x = \frac{f_x}{E} - \frac{\sigma f_z}{E}$$

$$= \frac{f_x}{E} - \frac{\sigma f_x}{2E}$$

$$= \frac{f_x}{E}\left(1 - \frac{\sigma}{2}\right).$$

Putting $f_x = \dfrac{pd}{2t}$ we obtain

$$e_x = \frac{pd}{2tE}\left(1 - \frac{\sigma}{2}\right).$$

This is the circumferential strain and the diametral strain is the same as this, since diameter and circumference increase in the same ratio. Further, the diameter increases in two mutually perpendicular directions, so that

Volumetric strain, $e_v = e_z + 2e_x$

$$= \frac{pd}{4tE}(1-2\sigma) + 2\left[\frac{pd}{2tE}\left(1-\frac{\sigma}{2}\right)\right]$$

which reduces to $e_v = \frac{pd}{tE}(1\cdot25 - \sigma).$

Change in volume = Volumetric strain × Original volume

$$= e_v V$$

$$= \frac{Vpd}{tE}(1\cdot25 - \sigma).$$

For the case considered, this is evidently an increase.

EXAMPLE. A storage tank is made of material 25 mm thick and consists of a cylinder 2·5m long, 1·25 m inside diameter, the ends of which are closed by hemispheres. If, initially, it is full of water at atmospheric pressure, determine how much additional water must be pumped in to bring about a test pressure of 3·5 MPa. Assume E and σ for the tank material to be 206 000 MPa and 0·287 respectively and assume the circumferential strains at the junction of cylinder and hemisphere to be the same. For water assume $K = 2200$ MPa.

Solution. Considering the cylindrical part first and assuming the hoop stress to be uniform:

Hoop stress, $\quad f_x = \dfrac{pd}{2t} = \dfrac{(3\cdot5 \times 10^6)1\cdot25}{2 \times 0\cdot025} \times \dfrac{1}{10^6} = 87\cdot5$ MPa

Longitudinal stress, $\quad f_z = \dfrac{pd}{4t} = 43\cdot75$ MPa.

Assuming the radial stress, f_y, to be zero, we have

Hoop strain, $\quad e_x = \dfrac{f_x}{E} - \dfrac{\sigma f_z}{E}$

$$= \frac{1}{E}[(87\cdot5 \times 10^6) - 0\cdot287(43\cdot75 \times 10^6)]$$

$$= \frac{75 \times 10^6}{E}$$

Longitudinal strain, $\quad e_z = \dfrac{f_z}{E} - \dfrac{\sigma f_x}{E}$

$$= \frac{1}{E}[(43\cdot75 \times 10^6) - 0\cdot287(87\cdot5 \times 10^6)]$$

$$= \frac{18\cdot2 \times 10^6}{E}$$

Volumetric strain, $\quad e_v = e_z + 2e_x = \dfrac{168 \times 10^6}{E}$

Increase in volume $= e_v V = \dfrac{168\times 10^6}{206\,000\times 10^6}\left(\dfrac{\pi}{4}\times 1\cdot 25^2\times 2\cdot 5\right)$

$= 0\cdot 0025 \text{ m}^3.$

(This result may also be obtained by substitution in the formula previously derived.)
Considering next the hemispherical ends:

Volumetric strain, $\quad e_v = 3e_x \quad$ (where e_x has the value found above),

i.e. $\quad e_x = \dfrac{225\times 10^6}{E}$

Increase in volume $= e_v V = \dfrac{225\times 10^6}{206\,000\times 10^6}\left[\dfrac{4}{3}\pi\left(\dfrac{1\cdot 25}{2}\right)^3\right]$

$= 0\cdot 00113 \text{ m}^3.$

Reduction in volume of original water $= p/K\times$ volume

$= \dfrac{3\cdot 5\times 10^6}{2200\times 10^6}\left[\left(\dfrac{\pi}{4}\times 1\cdot 25^2\times 2\cdot 5\right)+\left(\dfrac{4}{3}\pi\times 0\cdot 625^3\right)\right]$

$= 0\cdot 0065 \text{ m}^3.$

Additional water necessary $= 0\cdot 0025 + 0\cdot 0011 + 0\cdot 0065$

$= 0\cdot 0101 \text{ m}^3 \quad$ (about 10·1 litres).

EXAMPLES 4

1. Assume a factor of safety of 10 and compare the safe loads of two struts, 50×50×2540 mm and 50×50×760 mm, given that, in each case, one end is pin-jointed and one fixed. Take $f_c = 325$ MPa and $E = 208\,000$ MPa ($P_e = 4390$ N, $P_r = 34\,300$ N).
2. A strut 3·65 m long has a circular section and rounded ends. When freely supported at the ends in a horizontal position, a load of 40 N at the centre produced there a deflection of 9·65 mm. Determine the Euler critical load (3160 N). (It is not necessary to know the value of E.)
3. Two T-sections are riveted back to back to form a cruciform section 150×222 mm overall. Each section is 150×16×112 mm, the effective length is 6·1 m and the ends are direction-fixed. Find the safe load using a factor of 5 given that $f_c = 325$ MPa and that $a = 1/30\,000$ in the Rankine–Gordon formula (250 kN).
4. A link in a mechanism is to be of I-section steel 24×12 mm overall, the web being 2·5 mm thick and the flanges 3·2 mm thick. The bearing centres are to be 255 mm apart, the axes of the pins being normal to the link and parallel to the web. Take $f_c = 325$ MPa in the Rankine–Gordon formula and compare the crippling load in the plane of oscillation with that in the plane normal to it. Hence find the safe thrust using a factor of 10 (3·35 kN). (Note that the constant a has different values in the two cases.)
5. A tubular strut 1·5 m long 50 mm external diameter is to carry 20 kN axially with a factor of safety of 6 based on the Euler critical load for pin-jointed ends. Estimate the thickness required assuming $E = 206\,000$ MPa. Estimate also the factor of safety according to the Rankine–Gordon formula, assuming the constants to be 325 MPa and 1/7500 (3·2 mm, 3·7).
6. A block and tackle is suspended from the apex of a tripod, each leg of which makes an angle of 60 deg with the ground and consists of a pipe 64 mm external diameter, 50 mm internal diameter, 3 m long. Take the Rankine–Gordon constants as 1/7500 and 325 MPa and estimate what load may be raised assuming a safety factor of 10 (24 kN).
7. The force required at the end of a clutch operating lever is 250 N and is transmitted by means of a pin-ended steel rod 255 mm long. Take $E = 206\,000$ MPa and determine a suitable diameter given that the factor of safety based on the Euler formula is to be 10 (6·5 mm).

8. The section of a pin-ended strut is to be 9·5×3·2 mm. If it is to fail at an axial load of 1600 N, find the required length given that $E = 206\,000$ MPa (255 mm).
9. A cylinder 200 mm outside diameter, 100 mm inside diameter is subjected to an internal pressure of 210 MPa. Calculate the longitudinal stress, assuming this to be uniform (7 MPa).
10. Find the greatest and least hoop stresses in the previous example (35 and 14 MPa).
11. A hollow forged boiler drum 1·94 m outside diameter and 132 mm thick was tested before entering service to a pressure of 20 MPa. Calculate the greatest and least hoop stress under test (140 and 120 MPa).
12. Find the longitudinal and hoop stresses at the outer surface of a hydraulic pipe 50 mm bore, 100 mm outside diameter, at the operating pressure of 140 MPa. Estimate the increase in the outside diameter when under load. The pipe is of steel for which $\sigma = 0·26$ and $E = 206\,000$ MPa ($f_z = 46$ MPa, $f_x = 92·5$ MPa, 0·04 mm).
13. A hydraulic press is to have a ram diameter of 200 mm and operate at an internal pressure of 10·4 MPa. Estimate the wall thickness needed to limit the hoop stress to 25 MPa. Find the increase in the internal diameter of the cylinder when under load, taking $\sigma = 0·25$ and $E = 116\,000$ MPa (57 mm, 0·045 mm).
14. Determine the interference fit per millimetre of diameter necessary to produce a radial pressure of 185 MPa at the common surface of a compound tube. The inner and outer diameters are 100 mm and 255 mm and the inner tube has an outer diameter initially of 200 mm. What must be the initial inner diameter of the outer tube? $E = 206\,000$ MPa (198·88 mm).
15. A torpedo shell 510 mm diameter is 9·5 mm nominal thickness. Estimate the hoop stress induced by an internal pressure of 14 MPa, assuming the stress to be uniform (373 MPa).
16. The flywheel of a stamping machine is to have a moment of inertia of 270 kg m² and to be made from material of density 7250 kg/m³. Determine from first principles the outside diameter and radial thickness of the rim given that the stress due to rotation is not to exceed 4·14 MPa at 300 rev/min. Neglect the effect of the spokes and assume an axial width of 150 mm (1·52 m, 110 m).
17. Estimate the increase in the capacity of a shell 4·9 m long, 2·13 m diameter when subjected to an internal pressure of 0·83 MPa gauge, assuming $E = 206\,000$ MPa, $\sigma = 0·287$ and that the skin is nominally 22 mm thick (0·0068 m³).
18. An internal pressure of 5·2 MPa is applied to a steel tube 150 mm diameter, 3·2 mm thick, 610 mm long, the ends of which are plugged. If $E = 206\,000$ MPa and $\sigma = 0·28$, find:

    (a) hoop stress (125 MPa);
    (b) axial stress (62·5 MPa);
    (c) increase in capacity (13×10^{-6} m³);
    (d) increase in volume of tube material ($0·35 \times 10^{-6}$ m³).

19. An iron pipe 200 mm outside diameter 25 mm thick, failed under test at an internal pressure of 48 MPa. Use a factor of safety of 4 and find the safe pressure for a pipe of the same material and inside diameter but 38 mm thick. Assume maximum hoop stress to be the criterion of failure (16·5 MPa).

5

DYNAMICS

FUNDAMENTALS

Equations of Linear Motion

For uniform linear acceleration these are:

$$s = u_1 t + \tfrac{1}{2} f t^2, \quad u_2 = u_1 + ft \quad \text{and} \quad u_2^2 = u_1^2 + 2fs.$$

Equations of Angular Motion

For uniform angular acceleration these are:

$$\theta = \omega_1 t + \tfrac{1}{2} \alpha t^2, \quad \omega_2 = \omega_1 + \alpha t \quad \text{and} \quad \omega_2^2 = \omega_1^2 + 2\alpha\theta.$$

Relation between Linear and Angular Quantities

The instantaneous linear (tangential) velocity of a point rotating uniformly at radius r is given by

$$u = \omega r$$

where

$$\omega = \frac{2\pi}{60} \text{ (speed in rev/min)}$$

If ω is not constant, the instantaneous linear (tangential) acceleration of the point is given by

$$f = \alpha r$$

where

$$\alpha = (\omega_2 - \omega_1) \frac{1}{t}.$$

Force and Change in Motion

Force imparts, or tends to impart, motion to a stationary body. Force also changes the linear velocity of a body already in motion, i.e. brings about acceleration. The relevant equation is

$$F = Mf,$$

where $M = $ mass.

When the force is applied normal to and at the end of an arm mounted on a shaft, a torque is said to exist at the shaft axis, such torque being defined as the product of force and radius. Torque imparts, or tends to impart, angular motion to a stationary body. Torque also changes the angular velocity of a rotating body, i.e. brings about angular acceleration. The relevant equation is

$$\underline{T = I\alpha,}$$

where I = moment of inertia.

Kinetic Energy

This is the energy of motion and increases as the square of the speed.

For linear motion, K.E. $= \frac{1}{2}Mu^2$ (u = linear velocity)

For angular motion, K.E. $= \frac{1}{2}I\omega^2$ (ω = angular velocity).

Momentum

This is defined as the product of mass and velocity.

For linear motion, $M = Mu$

For angular motion, $M = I\omega$.

If mass (and therefore moment of inertia) is constant, a change in momentum requires a change in velocity, i.e. an acceleration. Linear momentum can be changed therefore only by the application of a force while angular momentum can be changed only by the application of a torque.

Centripetal Acceleration and Force

When a point moves in a circular path, the continuous change in the direction of its (tangential) velocity, ωr (the direction of such change being radial), requires the presence of a radial, i.e. centripetal, acceleration, $\omega^2 r$. If such a point is replaced by a mass, this acceleration requires a force to produce it, the equation being

$$\underline{F = M\omega^2 r} \quad \left(\text{or } M\frac{u^2}{r}\right).$$

A twirled bunch of keys is kept on its circular path by the centripetal force exerted on it by the chain. The chain exerts an equal and opposite force (centrifugal reaction) on the hand. The moon is maintained in orbit by the attraction (gravitational force) of the earth, the centrifugal reaction being the cause of tidal phenomena. If the coefficient of friction is sufficient, a vehicle is prevented from leaving (tangentially) its path round a bend by the centripetal force exerted on its tyres by the road. The reaction (unavoidably) acts at the mass centre of the vehicle so that an overturning moment exists. The centripetal force on the driver is provided by the seat, the reaction (acting at his mass centre) tending to make him lean outward. The centrifugal force in all cases is the resistance offered to the (directional) change in motion and is therefore an inertia force.

DYNAMICS

By reversing the centripetal force and equating it (i.e. the centrifugal force) to the resultant of the applied forces, the equivalent static condition may be simulated and a force polygon drawn. This concept—known as D'Alembert's Principle—is of use in obtaining balance in rotating systems.

Work and Power

If a force does work directly, by moving in a straight line, the energy expended is the product of force and distance, i.e. Fs.

If the force moves at uniform linear velocity the work done per unit time (power) is Fu.

If the force is applied at the end of an arm, the energy expended is the product of torque and angle turned through, i.e. $Fr\theta$ or $T\theta$.

If the arm moves at uniform angular velocity, the work done by the torque per unit time (power) is $T\omega$.

GYROSCOPIC EFFECTS

The angular momentum of a disk rotating about its polar axis, as shown in Figure 96, is given by

$$M = I\omega_s$$

where I = polar moment of inertia

and ω_s = angular velocity of spin.

This momentum may be represented by a vector Oa drawn from (Fig. 97) any point O on the axis, in a direction as travelled by a right-hand screw when rotating clockwise. If the axis XX is made to rotate at ω_p about a vertical through point O, then, after it has

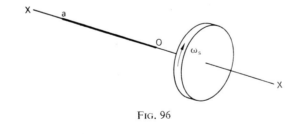

FIG. 96

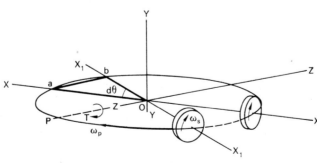

FIG. 97

103

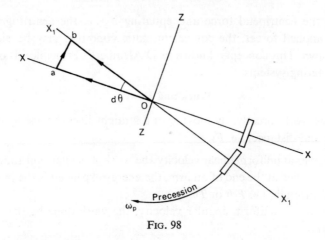

FIG. 98

described an angle $d\theta$, the angular momentum will be represented by the vector Ob. From Figure 98 (which is a plan view of Fig. 97) it is evident that Ob is the vector sum of Oa and ab where $ab = Oa \cdot d\theta$ and is an addition to the momentum.

If this change occurs in time dt and is represented by dM, then

$$\text{Rate of change in momentum} = \frac{dM}{dt}$$

$$= Oa \frac{d\theta}{dt}$$

$$= M\omega_p$$

$$= I\omega_s\omega_p.$$

But a change in momentum cannot take place in the absence of a torque so that to cause the rotation of XX about YY

Torque required = Rate of change in momentum

$$\underline{T = I\omega_s\omega_p.}$$

By applying the screw convention to the vector (ab) representing the momentum change (which is parallel to ZZ when $d\theta$ is infinitely small) the direction of this torque is seen to be clockwise when viewed from point P, Figure 97.

The rotation of the axis of spin XX about the axis YY when torque is applied to the axis ZZ is called *precession*, and results, usually, from the application of forces to the bearings as shown in Figure 99.

Equal and opposite forces are exerted by the shaft on the bearings as a result of the *gyroscopic reaction couple* which is of opposite sense to the applied torque. These forces are transmitted to whatever houses the bearings, the actual (equal) values being obtained by dividing the gyroscopic couple by the distance between the bearings.

If either the applied torque or the direction of spin is reversed, precession will also reverse in direction, but if both are reversed simultaneously there will be no change in the precession.

DYNAMICS

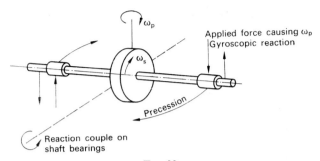

FIG. 99

EXAMPLE. The wheelbase and track of a vintage motor car weighing 12 900 N are respectively 3·13 m and 1·42 m. Each wheel has an effective diameter of 0·762 m and a moment of inertia of 2·1 kg m². In top gear the transmission rotates at 4·1 times the roadwheel speed in a clockwise direction when viewed from the front and is equivalent to a mass of 63·5 kg having a radius of gyration of 0·0915 m. If the front wheels together carry 0·6 of the weight when the car is stationary, determine the load on each wheel while traversing a horizontal right hand bend of 60 m radius at 20 m/s given that the mass centre is 0·61 m above the ground.

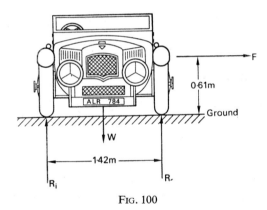

FIG. 100

Solution.

Centrifugal force, $\quad F = M\dfrac{u^2}{r} = \dfrac{12\,900}{9\cdot 81} \times \dfrac{20^2}{60} = 8760 \text{ N}$

When stationary:

$\quad$ Load/front wheel $= \frac{1}{2}(0\cdot 6 \times 12\,900) = 3870$ N

$\quad$ Load/rear wheel $= \frac{1}{2}(0\cdot 4 \times 12\,900) = 2580$ N

or, $\quad$ Load/side $\quad = 3870 + 2580 = 6450$ N

When in motion:

$\quad$ Moments about R_o give: $\quad (R_i \times 1\cdot 42) + (F \times 0\cdot 61) = W\left(\dfrac{1\cdot 42}{2}\right)$

105

Load on inner side, $R_i = \dfrac{(12\,900 \times 0\cdot71) - (8760 \times 0\cdot61)}{1\cdot42}$

$= 2680$ N

$\therefore \qquad R_o = 12\,900 - 2680$

$= 10\,220$ N

$\therefore$ Change/side $= 6450 - 2680 = 3770$ N (or $10\,220 - 6450$)
$\therefore$ Change/wheel $= 3770/2 = 1885$ N.

Hence, neglecting gyroscopic effects:

Load on inner front wheel $= 3870 - 1885 = 1985$ N

Load on inner rear wheel $= 2580 - 1885 = 695$ N

Load on outer front wheel $= 3870 + 1885 = 5755$ N

Load on outer rear wheel $= 2580 + 1885 = 4465$ N

For wheels:

Angular velocity of spin, $\omega_s = \dfrac{u}{r} = \dfrac{20}{0\cdot381} = 52\cdot5$ rad/s

Angular velocity of precession, $\omega_p = \dfrac{u}{R} = \dfrac{20}{60} = 0\cdot333$ rad/s

$\therefore$ Gyroscopic torque, $T = 4(2\cdot1 \times 52\cdot5 \times 0\cdot333) = 147$ Nm.

The reaction to this reduces the load on the inside wheels.

Gyroscopic force per side $= \dfrac{\text{Couple}}{\text{Arm}} = \dfrac{147}{1\cdot42} = 103\cdot5$ N

$\therefore$ Change in load per wheel $= \dfrac{103\cdot5}{2} = 52$ N.

This figure must be added to the load on the outer wheels and subtracted from the load on the inner wheels. Figure 101a shows the momentum vector triangle:

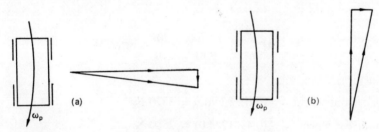

Fig. 101

For transmission:

Moment of inertia, $I = 63\cdot5 \times 0\cdot0915^2 = 0\cdot531$ kg m²
Angular velocity of spin, $\omega_s = 52\cdot5 \times 4\cdot1 = 215$ rad/s
Angular velocity of precession, $\omega_p = 0\cdot333$ rad/s (same)

∴ Gyroscopic torque, $T = 0.531 \times 215 \times 0.333 = 38$ Nm.
The reaction to this reduces the load on the front wheels.

$$\text{Gyroscopic force per end} = \frac{\text{Couple}}{\text{Arm}} = \frac{38}{3.13} = 12 \text{ N}$$

∴ Change in load per wheel $= \frac{12}{2} = 6$ N.

This figure must be added to the load on the rear wheels and subtracted from the load on the front wheels. Figure 101b shows the momentum vector triangle, while the resultant load distribution is as shown in Figure 102.

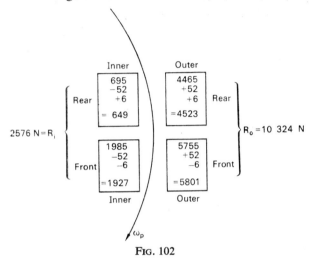

FIG. 102

ACCELERATION OF A TRAIN OF GEARS

For a wheel and pinion rotating at ω_2 and ω_1 respectively as shown in Figure 103,

Gear ratio, $$n = \frac{\omega_1}{\omega_2} = \frac{r_2}{r_1}.$$

If α_1 and α_2 are the respective angular accelerations resulting from the application of a torque T at the pinion axis, the linear acceleration of the pitch point (point of contact between

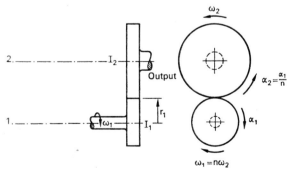

FIG. 103

equivalent friction circles) is given by

$$f = \alpha_2 r_2 = \alpha_1 r_1$$

so that
$$\alpha_2 = \left(\frac{r_1}{r_2}\right)\alpha_1 = \frac{\alpha_1}{n}.$$

If I_1 and I_2 are the respective moments of inertia and there are no losses, then, since torque is inversely proportional to speed,

$$\begin{aligned}\text{Torque required on pinion} \atop \text{to accelerate wheel} &= \frac{1}{n}\left({\text{Torque required on wheel} \atop \text{to accelerate wheel}}\right) \\ &= \frac{1}{n}I_2\alpha_2 \quad \text{where} \quad \alpha_2 = \frac{\alpha_1}{n} \\ &= \frac{I_2}{n^2}\alpha_1.\end{aligned}$$

$$\text{Torque required on pinion} \atop \text{to accelerate pinion} = I_1\alpha_1.$$

Hence, in addition to any torque required to overcome load and friction torques, the accelerating torque for the gears themselves is given by

$$T = \left(I_1 + \frac{I_2}{n^2}\right)\alpha_1.$$

The quantity in the brackets is known as the *equivalent moment of inertia* of the system referred to the input axis.

Thus,
$$I_{eq} = I_1 + \frac{I_2}{n^2}.$$

If there are three gears in series having respectively moments of inertia I_1, I_2 and I_3, the successive reductions being n_1 and n_2, then

$$I_{eq} = I_1 + \frac{I_2}{n_1^2} + \frac{I_3}{n_1^2 n_2^2}.$$

In the case of four gears there will be a fourth term $I_4/n_1^2 n_2^2 n_3^2$. Note that $n_1 n_2 n_3 = n =$ overall gear ratio.

If there are losses and the efficiency at each reduction is assumed to be the same and denoted by η, then

$$I_{eq} = I_1 + \frac{I_2}{\eta n_1^2} + \frac{I_3}{\eta^2 n_1^2 n_2^2} \quad \text{and so on.}$$

* If the axis of the second gear in a train is carried on an arm (as shown in Fig. 104) and this arm is able to rotate about the input axis, the train is said to be *epicyclic*, the input and second wheels being called respectively *sun* and *planet*. The third internal gear is also co-axial with the input and is called the *annulus*. (The method of determining the ratio n (i.e. ω_1/ω_3) is given later.)

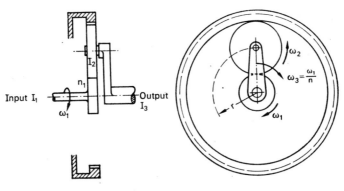

FIG. 104

Now, referred to the input axis, the moment of inertia of the output shaft is I_3/n^2 where $n = \omega_1/\omega_3$.

If the planet axis—and hence the output shaft—is stationary, the moment of inertia of the planet is I_2 about its own axis and I_2/n_1^2 about the input axis, but when the planet axis is rotating at output speed the parallel axes theorem requires the addition of the term

$$\text{Mass} \times (\text{Distance between axes})^2.$$

This addition, *referred to the output shaft*, is $M_2 r^2$ where M_2 is the mass of the planet and r is the radius of the path of the planet axis. *Referred to the input* this addition becomes

$$\frac{M_2 r^2}{n^2}$$

where $$n = \frac{\omega_1}{\omega_3}.$$

Thus, referred to the input shaft, the equivalent moment of inertia of a planet wheel is

$$\frac{I_2}{n_1^2} + \frac{M_2 r^2}{n^2}$$

where $$n_1 = \frac{\omega_1}{\omega_2}.$$

Note that ω_2 is now the speed of the planet wheel *in space* and not about its own axis.

Hence, for the train comprising sun, planet and arm, the equivalent moment of inertia referred to the input (the annulus being stationary) is given by

$$I_{eq} = I_1 + \left[\frac{I_2}{n_1^2} + \frac{M_2 r^2}{n^2}\right] + \frac{I_3}{n^2}$$

where n = input speed/output speed,
and n_1 = input speed/speed of planet in space.

For reasons of balance there are usually three equi-spaced planets so that the quantity in the brackets must be multiplied by three.

EXAMPLE. Figure 105 shows a double reduction gear in which each stage gives a ratio of 3·5 and has an efficiency of 0·96. Find the input torque required to raise the output speed from rest to 40 rev/min in 0·5 sec against a constant load torque of 80 Nm given that the polar moments of inertia of input, compound gear and output are respectively 0·002, 0·16 and 0·20 kg m².

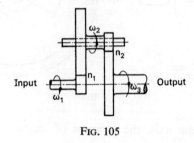

FIG. 105

Solution.

$$n_1 = n_2 = 3·5$$
$$n = n_1 n_2 = 12·25.$$

Referred to the input,

$$I_{eq} = 0·002 + \frac{0·16}{0·96 \times 3·5^2} + \frac{0·20}{0·96^2 \times 12·25^2} = 0·0162 \text{ kg m}^2$$

Final output, speed, $\omega_3 = \dfrac{2\pi \times 40}{60} = 4·19$ rad/s

Final input speed, $\omega_1 = 4·19 \times 12·25 = 51·4$ rad/s

Angular acceleration of input, $\alpha_1 = \dfrac{51·4}{0·5} = 102·8$ rad/s²

Input torque to accelerate parts $= 0·0162 \times 102·8$
$= 1·67$ Nm

Input torque to overcome load torque $= \dfrac{80}{12·25} \times \dfrac{1}{0·96^2}$

$= 7·11$ Nm

Required input torque, $T = 1·67 + 7·11 = 8·78$ Nm.

(Note that the torque required to accelerate the input alone is not affected by the efficiency.)

EXAMPLE. Figure 106 shows an epicyclic reduction gear. When the annulus is fixed, the output speed is one sixth of the input speed and in the same direction, while the planet speed in space is one quarter of the input speed and in the opposite direction. Calculate the equivalent moment of inertia of the gear train referred to the input, given that

(a) the polar moments for *A*, *B* and *C* are respectively

0·005, 0·046 and 0·115 kg m²,

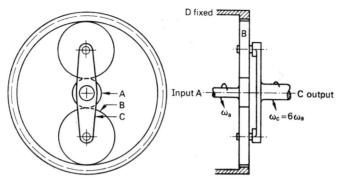

FIG. 106

(b) the mass of each planet is 5·5 kg,
(c) the distance between the input and planet axes is 200 mm. Find also the effective torque on a planet wheel when a torque of 3·5 Nm is applied to the input. Neglect losses.

Solution.

$$I_{eq} = 0.005 + 2\left[\frac{0.046}{4^2} + \frac{5.5 \times 0.2^2}{6^2}\right] + \frac{0.115}{6^2}$$

$$= 0.0247 \text{ kg m}^2.$$

Denoting the acceleration of the input by α_a we have:

Input torque, $\qquad T_a = I_{eq}\alpha_a$

$\qquad\qquad\qquad 3.5 = 0.0247\,\alpha_a$

whence $\qquad\qquad \alpha_a = 142 \text{ rad/s}^2$

∴ Acceleration of planet, $\qquad \alpha_b = \dfrac{142}{4} = 35.4 \text{ rad/s}^2$

i.e. Net torque on planet, $\qquad T_b = I_b\alpha_b$

$\qquad\qquad\qquad = 0.046 \times 35.4$

$\qquad\qquad\qquad = 1.63 \text{ Nm}.$

✳ EPICYCLIC GEARS ✳

The Epicycloid

An epicycloid is the path traced out (*APB*, Fig. 107) by a point such as *P* on the circumference of a circle centre C_1 when the circle rolls without slipping on the outside of a second circle centre C_2. Teeth on a rotating planet wheel follow an epicyclic path when the sun wheel is stationary.

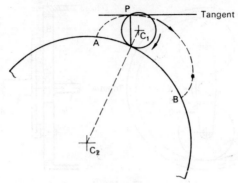

Fig. 107

Epicyclic Gears

The description *epicyclic* is given to a train of gears in which one or more of the wheel axes is given a motion relative to the housing. In addition to sun wheel and annulus, the train shown in Figure 108 contains three compound planets, BC, mounted on a three-armed spider, E. Any one of the three components A, D and E may be fixed, so fixing the velocity ratio between the other two.

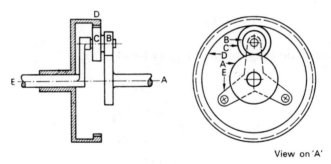

Fig. 108

Determination of Gear Ratio

Let E (Fig. 109) be one arm of the spider shown in Figure 108 and suppose one of the planets, C, to be fixed to it so that it cannot rotate about its own axis. Suppose also that C has a radial arrow marked on it and that the rest of the assembly is removed.

If the spider is rotated clockwise through 120° it is evident that, although the planet has not turned about its own axis, it too has made one third of a revolution *in space*. Thus one revolution of the spider involves one revolution of the planet in the same direction. If the planet is rotating clockwise about its own axis while the spider is rotating in the same direction, then the total revolutions made by the planet *in space* is the sum of the two motions. Similarly, when planet and arm are rotating in opposite directions, the net motion of the planet in space is given by the difference.

It is usual to distinguish between the two possible directions of motion by calling one direction positive and the reverse direction negative.

DYNAMICS

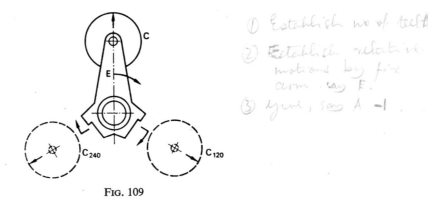

FIG. 109

Referring to Figure 108, suppose that the numbers of teeth are: $A = 36, B = 12, C = 18$. Then, for the same pitch,

$$D = A+B+C = 66 \text{ teeth.}$$

The relative motions may be established by holding the spider, E (so fixing all the axes and making an ordinary train), and giving A, say, one anti-clockwise revolution. If this is denoted by -1, then

Revolution of $BC = +\frac{36}{12} = +3$, i.e. clockwise.

Revolution of $D = +\frac{36}{12} \times \frac{18}{66} = +\frac{9}{11}$, i.e. in the same direction as C.

This information is best tabulated as below:

Action	Sun wheel A	Compound planet BC	Annulus D	Spider E
Fix E Give $A-1$	-1	$+\frac{36}{12} = +3$	$+\frac{36}{12} \times \frac{18}{66}$ $= +\frac{9}{11}$	0

These relative motions will not be altered by any motion the components may have in common so that if the spider is now given, say, $+1$ (i.e. clockwise) revolution about the axis of A, then $+1$ must be added to each of the above totals. This has the effect of fixing A as shown in the next line of the table:

Fix A by adding $+1$ throughout	$-1+1 = 0$	$+3+1 = +4$	$+\frac{9}{11}+1 = +\frac{20}{11}$	$0+1 = +1$

Thus if A is held stationary and E is driven at $+1$ rev/min, the annulus D will rotate at $+20/11$ rev/min in the same direction, i.e.

$$\text{Gear ratio,} \quad n = \frac{\text{Input speed}}{\text{output speed}} = \frac{1}{20/11} = \frac{11}{20} \quad \text{or} \quad 11:20.$$

MECHANICAL TECHNOLOGY FOR HIGHER ENGINEERING TECHNICIANS

This represents a "step-up". Evidently, if the speed of the spider is increased, the other speeds are raised in the same proportion so that, if A is held as before and E is driven at $+11$ rev/min we have:

Multiply previous line by 11	$0\times 11 = 0$	$+4\times 11 = +44$	$+\frac{20}{11}\times 11 = +20$	$+1\times 11 = +11$

Thus the speed of D is now $+20$ rev/min, i.e. clockwise.

The foregoing demonstrates that, once the relative motions have been established,

(1) any number may be added to all results in a line,

(2) all results in a line may be multiplied by a given number.

[*Note.* If the exact gear ratio is required, the ratios obtained when finding the relative motions must be kept in the form of fractions. This often simplifies the calculation.]

EXAMPLE. Find the ratio of the reduction gear shown in Figure 110 given that the input, A, has 20 teeth, the compound planet B/C has 24/16 teeth, the annulus E is fixed and that all gears are of the same pitch.

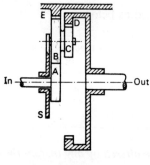

FIG. 110

Solution.

Teeth on E, $T_e = T_a + 2T_b$

$\qquad = 20 + (2\times 24)$

$\qquad = 68.$

Teeth on D, $T_d = T_a + 2\left[\dfrac{T_b}{2}\right] + 2\left[\dfrac{T_c}{2}\right]$

$\qquad = 20 + 24 + 16$

$\qquad = 60.$

Tabulating:

Action	A	B/C	E	D	S
Fix spider S Give $A+1$	$+1$	$-\frac{20}{24} = -\frac{5}{6}$	$-\frac{5}{6} \times \frac{24}{68} = -\frac{5}{17}$	$-\frac{5}{8} \times \frac{16}{60} = -\frac{2}{9}$	0
Add 5/17 to make E zero	$+1+\frac{5}{17} = +\frac{22}{17}$	$-\frac{5}{6}+\frac{5}{17} = -\frac{55}{102}$	$-\frac{5}{17}+\frac{5}{17} = 0$	$-\frac{2}{9}+\frac{5}{17} = +\frac{11}{153}$	$+\frac{5}{17}$
$\times\frac{17}{22}$ to bring A back to $+1$	$+\frac{22}{17}\times\frac{17}{22} = +1$		0	$+\frac{11}{153}\times\frac{17}{22} = +\frac{1}{18}$	
Multiply by 18	$+18$		0	$+1$	

Hence, Gear ratio, $n = \dfrac{\text{Input speed}}{\text{Output speed}} = \dfrac{18}{1}$ or $\underline{n = 18}$.

Transmission of Torque

Consider the reduction gearbox represented by Figure 111. Assuming that there is no loss and that $N_1 > N_2$ we have

$$\text{Output} = \text{Input}$$

i.e. $$T_2\omega_2 = T_1\omega_1$$

or $$N_2 T_2 = N_1 T_1$$

so that $$T_2 = \frac{N_1}{N_2} T_1.$$

If N_2 is in the same direction as N_1 then this output (resisting) torque will be in the opposite sense to T_1. Moreover, since $N_2 < N_1$ it follows that $T_2 > T_1$ so that a torque $T_2 - T_1$ in the same direction as T_1 will be required on the housing to prevent it from rotating in the direction of T_2. (This will be so whether the train is simple or epicyclic.) This is called a "fixing" torque and is provided by the holding down bolts. The various torques are shown in Figure 112, the net torque about the axis being zero. If the rotation of the output is reversed, it is evident that the fixing torque required on the housing will be $T_1 + T_2$ in the opposite direc-

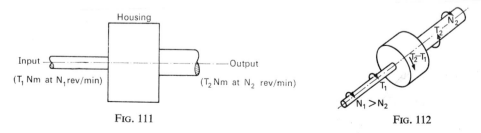

Fig. 111

Fig. 112

tion. If losses are taken into account we have, writing η for the efficiency,

Output torque
$$T_2 = \eta\left(\frac{N_1}{N_2}\right)T_1.$$

EXAMPLE. In the gear shown in Fig. 113 the spider S carries compound planets QR, the numbers of teeth being as follows: $P = 21, Q = 28, R = 14, A = 84$. If A is held stationary while 4 kW is delivered to P at 900 rev/min, find the speed of the output S and the holding torque required on A assuming an efficiency of 0·96. If the annulus is permitted to rotate at 90 rev/min in the same direction as P, find the new speed of S.

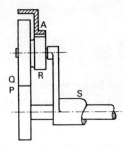

FIG. 113

Solution.

$$\text{Input speed, } \omega_p = \frac{2\pi \times 900}{60} = 94\cdot 2 \text{ rad/s}$$

$$\text{Input power} = T_p \omega_p = 4 \times 1000 = 4000 \text{ W}$$

$$\therefore \quad \text{Input torque, } T_p = \frac{4000}{94\cdot 2} = 42\cdot 5 \text{ Nm}.$$

Tabulating:

Action	S	P	QR	A
Hold S Give A -1	0	$+\frac{84}{14} \times \frac{28}{21} = +8$	$-\frac{84}{14}$	-1
Fix A by adding $+1$	$+1$	$+8+1 = +9$		$-1+1 = 0$
Make $P = 900$	$+100$	$+9 \times 100 = +900$		0

Hence, speed of output = 100 rev/min, i.e. $n = 9$.

$$\text{Output torque, } T_s = 0\cdot 96(42\cdot 5 \times 9)$$
$$= 367 \text{ Nm in a direction opposite to that of } T_p.$$

Required holding torque, $T_a = 367 - 42$
$= 325$ Nm in the same direction as T_s.

Tabulating again:

Action	S	P	QR	A
Add x to values in first line	x	$8+x$		$-1+x =$ $= x-1$
Multiply by y	xy	$y(8+x)$		$y(x-1)$

Then $y(x-1) = 90$ when $y(x+8) = 900$

so that
$$\frac{y(8+x)}{y(x-1)} = \frac{900}{90} = 10$$

i.e. $8+x = 10x-10$

whence $x = 2$.

Thus $y(8+2) = 900$

whence $y = 90$.

Hence,   New speed of $S = xy = 180$ rev/min.

BALANCING

Imbalance due to Centrifugal Force

Let a body of mass M_1 rotate as shown in Figure 114 at the end of a weightless link OM_1 of length r_1 about a fixed point O with constant angular velocity ω. To produce the required centripetal acceleration ($\omega^2 r_1$) a centripetal force $F_1 (= M_1 \omega^2 r_1)$ is required and the (centrifugal) reaction to this has the same value and acts radially outwards from point O, its direction changing as the link rotates. The state of point O is one of *imbalance* or *out of balance*.

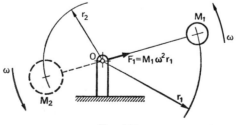

Fig. 114

Equilibrium may be restored to point O by introducing a balance mass M_2 (shown dotted) opposite to M_1 at some convenient radius r_2 also rotating at ω so that the centrifugal reaction F_2 due to it is equal and opposite to F_1 in all positions, i.e. so that

$$M_2\omega^2 r_2 = M_1\omega^2 r_1$$

or

$$\underline{M_2 r_2 = M_1 r_1.}$$

Consider the system of bodies shown in Figure 115(a), all in the same plane and all rotating at the same angular velocity.

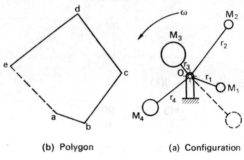

(b) Polygon   (a) Configuration

FIG. 115

The centrifugal forces will be proportional respectively to $M_1 r_1$, $M_2 r_2$ and so on (i.e. to the so-called Mr value) and can be represented by vectors ab, bc, cd and so on, drawn for a given position parallel to the relevant radius (to some suitable scale) to form what is known as a centrifugal force polygon.

Thus the polygon of Figure 115(b) is formed by drawing

ab parallel to OM_1 proportional to $M_1 r_1$

bc parallel to OM_2 proportional to $M_2 r_2$

cd parallel to OM_3 proportional to $M_3 r_3$

de parallel to OM_3 proportional to $M_4 r_4$.

The sum of these vectors (ΣMr) is ae and this represents (to the chosen scale) the resultant unbalanced Mr value in magnitude and direction. (The corresponding unbalanced force on the pivot O is obtained by multiplying this quantity by ω^2.) The equilibriant vector of the polygon is ea and the equilibriant force may be supplied by placing a balance weight M_b at some convenient radius r_b (shown dotted) so that $M_b r_b$ is proportional to ea. When M_b is in position the polygon will close, since then $\Sigma Mr = 0$. Evidently the system will then remain in any position when at rest, i.e. it will be in *static balance*.

EXAMPLE. Find the resultant out of balance force at the pivot O when the system shown in Figure 116(a) rotates at 10 rev/min and state its direction in the position shown. What value of balance mass would be required at 3 m radius and in what position?

DYNAMICS

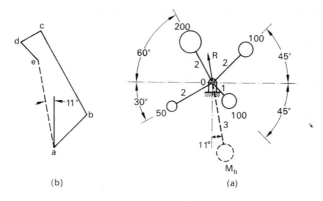

Fig. 116

Solution. At 10 rev/min, Angular velocity of system, $\omega = \dfrac{2\pi \times 10}{60}$

$$= 1{\cdot}047 \text{ rad/s.}$$

Tabulating the given information:

M	r	Mr	Vector
100	2	200	ab
200	2	400	bc
50	2	100	cd
100	1	100	de
M_b	3	$3M_b$	ea

The force polygon is shown in Figure 116(b) and from it, to scale, the resultant $ae = 375$ and is at 11° to the vertical.

∴ Unbalanced force, $\qquad R = 375 \times \omega^2$

$\qquad\qquad\qquad\qquad\qquad = 375 \times 1{\cdot}047^2$

$\qquad\qquad\qquad\qquad\qquad = 410$ N in the direction of the arrow, Figure 116(a).

From the table, $\qquad 375 = 3M_b$ so that $M_b = 125$ kg.

This must be placed so that the centrifugal reaction due to it is in opposition to R, i.e. so that M_b is in the dotted position.

Imbalance due to Centrifugal Couple

If the parts of a system rotate in different planes, the centrifugal forces, in addition to being out of balance, form *couples* which must also be eliminated if dynamic equilibrium is to be achieved.

The first step is to transfer each centrifugal force to a suitably chosen datum or *plane of reference* and the second is to draw the centrifugal force (Mr) polygon in that plane. Suppose M to rotate as shown in Figure 117(a). The centrifugal force, F, has a moment Fa about the bearing O which bends the shaft in a direction which is changing continuously.

Now imagine M_1 and M_2 to be attached at point O (as shown in Fig. 117(b)) so that their centrifugal forces are not only equal and opposite but are also equal to that at point Q,

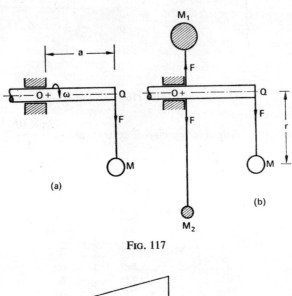

Fig. 117

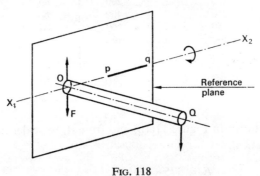

Fig. 118

i.e. are equal to F. The addition does not affect the equilibrium of the shaft and results in a pure couple, Fa, combined with a downward unbalanced force, F, acting at point O. This force F has, in effect, been "transferred" from Q to O.

The centrifugal force due to each of any number of components along a shaft may be transferred to the plane through point O in the same way and a polygon drawn for what are now co-planar forces. If the polygon does not close (i.e. if $\Sigma Mr \neq 0$) then the unbalanced Mr value is proportional to the vector sum while the closing vector (equilibriant) represents —in magnitude and position—the Mr value required for force equilibrium. The plane to which the various centrifugal forces are "transferred" is called a *plane of reference*.

Referring to Figure 118 it can be seen that $X_1 X_2$ is the neutral axis of the shaft when deflected by the couple Fa. The Mra value (to which the couple is proportional) may there-

DYNAMICS

fore be represented—to some suitable scale—by a vector *pq* drawn from some point *p* on this axis in a direction as would be travelled by a right-hand screw.

The *Mra* value due to the imbalance of each of any number of parts spaced along the shaft may be represented in the reference plane in the same way, so enabling a second or *couple polygon* to be drawn in the plane. Note that, if a part is on the far side of the plane chosen as reference, the effect of its couple is reversed, so that the vector must also be reversed (treated as negative) in the polygon.

In practice it is convenient to turn the *Mra* polygon clockwise through 90° so that each couple vector is parallel to the corresponding force vector, e.g. in Figure 118 the vector *pq* would in fact be drawn vertically downwards, parallel to *OF* instead of as shown. If the polygon does not close (i.e. if $\Sigma Mra \neq 0$) then the unbalanced *Mra* value is proportional to the vector sum while the closing vector represents—in direction and magnitude—the *Mra* value required for couple equilibrium. The actual value of the residual couple before balancing is obtained by multiplying the unbalanced *Mra* value by ω^2.

For complete (dynamic) balance, both the resultant force and the resultant couple must be zero,

i.e. $\quad\quad\quad\quad \Sigma Mr = 0 \quad \text{and} \quad \Sigma Mra = 0.$

Although the force polygon may be closed by placing a mass in the reference plane, such a mass will have no effect on any unbalanced couple so that, in general, masses must be provided in two planes. Traditionally such masses are called "balance weights".

Suppose the inertia of a rotating system to be represented by the five masses of Figure 119. (This is not a practical arrangement but will serve to show the method of analysis.)
Suppose it is required to determine, for a speed of 75 rev/min:

(a) the value of the resultant unbalanced couple,
(b) the additional load on each bearing due to (a) above,

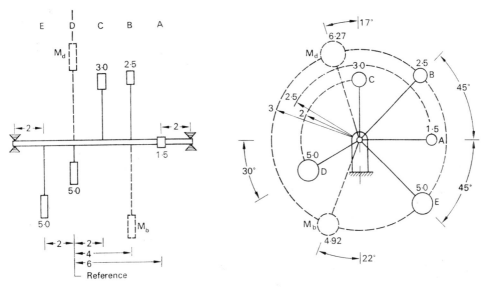

Fig. 119

121

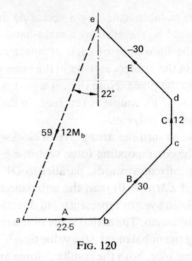

Fig. 120

(c) the sizes of the balance masses required at 3 m radius in planes B and D for dynamic balance,
(d) the resultant unbalanced force before balancing.

If one of the balance weight planes is chosen as reference, the couples due to the masses in that plane are zero and the equilibriant (closing vector) of the couple polygon can be equated to the Mra value required from the other balance weight. Taking plane D as reference and tabulating:

Plane	M (kg)	r (m)	Mr	a (m)	Mra
A	1·5	2·5	3·75	6	22·5
B	2·5	3	7·50	4	30·0
	M_b		$3M_b$		$12M_b$
	(= 4·92)	3	(= 14·76)	4	(= 59)
C	3	2	6	2	12
D	5	2	10	0	0
	M_d		$3M_d$		
	(= 6·27)	3	(= 18·8)	0	0
E	5	3	15	−2	−30

The Mra polygon is shown in Figure 120 and was constructed by drawing

ab proportional to 22·5 parallel to OA in a direction from O to A
bc proportional to 30 parallel to OB in a direction from O to B
cd proportional to 12 parallel to OC in a direction from O to C
de proportional to −30 parallel to OE in a direction from E to O.

Resultant Mra value, $ae = 59$ to scale.

Angular velocity, $\omega = \dfrac{2\pi \times 75}{60} = 7\cdot 85$ rad/s

$\therefore$ Unbalanced couple $= M\omega^2 r \times a$
$= Mra \times \omega^2$
$= 59 \times 7\cdot 85^2$
$= 3640$ Nm.

This couple introduces an additional load on each bearing which rotates with the system, the value being found by dividing the above figure by the "arm" of the couple, i.e. by the distance between the bearings. Thus

Additional load per bearing $= \dfrac{3640}{12} = 303$ N.

The Mra value required for equilibrium is represented by the equilibriant, ea, and, from the Mra column in the table, this must be equal to $12M_b$ where M_b is the balance mass required in plane B.

Hence $\qquad\qquad\qquad\qquad 12\, M_b = 59$

i.e. $\qquad\qquad\qquad\qquad M_b = 4\cdot 92$ kg.

This may now be inserted in the system at 3 m radius, this radius being drawn from point O (as shown dotted) parallel to ea, i.e. at 22° to the vertical. Also, the value of $3M_b$ in the Mr column ($= 14\cdot 76$) may now be inserted and, since there is now only one unknown, the Mr polygon (Fig. 121) can be constructed by drawing

pq proportional to $\ 3\cdot 75$ parallel to OA
qr proportional to $\ 7\cdot 5\ $ parallel to OB
rs proportional to $\ 6\cdot 0\ $ parallel to OC
st proportional to $10\cdot 0$ parallel to OD
tu proportional to $15\cdot 0$ parallel to OE
uv proportional to $14\cdot 76$ parallel to OM_b.

Resultant Mr value, $pv = 18\cdot 8$ to scale.

The Mr value required for equilibrium is the equilibriant, vp, and, from the Mr column in the table, this must be equal to $3M_d$ where M_d is the balance mass required in plane D.

Hence, $\qquad\qquad\qquad\qquad 3M_d = 18\cdot 8$

i.e. $\qquad\qquad\qquad\qquad M_d = 6\cdot 27$ kg.

This may also be inserted in the system at 3 m radius, this radius being drawn from point O (as shown dotted) parallel to vp, i.e. at 17° to the vertical.

Note that, after M_b has been added to the system (and the couple eliminated) but before M_d is inserted, the actual unbalanced force is given by

$$F = M_d(\omega^2 r) = M_d r \times \omega^2 = 18\cdot 8 \times 7\cdot 85^2 = 1160 \text{ N}.$$

MECHANICAL TECHNOLOGY FOR HIGHER ENGINEERING TECHNICIANS

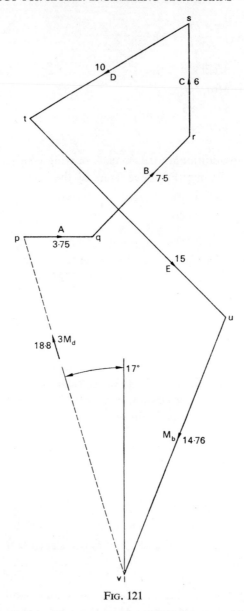

FIG. 121

With neither M_b nor M_d in position, the vector sum of the Mr values is pu which, to scale, is 12. Thus, originally, in addition to the unbalanced couple of 3640 Nm, there was an unbalanced force of $12 \times 7 \cdot 85^2$ or 740 N.

EXAMPLES 5

1. The rim of a rope brake forms a cooling trough of least diameter 1·25 m. Estimate the lowest speed at which it can retain coolant (38 rev/min approx.).
2. A motor vehicle crankshaft, flywheel and transmission assembly together weigh 625 N and have a radius of gyration of 90 mm. Find the torque required to bring the combination to rest in 6 s from 4000 rev/min and the revolutions made during the deceleration period (36 Nm, 200).

DYNAMICS

3. An underground train weighing 1·5 MN completes a route containing 20 stations in 50 min. Estimate the cost per hour of stopping the train assuming the brakes to be applied at 40 km/h and taking the cost of 1·0 MJ as 2 pence (455 pence).

4. A turntable is driven clockwise when viewed from above by two identical motors mounted on the underside at opposite ends of a diameter with their axes in a radial direction. Each armature assembly weighs 2220 N, has a radius of gyration of 152 mm and rotates at 1480 rev/min clockwise when viewed from the table centre. If the reduction between motors and table is 200 : 1, estimate the change in load on each motor bearing due to gyroscopic effect given that the bearings are 533 mm apart (1175 N transferred from inner to outer bearing).

5. Each of the two wheels of a river boat weighs 10 700 N and has a radius of gyration of 1·37 m. If the boat rounds a right hand bend in the river at 15 km/h and this corresponds to a paddle speed of 48 rev/min, calculate the magnitude of the gyroscopic couple and show, by means of a vector diagram, the effect on the boat of its reaction. The radius of the bend is 185 m (465 Nm. Boat heels to port).

6. An aircraft used for crop spraying has a single propeller weighing 286 N and having a radius of gyration of 0·915 m, the direction of rotation being clockwise when viewed from the front. Find the gyroscopic reaction couple during a turn of 760 m radius executed at 200 km/h if the propeller speed is then 1800 rev/min (340 Nm).
State the effect of this couple on the airframe if the turn is (a) to the left, (b) upward, i.e. a "loop." [(a) nose drops, (b) aircraft swings to left].

7. A car of total weight 8900 N has four wheels each having a moment of inertia of 0·95 kg m² and an effective diameter of 0·61 m. The axis of the motor and transmission parts is parallel to the direction of travel while the parts have a moment of inertia of 0·406 kg m² and rotate clockwise when viewed from the front. The wheelbase and track are 2·75 m and 1·37 m respectively and, when the car is stationary, its weight may be assumed equally distributed between the wheels. Adhesion is lost on a straight, level, ice-covered road when travelling at 15·25 m/s and in 1·5 s the vehicle executes two complete revolutions in a horizontal plane and in a clockwise direction when viewed from above. If the speed ratio between motor and wheels is 8·5:1, estimate the vertical load on each wheel during the skid. Justify your deductions with the aid of suitable vector diagrams (RH front 1385 N, LH front 2540 N, RH rear 1910 N, LH rear 3065 N).

8. The rotating parts of a machine tool have a moment of inertia of 38 kg m² and require a constant torque of 103 Nm. They are driven by an electric motor via a 9 : 1 reduction gear the efficiency of which is 0·95. If the motor has a starting torque of 27·1 Nm and its armature and pinion have a combined moment of inertia of 0·515 kg m², calculate:
(a) the motor output required for a machine operating speed of 200 rev/min (2·24 kW);
(b) the time taken for the machine to attain half speed (6·25 s).

9. The rotor of an auto pilot weighs 5·34 N and has a radius of gyration of 38·1 mm. It is driven at 30 000 rev/min anticlockwise when viewed from the front. If the aircraft in which it is mounted executes a right-hand turn of 1220 m radius when travelling at 2250 km/h, calculate the change in load on each of the two supporting bearings, if these are 127 mm apart. Illustrate your answer by means of a suitable vector diagram and state the directions of the changes in load. (10 N added to the load on the front bearing and subtracted from the load on the rear bearing.)

10. The armature of a motor has a moment of inertia of 2·03 kg m² and drives a machine via a 2:1 reduction gear. The machine requires constant torque, absorbs 3·73 kW at its operating speed of 500 rev/min and has rotating parts which are the equivalent of 90·5 kg at 0·38 m radius. If the gear efficiency is 0·85 determine the input torque required to accelerate the system from test to fullspeed in 20 s (72·6 Nm).

11. The sun wheel of a simple epicyclic train has 60 teeth and when this is fixed and the spider is driven at 240 rev/min, the annulus is required to rotate at 312 rev/min. Find the numbers of teeth needed by annulus and planets (200, 70).

12. Figure 122 shows a reduction gear in which the input sun wheel has 24 teeth and the fixed annulus has 90 teeth. If the moving annulus has 100 teeth and the sun wheel integral with it has 30 teeth, find the speed of the output when the input is driven at 1150 rev/min (100 rev/min).

13. Figure 123 shows a gear in which $A = C = 50$ teeth, $B = E = 30$ teeth. If A and E are driven in the same direction at 148 rev/min and 100 rev/min respectively, find what happens to D (175 rev/min, same direction as A).

14. Figure 124 represents a gear in which R = annulus having teeth of 216 mm p.c.d., 4·5 mm module, S = sun wheel (driven), Q = spider carrying three equi-spaced planet wheels, P.
If, when R is stationary, Q is to make one revolution for every five revolutions made by S, determine suitable numbers of teeth. Show that it is a condition of assembly that the numbers of teeth on S and R shall be divisible by 3 ($R = 48$, $S = 12$, $P = 18$).

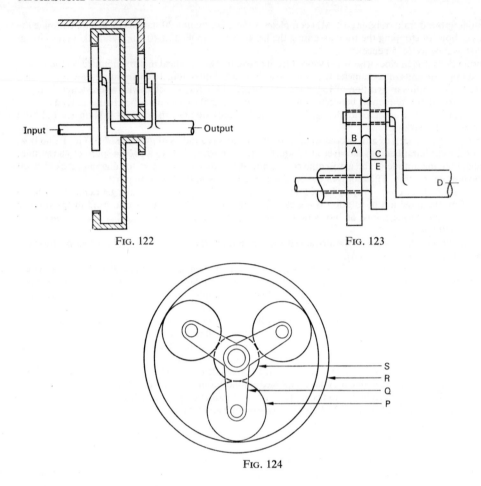

Fig. 122

Fig. 123

Fig. 124

15. Determine the size and position of the mass required at 150 mm radius to give balance to the following coplanar system:
    12 kg at 100 mm radius,
    24 kg at 75 mm radius 90° anticlockwise from the first radius,
    30 kg at 100 mm radius 240° anticlockwise from the first radius (5·77 kg at 70° anticlockwise from first radius).
    Find also the size of each of the two balance masses which could be substituted for the single one already found if these are to be at 75 mm radius and situated respectively at 30° and 150° anticlockwise from the 12 kg load (13 kg and 8·9 kg).

16. Four cranks each having a throw of 50 mm and numbered 1 to 4 carry respectively masses of 20, 12·5, 100 and M kg. The spacing between the first and second cranks is 125 mm, between the second and third is 100 mm, and between the third and fourth is 50 mm. Find the value of M and the relative angular positions to give dynamic balance (82 kg at 16° to number 1).

17. Part of a machine is 1·5 m long and its imbalance is equivalent to:
    (a) 10 kg at 0·04 m radius at the right-hand end, vertically up from the axis;
    (b) 10 kg at 0·06 m radius at the left-hand end, 120° anticlockwise from the first when viewed from the right-hand end.
    Take the left-hand end as reference and find, for static balance, the mass required at 0·5 m radius and its angular position relative to the first (1·06 kg at 260° anticlockwise from (a)). Find also
    (c) the axial position of the balance mass which makes the residual couple a minimum (0·22 m from LH end);
    (d) the value of this couple when the system is driven at 500 rev/min (412 Nm).

6

MECHANISMS

THE VELOCITY DIAGRAM

Configuration

This is the name given to a diagram of the mechanism, to some convenient scale, showing the instantaneous relative positions of the links, which one is driving and which of the pivots is fixed. Details of bearings, method of lubrication and so on are omitted.

Since it is seldom practicable to calculate the instantaneous velocity of each point in a mechanism, some salient point is chosen (e.g. the pin of the driving crank) and a graphical method used, based on the configuration, to obtain a second diagram from which can be found the velocity of any point relative to the chosen one.

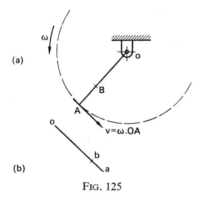

Fig. 125

Figure 125(a) shows a rigid link OA being driven at ω rad/s about a fixed point O. Instantaneously the velocity of point A is $\omega.OA$ in the direction of the tangent, i.e. in a direction at right angles to the link. This velocity may be represented by the vector oa, shown in Figure 125(b), drawn parallel to the tangent to some convenient scale. It is now possible to state that

$$\text{Velocity of } A \text{ relative to } O, \quad oa = \omega.OA.$$

For any point B between O and A we can also state that

$$\text{Velocity of } B \text{ relative to } O, \quad ob = \omega.OB.$$

MECHANICAL TECHNOLOGY FOR HIGHER ENGINEERING TECHNICIANS

Since velocity is proportional to radius we can write also

$$\frac{ob}{oa} = \frac{OB}{OA}.$$

Note that points in the configuration (Fig. 125(a)) are denoted by capital letters, while points in Figure 125(b)—which is called a *velocity diagram*—are denoted by small letters. Note also that the letter at the start of a velocity vector corresponds with the point about which the motion is relative—in this case the fixed point, O. Since this pivot has an equal and opposite velocity relative to the moving point, A, we can write

Velocity of O relative to A, $\underline{ao = \omega.AO.}$

The single vector thus represents two equal and opposite motions, depending on the order of the letters. For this reason arrows should not be inserted.

If the length of the link is known and the velocity of one end relative to the other is found from the velocity diagram, the angular velocity of the link can be obtained. Thus, in the case considered

Angular velocity of link, $\omega = \dfrac{ao}{AO}.$

Thus if the moment of inertia of the link, I, is known, the instantaneous kinetic energy of the link can be found from the relation

$$\text{K.E.} = \tfrac{1}{2}I\omega^2.$$

The foregoing will now be applied to one or two common mechanisms.

The Four-bar Chain

This is the basis of many mechanisms and consists of three movable links and one fixed link, the configuration being as shown in Figure 126.

If the link AB is driven at ω_1 the links BC and CD will, respectively, have instantaneous velocities ω_2 and ω_3. Then

Velocity of B relative to A, $ab = \omega_1 AB.$

This may be represented, to some convenient scale, by the vector ab drawn normal to AB as shown in Figure 127(a). Also,

Velocity of C relative to B, $bc = \omega_2 BC.$

This may be represented by the vector bc drawn from point b normal to BC. Finally,

Velocity of C relative to D, $dc = \omega_3 DC.$

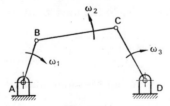

Fig. 126

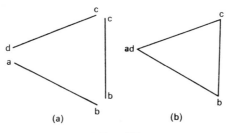

FIG. 127

This may be represented by the vector *cd* drawn from point *c* normal to *DC*. Evidently the three vectors of Figure 127(a) may be combined to form a triangular velocity diagram as shown in Figure 127(b), points *a* and *d* being the same since both *A* and *D* are fixed. The vector *ac* is seen to be the vector sum of *ab* and *bc*, i.e. the velocity of *C* relative to *A* is the vector sum of its velocity relative to *B* and the velocity of *B* itself relative to the housing *AD*.

EXAMPLE. Figure 128 shows a four-bar chain in which $AB = 0.417$ m, $BC = CD = 0.583$m and the distance between the fixed pivots *A* and *D* is 0.834 m. Find the velocity of point *C* and the angular velocity of the link *BC* when link *AB* is driven clockwise at 50 rev/min and makes with *AD* an angle of (a) 70°, (b) 30°.

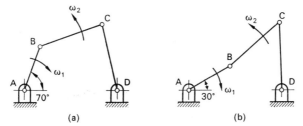

FIG. 128

Solution. Velocity of *B* relative to *A*,

$$ab = \omega_1 AB$$

$$= \left(\frac{2\pi \times 50}{60}\right) 0.417$$

$$= 2.18 \text{ m/s.}$$

The velocity diagrams for both cases are shown in Figure 129.

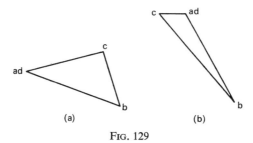

FIG. 129

MECHANICAL TECHNOLOGY FOR HIGHER ENGINEERING TECHNICIANS

In case (a): $ab = 2\cdot 18$
$\qquad bc = 1\cdot 25$ to scale
$\qquad ac = 1\cdot 72$ to scale

$$\omega_2 = \frac{bc}{BC}$$
$$= \frac{1\cdot 25}{0\cdot 583}$$
$$= 2\cdot 14 \text{ rad/s}$$

In case (b): $ab = 2\cdot 18$
$\qquad bc = 2\cdot 42$ to scale
$\qquad ac = 0\cdot 47$ to scale

$$\omega_2 = \frac{bc}{BC}$$
$$= \frac{2\cdot 42}{0\cdot 583}$$
$$= 4\cdot 14 \text{ rad/s}$$

The Reciprocating Mechanism

This is also known as the *slider crank chain* and is derived from the four-bar chain by making the link CD infinitely long relative to the link AB, as shown in Figure 130(a). Point C then lies on the circumference of a circle of infinite radius so that rotation of AB causes it to oscillate (i.e. to reciprocate) in a straight line. A guide may be introduced therefore as shown in Figure 130(b) and the link CD discarded; the remaining links AB and BC are then known respectively as the *crank* and the *connecting rod*.

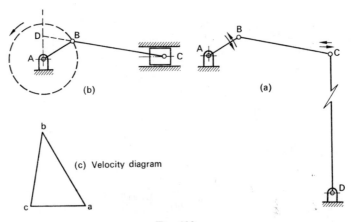

Fig. 130

For the rotation shown,

Velocity of B relative to A, $ab = \omega AB$ in a direction normal to AB.

Velocity of C relative to A, ac is horizontal but of unknown magnitude.

Velocity of C relative to B, bc is also unknown in magnitude but is in a direction normal to BC.

The velocity diagram abc is shown in Figure 130(c), point c being fixed by the intersection of ac and bc. The velocity of the slide C is obtained by measuring ac to scale.

Referring to Figure 130(b) it is evident that, if CB is produced to D, the triangle ABD

is similar to triangle abc so that

$$\frac{ac}{ab} = \frac{AD}{AB},$$

i.e. Slide velocity, $\quad ac = \dfrac{AD}{AB} \times ab \quad$ and $\quad ab = \omega AB$

$$= \frac{AD}{AB} \times \omega AB$$

or $\quad ac = \omega AD.$

The velocity of the slide in a given configuration may be found therefore by measuring AD to scale and multiplying the result by the angular velocity of the crank.

Relative Velocity between Pin and Link

A pin connecting two links may be attached to either, or it may "float", in which case a means of keeping it in position must be provided.

Let a link rotate about a stationary pin of radius r at ω_1 rad/s as shown in Figure 131(a). Neglecting the clearance,

Relative velocity between link and pin $= \omega_1 r.$

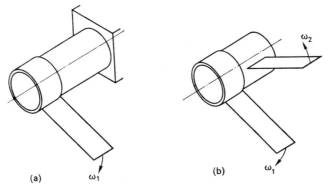

Fig. 131

If the pin is not stationary but is attached to a second link rotating at ω_2 rad/s in the opposite direction as shown in Figure 131b, then,

Relative velocity between link and pin $= (\omega_1 + \omega_2) r.$

If the direction of rotation of the pin is reversed, then,

Relative velocity between link and pin $= (\omega_1 - \omega_2) r.$

EXAMPLE. The motor of a powered cycle has a stroke of 72 mm, and the bearing centres of the connecting rod are 136 mm apart. If the diameters of piston pin and crankpin are respectively 12·7 and 38 mm, find, when the crank is at 45 deg after inner dead centre (i.d.c.) and rotating at 4000 rev/min anticlockwise,

(a) velocity of piston,
(b) angular velocity of connecting rod,
(c) relative velocities at connecting rod bearings.

Solution. The configuration is shown in Figure 132 and from it to scale: $AD = 30$ mm and $BD = 26$ mm.

Angular velocity of crank $= \dfrac{2\pi \times 4000}{60}$

i.e. $\omega_1 = 419$ rad/s.

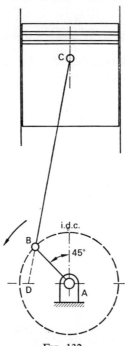

FIG. 132

Velocity of piston, (C), $= \omega_1 AD$
$= 419 \times 0.030$
$= 12.57$ m/s.

Velocity of C relative to $B = \omega_1 BD$
$= 419 \times 0.026$
$= 10.9$ m/s.

Angular velocity of $BC = \dfrac{10.9}{136}$

i.e. $\omega_2 = 80$ rad/s.

Assuming the piston pin to be fixed in the piston,

Relative velocity between BC and piston pin $= 80\left(\dfrac{0\cdot0127}{2}\right)$

$\qquad\qquad = 0\cdot508$ m/s.

Assuming the crankpin to be fixed in the crank (as is usual),

Relative velocity between BC and crankpin $= (419+80)\left(\dfrac{0\cdot038}{2}\right)$

$\qquad\qquad = 9\cdot5$ m/s.

Quick Return Mechanisms

These are often employed in machine tools of the reciprocating type to reduce the time taken by the return stroke of the tool. They consist basically of an oscillating lever AC, Figure 133, actuated by a crank AB via a sliding block B.

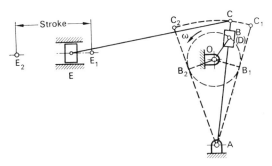

Fig. 133

The ends of the stroke, E_1 and E_2, correspond with the block positions B_1 and B_2 and the tool E performs the cutting stroke while the pin B traverses the arc B_1BB_2. The return is performed during the remaining arc and, if ω is uniform, we have

$$\frac{\text{Cutting time}}{\text{Return time}} = \frac{\text{Arc } B_1BB_2}{\text{Arc } B_2B_1}.$$

The velocity of the block relative to the housing (either O or A) is the vector sum of the velocity of rotation of the coincident point, D (on AC beneath B), and the sliding velocity of the block itself relative to this point. In other words, referring to Figure 134,

$$ab = ad + db$$

where
$\qquad ad =$ velocity of rotation of D about A

$\qquad db =$ velocity of sliding of B along AC away from A.

Since point C is also rotating about A we have

$$\frac{ac}{ad} = \frac{AC}{AD} = \frac{AC}{AB}.$$

i.e.
$$ac = \frac{AC}{AB} ad.$$

This enables point c to be fixed. Point e is now found by drawing a line through c normal to CE to cut the horizontal through point a. To scale, the velocity of cutting is then ae.

By arranging the lever to pivot about a point between C and B a version known as the "Whitworth" mechanism is obtained and this is more compact. The lever AC now rotates about A (instead of oscillating), the configuration being shown in Figure 135. The return stroke is performed while the block B traverses the arc B_2BB_1.

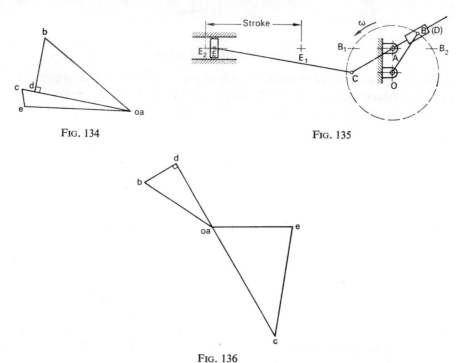

Fig. 134

Fig. 135

Fig. 136

As before, the velocity of the block relative to the housing (either O or A) is the vector sum of the velocity of rotation of the coincident point, D (beneath B on CA produced), and the sliding velocity of the block itself relative to this point. In other words, referring to Figure 136,
$$ab = ad + db$$
where $\quad ad =$ velocity of rotation of D about A
$\quad\quad\quad db =$ velocity of sliding of B along AC towards A.

Since point C is also rotating about A we have
$$ac = \frac{AC}{AB} ad.$$

This enables point c to be fixed. Point e is found by drawing a line through c normal to CE to cut the horizontal through A. To scale, the velocity of cutting is then ae.

EXAMPLE. The mechanism of a shaper is shown in Figure 137. Find the instantaneous velocity of cutting in the position shown when the machine is making 90 cutting strokes per minute. $OA = 75$ mm, $OB = AC = 150$ mm and $CE = 340$ mm.

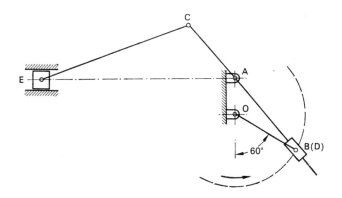

FIG. 137

Solution.

Velocity of B relative to O, $\quad ob = \left(\dfrac{2\pi \times 90}{60}\right) 0\cdot 15$

$\qquad\qquad\qquad\qquad\qquad\qquad = 1\cdot 413$ m/s

To scale, $\qquad\qquad\qquad\qquad ad = 1\cdot 332$ m/s

By measurement, $\qquad\qquad AB = 198$ mm

$\dfrac{ac}{ad} = \dfrac{AC}{AB}$ so that $\qquad ac = \dfrac{150}{198} \times 1\cdot 332 = 1\cdot 01$ m/s.

The velocity diagram is as shown in Figure 138 and from it, to scale,

Tool velocity, $\quad oe = 1\cdot 017$ m/s.

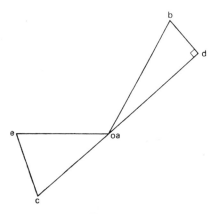

FIG. 138

The Principle of Work

Referring to Figure 139, let F be the instantaneous effort (applied force) required at point B in a direction normal to AB to produce rotation as shown against a load (resisting force) P applied at point C in a direction normal to CD.

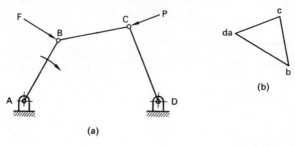

Fig. 139

Assuming that there is no loss and that the motion is uniform (i.e. that there is no inertia force) we have

$$\text{Work done/sec by effort} = \text{Work done/sec on load}$$
$$F \times \text{Velocity of } B = P \times \text{velocity of } C$$
$$F \times ab = P \times ac.$$

This is a statement of what is known generally as the "Principle of Work" and is useful when the mechanical advantage of linkwork is required to be found. Thus in this case, for a given value of P,

$$\text{Least effort required,} \quad F = \frac{ac}{ab} \times P.$$

Suppose the effort to be applied (by, say, gas pressure) to the piston of a reciprocating mechanism as shown in Figure 140. Then, applying the Principle of Work,

$$F \times ac = P \times ab.$$

But, since it has been shown that $\dfrac{ac}{ab} = \dfrac{AD}{AB}$ we have

$$F \times AD = P \times AB.$$

The right-hand side of this equation is evidently the instantaneous torque on the crankshaft so that this quantity may be found simply by multiplying the *effective* force on the piston by the intercept AD.

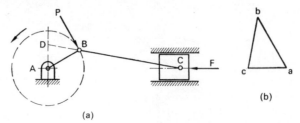

Fig. 140

Note that when D is below A, the direction of the torque is reversed unless the direction of F is reversed also. Note also that the effective force is the gas force less the inertia force opposing the motion and that some of the torque accelerates the connecting rod.

EXAMPLE. In the press mechanism shown in Figure 141(a) ABC is a 90° bell crank. Assume no loss and find the minimum force required at point A to overcome a resistance of 1500 N at the slide given that $BD = 254$ mm, $AB = 4BC = 305$ mm.

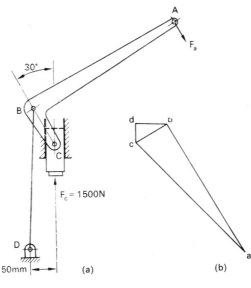

FIG. 141

Solution. For the force to be a minimum it must be applied at 90° to the link AB so that, letting the plunger, C, have unit velocity relative to the housing, D, the velocity diagram is as shown in Figure 141(b).

Since $dc = 1·0$ and $da = 8·68$ to scale, we have, applying the Principle of Work:

$$F_a \times da = F_c \times dc$$

where
$$F_c = 1500 \text{ N},$$

whence
$$F_a = 1500 \left(\frac{1·0}{8·68} \right) = \underline{173 \text{ N}}.$$

Instantaneous Centre

Consider the link BC of the four-bar chain shown in Figure 142(a).

Since point B is moving at right angles to the link AB it may be considered to be rotating instantaneously about any point on a line through AB. Similarly point C may be considered to be rotating instantaneously about any point on a line through DC. Since these lines intersect at point I, both B and C (and hence all points on the link) may be looked on as rotating momentarily about I at some angular velocity ω. This point is known therefore as

137

MECHANICAL TECHNOLOGY FOR HIGHER ENGINEERING TECHNICIANS

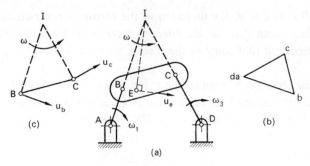

FIG. 142

the *instantaneous centre* of the link BC. Evidently the velocity of any point on the link is proportional to its distance from I and is in a direction normal to the line joining it to I. Considering for example point E,

Instantaneous velocity, $\qquad u_e = \omega EI$.

Now, Velocity of B, $\quad u_b = \omega BI \quad$ so that $\quad \omega = \dfrac{u_b}{BI}$

and Velocity of C, $\quad u_c = \omega CI \quad$ so that $\quad \omega = \dfrac{u_c}{CI} \quad$ also

i.e. $\qquad \dfrac{u_b}{BI} = \dfrac{u_c}{CI} \quad$ (equating expressions for ω)

so that $\qquad \underline{\dfrac{u_c}{u_b} = \dfrac{CI}{BI}}.$

But, from the velocity diagram, Figure 142(b),

$$\dfrac{u_c}{u_b} = \dfrac{dc}{ab} = \dfrac{ac}{ab} \quad \text{so that} \quad \underline{\dfrac{ac}{ab} = \dfrac{CI}{BI}}.$$

Evidently the angles CIB and cab are equal so that triangles bac and BIC are similar, showing that BIC is in effect a velocity diagram.

THE ACCELERATION DIAGRAM

A body will move in a circular path only when given a centripetal acceleration, i.e. when a radially inward force is applied. If the tangential (peripheral) speed is u and the path radius is r then the centripetal acceleration is u^2/r. Since $u = \omega r$ this acceleration is also $\omega^2 r$.

Suppose the link AB of Figure 143(a) to be rotating at ω about point A and let the velocity of point B relative to A be u. Then the centripetal acceleration of point B is u^2/AB and may be represented, to some suitable scale, by a vector $a_1 b_a$ [Fig. 143(b)] drawn parallel to BA.*

---

* The notation adopted is that due to Dyson and is used in his book *Principles of Mechanism*.

138

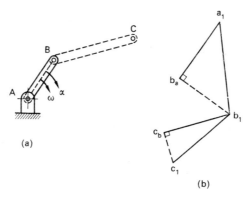

Fig. 143

If ω is not uniform but is changing (say increasing) at α rad/s², then u will be increasing at, say, f, i.e. there will be a tangential acceleration given by $f = \alpha \cdot AB$. This may be represented to the same scale by the vector $b_a b_1$ drawn from point b_a normal to $a_1 b_a$. If points a_1 and b_1 are joined, the vector $a_1 b_1$ will represent in magnitude and direction the resultant or total acceleration of B relative to A.

The acceleration of a third point, C, relative to A may be found by adding vectorially its *total* acceleration ($b_1 c_1$) relative to B to the total acceleration relative to A of B itself, so obtaining the vector sum $a_1 c_1$ (not shown).

Notes

1. Before constructing the acceleration diagram for a given mechanism, the centripetal acceleration of each salient point must be found using the expression u^2/r, the values of u being obtained from the velocity diagram and those of r from the configuration.
2. The letter at the start of centripetal and total acceleration vectors corresponds with the point to which the acceleration is relative and has the subscript ($_1$).
3. The letter at the end of tangential and total acceleration vectors corresponds with the relatively moving point and also has the subscript ($_1$).
4. The letter at the end of the centripetal acceleration vector also corresponds with the relatively moving point while its subscript denotes the "fixed" point.
5. If the path is straight (i.e. of infinite radius), the first (centripetal) vector is zero so that the letters denoting it coincide in the acceleration diagram. For example, in Figure 143(b) points a_1 and b_a would appear together.
6. If ω is constant the second (tangential) vector is zero. The letters denoting it will therefore coincide in the diagram.
7. When the tangential acceleration (f) of one end of a link has been found (from the acceleration diagram) the angular acceleration of the link itself can be found using the expression $\alpha = f/r$, where r is the length of the link. This enables calculation of the instantaneous torque ($I\alpha$) required to accelerate the link, and hence of the forces on the pins.

The foregoing will now be applied to one or two common mechanisms.

The Four-bar Chain

Suppose the configuration and velocity diagram to be as shown in Figure 144 and let the link AB have an angular acceleration, α.

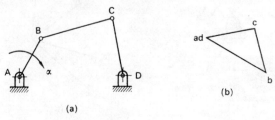

Fig. 144

Centripetal acceleration of C relative to D

$$= \frac{dc^2}{DC} \quad \text{represented by vector } d_1 c_d.$$

Centripetal acceleration of C relative to B

$$= \frac{bc^2}{BC} \quad \text{represented by vector } b_1 c_b.$$

Centripetal acceleration of B relative to A

$$= \frac{ab^2}{AB}. \quad \text{represented by vector } a_1 b_a.$$

Tangential acceleration of B relative to A

$$= \alpha . AB \quad \text{represented by vector } b_a b_1.$$

Construction

Draw, to some suitable scale, as shown in Figure 145,
$a_1 b_a$ parallel to BA proportional to ab^2/AB.
$b_a b_1$ normal to $a_1 b_a$ from b_a proportional to $\alpha . AB$.
(The resultant $a_1 b_1$ represents the total acceleration of point B relative to point A.)

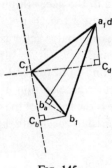

Fig. 145

From point b_1 draw b_1c_b parallel to CB proportional to bc^2/BC and from point c_b draw at right angles a vector of unknown length in both directions. Point c_1 lies somewhere on this vector.

Put in point d_1 coincident with point a_1.

Draw d_1c_d parallel to CD proportional to dc^2/DC and from point c_d draw at right angles a vector of unknown length in both directions. Point c_1 lies somewhere on this vector also so that the junction of this and the first unknown vector must fix point c_1. The total acceleration of point C relative to point B is b_1c_1 while the total acceleration of point C relative to point D (or A) is d_1c_1 (or a_1c_1) which is the vector sum of d_1b_1 and b_1c_1.

Finally, $\qquad$ Angular acceleration of $DC = \dfrac{c_d c_1}{DC}$

and $\qquad$ Angular acceleration of $BC = \dfrac{c_b c_1}{BC}$.

The Reciprocating Mechanism

As stated previously, this is derived from the four-bar chain by making the link CD infinitely long. The slide, C, moves therefore in a straight path and hence has zero centripetal acceleration. If the crank rotates at uniform angular velocity the crankpin, B, will have zero tangential acceleration.

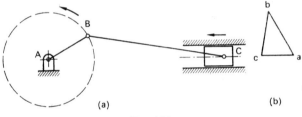

FIG. 146

Referring to Figure 146(a),

Centripetal acceleration of C relative to A, $\quad a_1 c_a = 0$

Centripetal acceleration of C relative to B, $\quad b_1 c_b = \dfrac{bc^2}{BC}$

Centripetal acceleration of B relative to A, $\quad a_1 b_a = \dfrac{ab^2}{AB}$

Tangential acceleration of B relative to A, $\quad b_a b_1 = 0$.

Construction

Draw, to some suitable scale, as shown in Figure 147, $a_1 b_a$ parallel to BA proportional to ab^2/AB.

Put in point b_1 coincident with point b_a, and from it draw $b_1 c_b$ parallel to CB proportional to bc^2/BC. Point c_1 lies somewhere on a normal through point c_b.

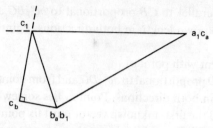

Fig. 147

Put in point c_a coincident with point a_1 and from it draw $c_a c_1$ parallel to CA to intersect the normal through point c_b and so fix point c_1.

Then, $\quad$ Angular acceleration of connecting rod $= \dfrac{c_b c_1}{BC}$.

Note that reversal of the direction of crank rotation has no effect on the acceleration diagram since the centripetal acceleration of B is always towards A.

Acceleration of a Block sliding along a Rotating Link

Let the link OA shown in Figure 148 rotate anticlockwise about the fixed pivot, O, with uniform angular velocity ω and let the block, B, slide radially outwards along OA with instantaneous velocity and acceleration v and f respectively. Let these directions be considered positive.

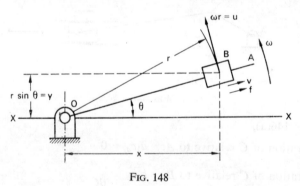

Fig. 148

When OA has turned through some angle, θ, from the datum XX, the vertical displacement of B from XX is given by

$$y = r \sin \theta.$$

Differentiating the right-hand side as a product (with respect to time) will give dy/dt, the vertical component of the velocity of B. Thus

$$\frac{dy}{dt} = r\left(\cos\theta \frac{d\theta}{dt}\right) + \sin\theta \left(\frac{dr}{dt}\right)$$

i.e. $\quad \dfrac{dy}{dt} = \omega r \cos\theta + v \sin\theta.$ $\hfill (1)$

(Alternatively, the velocity of B is the vector sum of its velocity normal to the link, ωr, and its velocity along the link, v, and the vertical components are respectively $\omega r \cos \theta$ and $v \sin \theta$.)

Differentiating eq. (1) will give the vertical component of the acceleration of B. Thus

$$\frac{d^2y}{dt^2} = \left[\omega r\left(-\sin\theta \frac{d\theta}{dt}\right) + \cos\theta\left(\omega \frac{dr}{dt} + r\frac{d\omega}{dt}\right)\right]$$

$$+ \left[v\left(\cos\theta \frac{d\theta}{dt}\right) + \sin\theta\left(\frac{dv}{dt}\right)\right]$$

$$= -\omega r \sin\theta \frac{d\theta}{dt} + \omega \cos\theta \frac{dr}{dt} + r \cos\theta \frac{d\omega}{dt}$$

$$+ v \cos\theta \frac{d\theta}{dt} + \sin\theta \frac{dv}{dt}.$$

But $d\omega/dt = 0$ since the angular velocity is uniform, while $dv/dt = f$, the linear acceleration of the block relative to the link, so that

$$\frac{d^2y}{dt^2} = -\omega^2 r \sin\theta + \omega v \cos\theta + 0 + \omega v \cos\theta + f \sin\theta$$

$$= -\omega^2 r \sin\theta + 2\omega v + f \sin\theta$$

or $$\frac{d^2y}{dt^2} = (f-\omega^2 r) \sin\theta + 2\omega v \cos\theta. \qquad (2)$$

Now, when $\theta = 0$, $\cos\theta = 1$ and $\sin\theta = 0$ so that, when the block is in the position shown in Figure 149.

Horizontal component of acceleration of $B = f$

and Vertical component of acceleration of $B = \dfrac{d^2y}{dt^2} = 2\omega v.$

Fig. 149

Thus, whenever a pin-carrying block slides radially with velocity v along a link rotating with angular velocity ω there is an acceleration of $2\omega v$ in a direction *normal to the link*, this direction leading the direction of v by 90° in the direction of ω. It is evident from eq. (2) that this acceleration is present even when v is uniform, i.e. when f is zero. This part of the total acceleration of the block requires a lateral force on the block to be exerted by the link and is named after G. G. Coriolis (1792–1843) the Frenchman who first revealed its existence.

Referring again to eq. (2) it is evident that, since $\omega^2 r$ is the centripetal acceleration of the point on the link coincident with the block, the quantity in the brackets is the net radial component of the total acceleration. This is illustrated by Figure 150.

If the angular velocity of the link is changing (say increasing), i.e. there is an angular

143

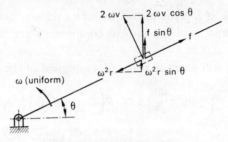

FIG. 150

acceleration, α, then, in addition to the Coriolis component, the block will have a tangential acceleration αr. Referring to Figure 151(a), the various velocities and accelerations are

ω = angular velocity of link OA,
α = angular acceleration of link OA,
$\omega^2 r$ = centripetal acceleration of point on OA coincident with B,
αr = tangential acceleration of point on OA coincident with B,
v = radial velocity of B relative to OA,
f = radial acceleration of B relative to OA,
$2\omega v$ = tangential acceleration of B relative to OA.

The last quantity ($2\omega v$) must be added to αr to obtain the tangential acceleration of B relative to O.

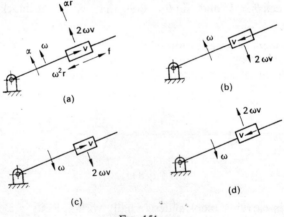

FIG. 151

From Figure 151 (b, c and d) it is evident that, since the direction of the Coriolis component is deduced by rotating the direction of v through 90° in the direction of ω, the direction of the component will be reversed if the direction of either ω or v is reversed, but will remain unchanged if the directions of both are reversed simultaneously.

EXAMPLE. The configuration shown in Figure 152(a) is that of a shaping machine. If the crank, OB, has a throw of 127 mm and is driven anticlockwise at 25 rev/min, find the angular velocity and acceleration of the link AC and the linear acceleration of the toolholder E given that $AC = 762$ mm, $CE = 610$ mm, and $OA = 406$ mm.

Solution.

$$\text{Angular velocity of } OB = \frac{2\pi \times 25}{60} = 2\cdot 62 \text{ rad/s.}$$

Let D be the point on the link AC corresponding with the instantaneous position of the block B. To scale, $AD = 505$ mm. (Hereafter D will be referred to as the *coincident point*.)

Velocity of B relative to O, $ob = 2\cdot 62 \times 0\cdot 127 = 0\cdot 333$ m/s.

Velocity of rotation of D about A, $ad = 0\cdot 273$ m/s.

(Point a is coincident with point o.)

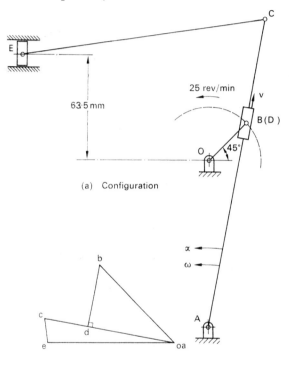

(a) Configuration

(b) Velocity diagram

Fig. 152

Velocity of sliding of B away from A, i.e. relative to D,

$$db = 0\cdot 19 \text{ m/s. (This is } v \text{ for the block.)}$$

Angular velocity of link $AC = \dfrac{ad}{AD} = \dfrac{0\cdot 273}{0\cdot 505} = 0\cdot 54$ rad/s. (This is ω for the link.)

Velocity of C relative to A,

$$ac = \frac{AC}{AD} \times ad$$

$$= \frac{0\cdot 762}{0\cdot 505} \times 0\cdot 273$$

$$= 0\cdot 412 \text{ m/s.}$$

Velocity of E relative to C, $ce = 0.076$ m/s to scale.
Velocity of E in guide, $ae = 0.394$ m/s to scale.

Coriolis component $= 2\omega v$

$$= 2 \times 0.54 \times 0.19$$

$$= 0.205 \text{ m/s}^2 \text{ at } 90° \text{ to } AC \text{ in direction of } \omega.$$

Centripetal acceleration of B relative to 0, $o_1 b_0 = \dfrac{ob^2}{OB}$

$$= \dfrac{0.333^2}{0.127}$$

$$= 0.873 \text{ m/s}^2.$$

Tangential acceleration of B relative to O, $b_0 b_1 = 0$ (because the angular velocity of the crank is uniform). Hence point b_1 is coincident with point b_0.

Centripetal acceleration of D relative to A, $a_1 d_a = \dfrac{ad^2}{AD}$

$$= \dfrac{0.273^2}{0.505}$$

$$= 0.147 \text{ m/s}^2.$$

Tangential acceleration of D relative to A, $d_a d_1 = \alpha . AD$, where α is the required angular acceleration of the link AC.

Construction of Acceleration Diagram

Draw $o_1 b_0$ (Fig. 153) parallel to BO and proportional to 0.873 and put in points b_1 and a_1.

Draw $a_1 d_a$ parallel to DA and proportional to 0.147. Point d_1 lies somewhere on the normal through d_a (shown dotted) and, when found, fixes the length of $a_1 d_1$, i.e. fixes the value of the total acceleration of the coincident point D relative to point A.

The total acceleration of B relative to O, i.e. to A ($a_1 b_1$ already known), must also be the result of adding to $a_1 d_1$ the vector sum of f and $2\omega v$, i.e. of $d_1 d_2$ and $d_2 b_1$.

Draw $d_1 b_1$ back from b_1 at $90°$ to AC and proportional to 0.205. Point d_1 lies on a line through point d_2 parallel to AC and is evidently the intersection of this line with the dotted line. Join $a_1 d_1$.

Sliding acceleration of block, $f = d_1 d_2 = 0.567$ m/s² to scale.
Tangential acceleration of D relative to A, $d_a d_1 = 0.292$ m/s² to scale.

∴ Angular acceleration of AC, $\alpha = \dfrac{d_a d_1}{AD} = \dfrac{0.292}{0.505} = 5.78$ rad/s².

Total acceleration of D relative to A, $a_1 d_1 = 0.33$ m/s² to scale.

Point c_1 is obtained by producing $a_1 d_1$ so that $\dfrac{a_1 c_1}{a_1 d_1} = \dfrac{AC}{AD}$.

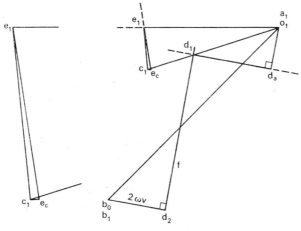

Fig. 153

Total acceleration of C relative to A, $a_1c_1 = 0 \cdot 33 \left(\dfrac{0 \cdot 762}{0 \cdot 505} \right) = 0 \cdot 497 \text{ m/s}^2$.

Centripetal acceleration of E relative to C, $c_1e_c = \dfrac{ce^2}{CE}$

$$= \dfrac{0 \cdot 076^2}{0 \cdot 610}$$

$$= 0 \cdot 0094 \text{ m/s}^2.$$

Point e is obtained by drawing a normal through point e_c to intersect a horizontal through point a_1. Then

Linear acceleration of E, $a_1e_1 = 0 \cdot 482 \text{ m/s}^2$ to scale.

The Inertia of a Link

Suppose a link BC, Figure 154(a), to be located by pins at points B and C and suppose f_g to be the instantaneous linear acceleration of its mass centre, G. The force required to impart this acceleration (i.e. to overcome the inertia) is $M \cdot f_g$ where M is the mass of the link. The equal and opposite reaction to this force, i.e. the inertia force, is denoted in the figure by F_i.

Suppose also that α is the instantaneous angular acceleration of the link about G. The torque required to impart this acceleration (i.e. to overcome the moment of the inertia) is

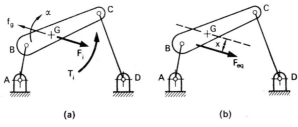

Fig. 154

MECHANICAL TECHNOLOGY FOR HIGHER ENGINEERING TECHNICIANS

$I\alpha$ where I is the moment of inertia about G. The equal and opposite reaction to this torque, i.e. the inertia torque, is denoted in the figure by T_i.

Now suppose the axis of the inertia force could be displaced in such a way as to make the moment of F_i about G equal to T_i and of the same sense. The situation would then be as shown in Figure 154(b) in which the displaced or *equivalent inertia force* (denoted by F_{eq}) replaces both F_i and T_i. Denoting this displacement by x we have, for equivalence,

$$F_{eq}x = F_i x = T_i$$

or

$$x = \frac{T_i}{F_i} = \frac{\text{Inertia torque}}{\text{Inertia force}} = \frac{I\alpha}{Mf_g}.$$

The values of f_g and α are obtainable from the acceleration diagram so that the value of the supposed displacement, x, can be calculated and an arrow representing F_{eq} inserted in the configuration as shown in Figure 154b. The forces which must be exerted at B and C to overcome F_{eq} can then be found by drawing a force polygon (triangle) for the link.

The method is illustrated by the following example.

EXAMPLE. Figure 155(a) shows a four-bar chain in which the link AB is 0·122 m long and is driven anticlockwise at 50 rev/min. The link CD is 0·229 m long. The link BC is 0·427 m

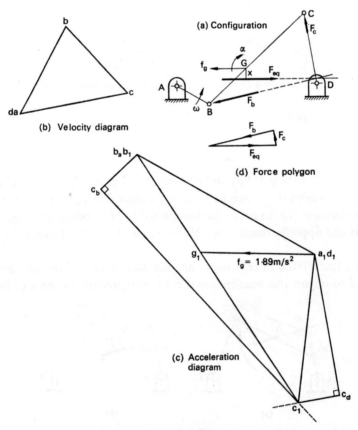

FIG. 155

long, has its mass centre, G, at a point 0·168 m from B and, about this point, has a moment of inertia of 0·02 kg m², while its mass is 3·4 kg. The bearings A and D are at the same level and are 0·458 m apart. Find the forces exerted on the link by the pins at B and C when $\widehat{BAD} = 30°$.

Solution.

Angular velocity of AB, $\quad \omega = \dfrac{2\pi \times 50}{60} = 5·24$ rad/s.

Velocity of B relative to A, $\quad ab = \omega \cdot AB = 5·24 \times 0·122 = 0·640$ m/s.

Velocity of C relative to D, $\quad dc = 0·732$ m/s to scale (Fig. 155b).

Velocity of C relative to B, $\quad bc = 0·595$ m/s to scale.

Centripetal acceleration of B relative to A, $\quad a_1 b_a = \dfrac{ab^2}{AB} = \dfrac{0·640^2}{0·122} = 3·35$ m/s².

Tangential acceleration of B relative to A, $\quad b_a b_1 = 0$.

Centripetal acceleration of C relative to B, $\quad b_1 c_b = \dfrac{bc^2}{BC} = \dfrac{0·595^2}{0·427} = 0·827$ m/s².

Centripetal acceleration of C relative to D, $\quad d_1 c_d = \dfrac{dc^2}{DC} = \dfrac{0·732^2}{0·229} = 2·34$ m/s².

The acceleration diagram is shown in Figure 155(c), point c_1 being fixed by the intersection of normals drawn through points c_b and c_d. Then

Total acceleration of C relative to B, $\quad b_1 c_1 = 4.85$ m/s² to scale.

Total acceleration of G relative to B, $\quad b_1 g_1 = \dfrac{BG}{BC} \times b_1 c_1$

$$= \left(\dfrac{0·168}{0·427}\right) 4·85$$

$$= 1·9 \text{ m/s}^2. \quad (f_g \text{ in figure.})$$

This fixes point g_1 and enables $a_1 g_1$ to be drawn.

Total acceleration of G relative to A, $\quad a_1 g_1 = 1·89$ m/s² to scale.

Inertia force, $\quad F_i = $ Mass $\times$ acceleration of mass centre

$$= 3·4 \times 1·89$$

$$= 6·42 \text{ N}.$$

This force is in a direction opposite to that of f_g.

Tangential acceleration of C relative to B, $\quad c_b c_1 = 4·79$ m/s² to scale.

Angular acceleration of $CB = \dfrac{c_b c_1}{CB} = \dfrac{4·79}{0·427} = 11·2$ rad/s² clockwise.

Referring to Figure 155(a) and denoting this by α, we obtain

Inertia torque, T_i = Moment of inertia × angular acceleration

$$= 0.02 \times 11.2$$
$$= 0.224 \text{ Nm.}$$

This torque is in a direction opposite to that of α, i.e. is anticlockwise.

The distance by which the inertia force would have to be displaced so as to provide the inertia torque as well is given therefore by

$$x = \frac{T_i}{F_i} = \frac{0.224}{6.42} = \underline{0.035 \text{ m. }} \quad (35 \text{ mm.})$$

The displaced F_i (i.e. the equivalent inertia force) may now be inserted as F_{eq} in the configuration, parallel and opposite to f_g and distant x from its line of action.

The forces required to overcome the effects of F_{eq} (i.e. required *on* the link for its dynamic equilibrium) are F_c at point C and F_b at point B and, because no torque acts at D, the force F_c must act along the link DC. To close the force triangle, Figure 155(d), the axis of F_b must pass through the point—just above the bearing D—at which the axis of F_{eq} intersects CD. Then, to scale,

$$F_b = 6.23 \text{ N} \quad \text{and} \quad F_c = 1.51 \text{ N.}$$

Note that the reaction to F_c puts the link CD in compression. Note also that the reaction to F_b has a radial component which puts the link AB in tension, and a tangential component which applies an anticlockwise torque to point A.

EXAMPLES 6

1. The crank and connecting rod of a reciprocating pump are respectively 0·038 m and 0·152 m. Find the velocity of the plunger when the crank is driven anticlockwise at 200 rev/min and has turned from the inner dead centre through an angle of (a) 60°, (b) 130°. Find also the corresponding velocities of a point on the connecting rod 0·03 m from the small end [(a) 0·781 and 0·766 m/s, (b) 0·488 and 0·530 m/s].
2. The configuration of a wrapping machine is shown in Figure 156.
   Find the velocities of the pushers, C and D, given that $OA = 25$ mm, $AB = BC = 50$ mm, $BD = 100$ mm (0·455 and 0·135 m/s).

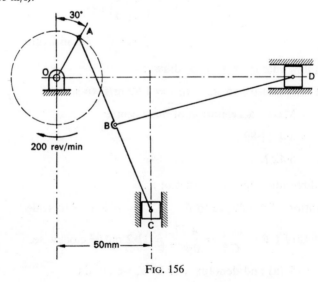

FIG. 156

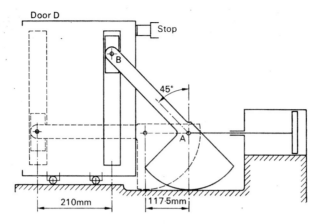

Fig. 157

Fig. 158

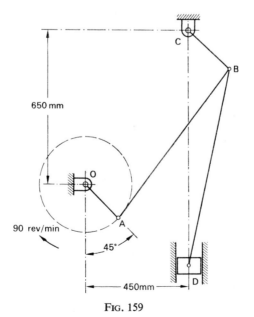

Fig. 159

MECHANICAL TECHNOLOGY FOR HIGHER ENGINEERNG TECHNICIANS

3. The configuration shown in Figure 157 represents one of a pair of pneumatically operated sliding doors in the open position. The arm AB is 300 mm long and is integral with the quadrant which is 150 mm radius and gears with the horizontal rack, E. If the piston is moving at 0·3 m/s at the end of the opening stroke, find the velocity of the door at the instant it strikes the stop. If the door is closed when AB is horizontal, find its total travel (0·723 m/s, 210 mm).

4. In the mechanism shown in Fig. 158, $AB = 25$ mm. Determine the stroke of the slide (52 mm). Find also, for the given configuration:

   (a) the time taken for the outstroke (0·105 s);
   (b) the time taken for the instroke (0·094 s);
   (c) the velocity and acceleration of the slide (3·35 m/s, 28·7 m/s²).

5. The configuration shown in Figure 159 is that of a moulding press. Determine the velocity and acceleration of the ram, D, given that $AB = 4AO = 800$ mm, $BC = 237$ mm and $BD = 850$ mm (0·136 m/s, 0·885 m/s²).

6. A reciprocating mechanism has a crankpin 75 mm diameter which rotates at 300 mm radius, while the connecting rod is equal in length to four cranks. Find the angular velocity of the connecting rod when the crank is 45° past the i.d.c. and rotating at 100 rev/min, and the relative velocity between it and the crankpin (1·85 rad/s, 0·463 m/s).

7. A mine cage is allowed to fall freely for the first 60 m of its descent after which its speed is maintained constant by the mine motor before slowing down. If the cage runs in guides and the mine is at the equator, find the maximum side thrust exerted by the cage on a man weighing 755 N and state its direction. (0·387 N in a direction opposite to that of the Coriolis component).

8. Two fixed pivots, O and A, are on a horizontal line and 0·3 m apart. The upper end of a link pivoted at O (which is to the left of A) is pinned to a block B which can slide along a second link pivoted at A. If $OB = 0·6$ m, the angle OAB is 30° and the link AB is driven anticlockwise at a uniform speed of 5 rad/s, find the resulting angular acceleration of the link OB (4·5 rad/s²).

9. In a certain four-bar chain $ABCD$, $AB = 50$ mm, $BC = 175$ mm and has a moment of inertia about its mass centre of 0·00665 kg /m², $CD = 100$ mm and $AD = 200$ mm. Points A and D are fixed. For the position of the linkage in which B and C are on opposite sides of AD and the angle BAD is 30°, find the inertia torque acting on the link BC when AB is driven uniformly at 150 rev/min anticlockwise (0·16 Nm).

7

FLUID MECHANICS

INTRODUCTION

The resistance to change in shape (flow) offered by an actual—as distinct from an ideal or perfect—fluid is the resistance to motion between adjacent layers, i.e. it is a shear resistance. It follows that when the motion has ceased and the fluid is at rest (stagnant, static or stationary) the shear stress is zero. The stress on each interface is then entirely normal and, at any point, is the same in all directions. This is known as the *pressure*. A fluid differs from a solid therefore in that it can offer resistance to change in shape only while that change is taking place. (On occasion it is convenient to assume that an ideal (non-viscous or inviscid) fluid can exist.)

A fluid may be gaseous and highly compressible or it may be liquid and relatively incompressible, e.g. for water the bulk modulus, K, has a value of 2070 MPa. (For this reason it is not usual to take compressibility into account until the pressure exceeds 30 MPa.) Hence a given mass of liquid may be assumed to occupy a fixed volume. This fixed volume will be forced by gravity to assume the shape of any solid body interposed between it and the earth. If the body is a container of sufficient size the boundary between the liquid and the atmosphere (or other gas if the container is closed) will be a free plane surface. On the other hand a gas expands to fill any enclosed space, while in a mixture of gases each constituent exerts a pressure on the container.

Definitions

The *specific weight* is the gravitational force acting on unit volume (weight per unit volume) and is denoted by w. At s.t.p. (see p. 193) the value of w for water is 9800 N/m³. The ratio of the specific weight of any other substance—fluid or solid—to this value is called the *specific gravity* and is denoted by s. Specific gravity (which is sometimes referred to as *relative density*) is measured by means of a calibrated float or *hydrometer*. Provided that the specific weight of a fluid is constant, the weight of a known volume, V, is wV. Specific weight varies with temperature and with pressure and, in the case of a liquid, will be changed if a salt is dissolved in it, if a powder is suspended in it or if it is made to form an emulsion with a second liquid.

The specific volume is the volume on which the earth exerts unit gravitational force (volume occupied by unit weight) and is denoted by v. Thus $v = 1/w$ and the value for water at s.t.p. is $1/9800$ or 102×10^{-6} m³/N approx. The corresponding value of v for air is 0.0833 m³/N. (In fluid mechanics the symbol for velocity is u.)

Effect of Depth on Pressure

Figure 160(a) shows a column of liquid of specific weight w, section dA and length L. If the downward pressure on the free surface is p_a the resultant downward force is

$$p_a\, dA + w(dA.L).$$

For vertical equilibrium the upward force on the underside of the column must be equal and opposite so that, if the upward pressure on this surface is p,

$$p.dA = p_a\, dA + w(dA.L)$$

i.e.
$$p = p_a + wL.$$

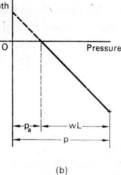

(a) (b)

Fig. 160

Thus if w is constant, the graph of p against L is a straight line as shown in Figure 160(b). The pressure on the free surface, p_a, is usually atmospheric. The additional pressure, wL, is the amount by which the pressure exceeds the atmospheric value and is that measured (usually) by a gauge. Hence the term "gauge pressure".

From the foregoing it follows that, in a static fluid, the pressure is constant on any horizontal plane; it also follows that a change in p_a will be transmitted throughout the fluid, a principle made use of by pumps and other machinery.

Force on a Submerged Plane Surface

Figure 161 shows an irregular plane surface of negligible thickness and area A submerged in a liquid of specific weight w and inclined to the free surface at an angle θ. The position of the centroid is denoted by G and its depth by $\bar{x}$, while S represents the intersection of the free surface by the inclined plane.

The depth of an elemental strip of length B will vary between x and $(x+dx)$ so that the distance from S of points in the strip will vary between $x/\sin\theta$ and $(x/\sin\theta + dx/\sin\theta)$. The area of the strip is $B(dx/\sin\theta)$ and the pressure at depth x is wx, so that

$$\text{Force on strip} = wx\left(B\frac{dx}{\sin\theta}\right)$$

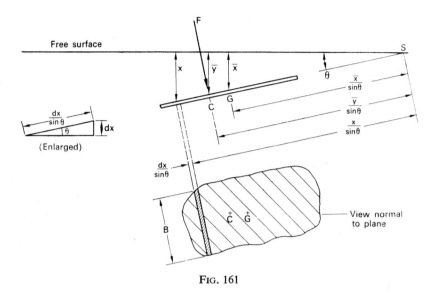

Fig. 161

$$\therefore \text{Total force on one side} = w \int \frac{B.dx}{\sin \theta} x$$

$$= w \sin \theta \int \frac{B.dx}{\sin \theta} \frac{x}{\sin \theta}$$

$$= w \sin \theta \, (\text{1st Moment of Area about } S)$$

$$= w \sin \theta \times A \left(\frac{\bar{x}}{\sin \theta}\right)$$

i.e. $$\underline{F = A(w\bar{x}).}$$

Thus the resultant force on one side is the product of the area and the pressure at its centroid, whatever the value of θ. Because multiplying by A gives the resultant force, the quantity $w\bar{x}$ must be the average pressure on the surface. (Note that $\bar{x}$ will only be the mean depth of the surface if B is constant.)

Centre of Pressure

Since the pressures on the elemental strips below G (Fig. 161) are greater than those above G, the resultant force on one side, F, will not act at G but at some point, C, at a greater depth, say $\bar{y}$. As before

$$\text{Force on one side of strip} = wx\left(B\frac{dx}{\sin \theta}\right)$$

$$\text{Moment of this about } S = wx\left(B\frac{dx}{\sin \theta}\right) \times \frac{x}{\sin \theta}$$

$$= w \sin \theta \left(B\frac{dx}{\sin \theta}\right)\left(\frac{x}{\sin \theta}\right)^2$$

$$\text{Total moment about } S = w \sin \theta \int \left(B \frac{dx}{\sin \theta}\right)\left(\frac{x}{\sin \theta}\right)^2$$

$$= w \sin \theta \,(\text{2nd Moment of Area about } S)$$
$$= w \sin \theta \,(I_s).$$

But, Total moment about $S = F\left(\dfrac{\bar{y}}{\sin \theta}\right)$ and $F = Aw\bar{x}$

so that
$$Aw\bar{x}\left(\frac{\bar{y}}{\sin \theta}\right) = I_s w \sin \theta$$

or
$$\bar{y} = \frac{I_s \sin^2 \theta}{A\bar{x}}.$$

Note that, since $A(\bar{x}/\sin \theta)$ is the 1st Moment of Area about S we have

$$\bar{y} = \left(\frac{\text{2nd Moment}}{\text{1st Moment}}\right) \sin \theta.$$

For the shapes encountered in engineering, the value of I about a horizontal axis through the centroid, I_g, is usually calculable. By applying the parallel axis theorem the value of I_s can be obtained by adding the quantity

$$A\,(\text{distance between axes})^2$$

i.e.
$$I_s = I_g + A\left(\frac{\bar{x}}{\sin \theta}\right)^2.$$

Since $I_g = Ak_g^2$, where k_g is the corresponding radius of gyration, we have

$$\bar{y} = \left(\frac{\text{2nd Moment}}{\text{1st Moment}}\right) \sin \theta$$

$$= \frac{Ak_g^2 + A\left(\dfrac{\bar{x}}{\sin \theta}\right)^2}{A\left(\dfrac{\bar{x}}{\sin \theta}\right)} \times \sin \theta$$

or
$$\bar{y} = \frac{k_g^2 + \left(\dfrac{\bar{x}}{\sin \theta}\right)}{\left(\dfrac{\bar{x}}{\sin \theta}\right)} \times \sin \theta.$$

For a vertical surface $\theta = 90$, i.e. $\sin \theta = 1$, so that this then reduces to

$$\bar{y} = \frac{k_g^2 + \bar{x}^2}{\bar{x}} = \frac{\text{2nd Moment}}{\text{1st Moment}}.$$

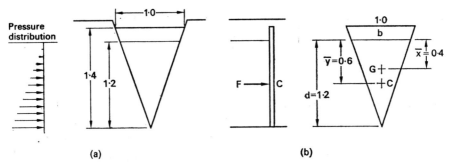

Fig. 162

EXAMPLE. The maximum depth of water in a concrete V-section channel is 1·2 m. If a triangular plate is inserted as shown in Figure 162(a), find:

(a) the width of water at the free surface,
(b) the force acting on the plate,
(c) the depth of the centre of pressure.

(Assume $w = 9800$ N/m³.)

Solution.

Width of surface, $b = \left(\dfrac{1 \cdot 2}{1 \cdot 4}\right) 1 \cdot 0 = 0 \cdot 857$ m.

Depth of centroid of submerged area, $\bar{x} = \dfrac{1 \cdot 2}{3} = 0 \cdot 4$ m.

Force on plate, $F = Aw\bar{x} = \dfrac{0 \cdot 857 \times 1 \cdot 2}{2}(9800 \times 0 \cdot 4) = 2010$ N.

For a triangle of height d base b,

$$I_g = \frac{bd^3}{36} = Ak_g^2$$

where

$$A = \frac{bd}{2}$$

∴

$$k_g^2 = \frac{bd^3}{36}\left(\frac{2}{bd}\right)$$

$$= \frac{d^2}{18} \quad \text{and} \quad d = 1 \cdot 2 \text{ m}$$

$$= \frac{1 \cdot 44}{18}$$

$$= 0 \cdot 08 \text{ m.}$$

MECHANICAL TECHNOLOGY FOR HIGHER ENGINEERING TECHNICIANS

∴ Depth of centre of pressure, $\bar{y} = \dfrac{k_g^2 + \bar{x}^2}{\bar{x}}$ (since plate is vertical)

$$= \dfrac{0\cdot 08 + 0\cdot 4^2}{0\cdot 4}$$

$$= 0\cdot 6 \text{ m}.$$

Note that this is half the greatest depth.

EXAMPLE. A circular opening 0·2 m diameter in the vertical side of a reservoir is closed by a concentric flap 0·25 m diameter and hinged at the top, the hinge being 0·975 m below the free surface. Estimate:

(a) the mean pressure on the flap,
(b) the thrust on the flap,
(c) the depth of the centre of pressure,
(d) the horizontal force required on the bottom of the flap to open it.

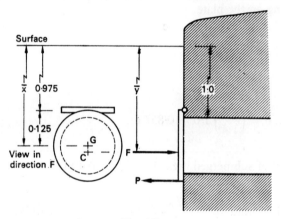

FIG. 163

Solution.

Depth of centroid, $\bar{x} = 0\cdot 975 + 0\cdot 125 = 1\cdot 1$ m

Mean pressure, $w\bar{x} = 9800 \times 1\cdot 1 = 10,780$ N/m² or Pa

Thrust, $F = Aw\bar{x} = \left(\dfrac{\pi}{4} \times 0\cdot 2^2\right) 10\,780 = 338$ N.

For a circle of diameter d, $I_g = \dfrac{\pi d^4}{64} = Ak_g^2$

where $A = \dfrac{\pi d^2}{4}.$

∴ $k_g^2 = \dfrac{\pi d^4}{64}\left(\dfrac{4}{\pi d^2}\right)$

$= \dfrac{d^2}{16}$ and $d = 0\cdot 2$

$= 0\cdot 0025.$

Depth of centre of pressure, $$\bar{y} = \frac{0.0025 + 1 \cdot 1^2}{1 \cdot 1} = 1.102 \text{ m}.$$

Thus C is 0·002 m (2 mm) below G, i.e. 0·127 m below the hinge. Moments about the hinge give
$$F \times 0.127 = P \times 0.250$$

Required force, $$P = 338 \left(\frac{0.127}{0.25}\right)$$
$$= \underline{172 \text{ N}}.$$

EXAMPLE. *Each of the gates to a dry dock is 12 m wide. Find, for one gate, the thrust on each side when the depth on one side is 6 m and on the other side is 15 m. Hence find the resultant thrust on each gate and the height at which it acts.*

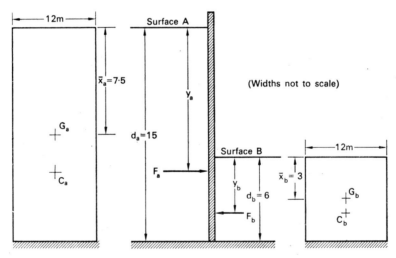

FIG. 164

Solution.

For a rectangle width b, depth d,
$$I_g = \frac{bd^3}{12} = (bd) k_g^2$$

$$\therefore \quad k_g^2 = \frac{bd^3}{12}\left(\frac{1}{bd}\right) = \frac{d^2}{12}.$$

For side A: $\quad \bar{x}_a = \frac{15}{2} = 7.5 \text{ m}$

Mean pressure $= 9800 + 7.5$

$= 73\,500 \text{ Pa}.$

∴ Thrust,
$$F_a = 73\,500 \times 15 \times 12$$
$$= 13 \cdot 2 \text{ MN}$$

$$k_g^2 = \frac{15^2}{12} = 18 \cdot 75.$$

∴
$$\bar{y}_a = \frac{18 \cdot 75 + 7 \cdot 5^2}{7 \cdot 5} = 10 \text{ m}.$$

Note that this is two thirds of the depth on side A.

For side B:
$$\bar{x}_b = \tfrac{6}{2} = 3 \cdot 0 \text{ m}$$

Mean pressure
$$= 9800 \times 3 \cdot 0$$
$$= 29\,400 \text{ Pa}.$$

∴ Thrust,
$$F_b = 29\,400 \times 6 \times 12$$
$$= 2 \cdot 12 \text{ MN}$$

$$\bar{y}_b = \tfrac{2}{3} \times 6 \cdot 0 = 4 \cdot 0 \text{ m}$$

Resultant moment about base
$$= (13 \cdot 2 \times 10^6)\,5 - (2 \cdot 12 \times 10^6)\,2$$
$$= 61 \cdot 76 \times 10^6 \text{ Nm}$$

Resultant thrust on gate
$$= (13 \cdot 20 - 2 \cdot 12)\,10^6$$
$$= 11 \cdot 08 \times 10^6 \text{ MN}$$

Height above base at which this acts $= \dfrac{61 \cdot 76 \times 10^6}{11 \cdot 08 \times 10^6} = \underline{5 \cdot 57 \text{ m}.}$

Force on a Submerged Curved Surface

The resultant force, F, on one side of a plane inclined surface such as shown in Figure 165(a), has been shown to act normal to the plane at a point C, called the centre of pressure, the depth of which is given by
$$\bar{y} = I_s \sin^2 \theta / A\bar{x}.$$

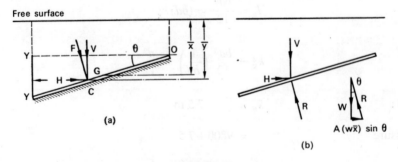

FIG. 165

FLUID MECHANICS

This force may be resolved into

1. A vertical component, V (= $F \cos \theta$), which is equal to the weight of liquid supported by the surface and acts through the mass centre of that liquid.

2. A horizontal component, H (= $F \sin \theta$), which is equal and opposite to the thrust on the projection YY, since the liquid OYY is in horizontal equilibrium.

The equilibriant, R, and the force triangle are shown in Figure 165(b).

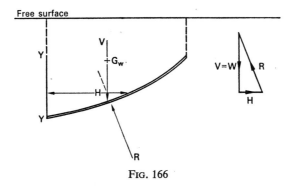

FIG. 166

If the surface is curved as shown in Figure 166, the vertical component of the unknown resultant force is again equal to the weight of liquid supported by the surface (and acts at G_w, the mass centre of that liquid) while the horizontal component is equal and opposite to the thrust on the projection YY and acts at the centre of pressure of YY. When V and H have been found, the equilibriant, R, is obtainable from the force triangle in the same way.

Note that V and H no longer intersect at a point on the submerged surface.

EXAMPLE. The section of a dam is represented in Figure 167 by $abcdef$, part bd of which is parabolic, the equation being $y = 0 \cdot 1x^2$. If the depth of water is 90 m, find from first principles, per metre width,

(a) the weight of water supported by bc,

(b) the horizontal distance from bg at which this weight acts,

(c) the horizontal thrust on bc and the height above ab at which it acts,

(d) the magnitude and direction of the resultant thrust.

Solution.

(a) The equation is $\qquad y = 0 \cdot 1x^2$

so that $\qquad x^2 = 10y$, i.e. $x = 3 \cdot 162 y^{0 \cdot 5}$.

161

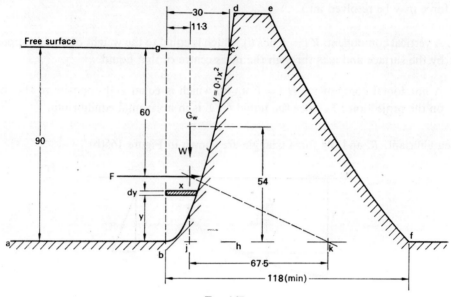

Fig. 167

Hence $(gc)^2 = 10 \times 90,$ i.e. $gc = 30$ m

$$\text{Area } bcg = \int_0^{90} x \, dy = 3 \cdot 162 \int_0^{90} y^{0 \cdot 5} \, dy$$

$$= 3 \cdot 162 \left. \frac{y^{1 \cdot 5}}{1 \cdot 5} \right|_0^{90}$$

$$= \frac{3 \cdot 162}{1 \cdot 5} (90)^{1 \cdot 5}$$

$$= \underline{1800 \text{ m}^2}.$$

(Alternatively, the area is 2/3 of rectangle $bhcg$.)

Hence, weight supported by bc per metre width $= (1800 \times 1)9800$

i.e. $$W = \underline{17 \cdot 64 \times 10^6 \text{ N}}.$$

(b) The 1st Moment of area bcg about $bg = \int_0^{90} x \, dy \left(\frac{x}{2}\right)$ and $x^2 = 10y$

$$= 5 \int_0^{90} y \, dy$$

∴ Distance from *bg* of the centroid G_w of this area

$$= \frac{\text{1st Moment}}{\text{Area}}$$

$$= \frac{5\int_0^{90} y\, dy}{3\cdot 162 \int_0^{90} y^{0\cdot 5}\, dy} = \frac{5\left[\frac{y^2}{2}\right]_0^{90}}{3\cdot 162 \left[\frac{y^{1\cdot 5}}{1\cdot 5}\right]_0^{90}}$$

$$= \frac{5}{3\cdot 162}\left(\frac{90^2}{2}\right)\left(\frac{1\cdot 5}{90^{1\cdot 5}}\right) = \frac{5\times 1\cdot 5}{3\cdot 162\times 2}(90^{0\cdot 5})$$

$$= 11\cdot 3 \text{ m}.$$

(It may be shown in similar fashion that G_w is 54 m above *ab*.)

(c) The horizontal thrust, F, on *bc* must be equal and opposite to that on the plane rectangular section *bg*, the centre of pressure of which is $\frac{2}{3}(90)$ or 60 m below the free surface.

Mean pressure on $bg = 9800(\frac{90}{2}) = 441\,000$ N/m²

∴ Per metre width, $F = 441\,000\,(90\times 1)$
$\qquad\qquad\qquad\quad = 39\cdot 7\times 10^6$ N at 30 m above *ab*.

(d) Referring to Figure 168(a),

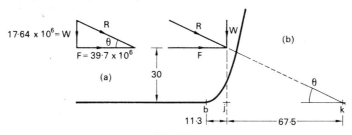

FIG. 168

Resultant thrust, $\quad R = \sqrt{(W^2 + F^2)}$
$\qquad\qquad\qquad\quad = 10^6 \sqrt{(17\cdot 64^2 + 39\cdot 7^2)}$
$\qquad\qquad\qquad\quad = 43\cdot 3\times 10^6$ N.

This is inclined to the horizontal at

$$\theta = \tan^{-1}\frac{17\cdot 64}{39\cdot 7} = \tan^{-1} 0\cdot 445.$$

From Figure 168 (b), $\quad 30 = jk \tan\theta$

so that $\qquad\qquad jk = \frac{30}{0\cdot 445} = 67\cdot 5$ m.

Hence $\qquad\qquad bk = 67\cdot 5 + 11\cdot 3 = 78\cdot 8$ m.

Hydrostatic Head and Energy

Consider the stationary element of fluid shown in Figure 169. The gravitational force on it is $w(dA.dz)$ and if it is assumed that the pressure is p at a height z *above* the datum and that a change, dz, corresponds to a pressure change dp, then, for vertical equilibrium,

$$(p+dp)\,dA + w(dA.dz) = p.dA$$

$$p + dp + w.dz = p$$

or
$$dp = -w.dz$$

so that
$$\frac{dp}{dz} = -w.$$

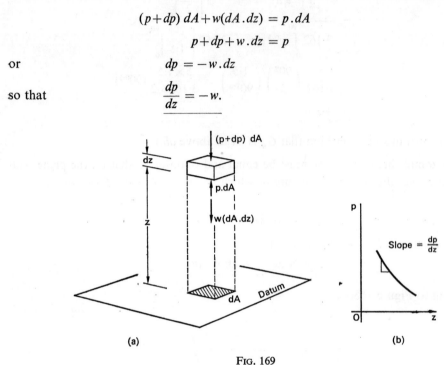

FIG. 169

Thus the rate of change in p with respect to z is negative so that the pressure falls as the height increases, the rate of fall depending alone on the specific weight at the point considered. Transposing gives

$$\frac{1}{w}.dp = -dz.$$

If the fluid is incompressible (i.e. if w is constant), then

$$\frac{1}{w}\int dp = -\int dz$$

or
$$\frac{p}{w} = -z + \text{constant}$$

i.e.
$$z + \frac{p}{w} = \text{constant}.$$

This constant is called the *hydrostatic head* and denoted by H, the components z and p/w

being called, respectively, the *datum head* and the *pressure head*.

Thus
$$z + \frac{p}{w} = H.$$

As already shown, at any depth, L, below the free surface of a stationary liquid subject to atmospheric pressure, p_a, the pressure is given by

$$p = p_a + wL$$

or
$$p = wL \text{ above atmospheric}$$

i.e.
$$\frac{p}{w} = L = \text{depth below surface.}$$

Hence,
$$z + L = H.$$

Thus, as shown in Figure 170, the hydrostatic head is the height of the free surface above the datum from which z is measured.

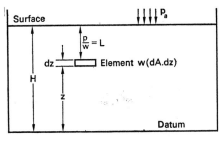

Fig. 170

Now, neglecting losses, the work which would have to be done (against gravity) to lift the element $w(dA.dz)$ from the datum to the surface is $w(dA.dz)H$ so that this quantity must be the hydrostatic (or potential) energy of the element when it is at the free surface. It follows that, at this surface, per unit weight, the hydrostatic energy is H. Now, whatever the depth, the upward and downward forces on any infinitely small static element are equal and opposite so that—in the absence of loss—no work would be required to reduce its height from H to, say, z. The hydrostatic energy per unit weight must, therefore, remain unaffected by such a change in height although the height energy is thereby reduced from H to z. From the equation above, the difference is p/w which is a compensating increase in the pressure energy.

Steady Flow and Kinetic Energy

Figure 171 represents an element of incompressible fluid of specific weight w, the velocity and pressure at the inlet section being constant and denoted respectively by u and p.

Suppose that, over an inclined length dx, the section increases from A to $(A+dA)$ and that the other quantities vary as shown. (Since the rate of flow is the same at all sections, it is evident that the change in velocity, du, is negative. The sign of the change in pressure is not, however, obvious.)

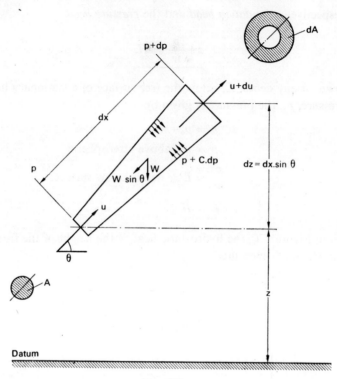

Fig. 171

Mean section of element $= \frac{1}{2}[A+(A+dA)] = A+\frac{dA}{2}$

∴ Volume of element $= \left(A+\dfrac{dA}{2}\right)dx$

and Weight of element $W = w\left(A+\dfrac{dA}{2}\right)dx$

Axial component of $W = w\left(A+\dfrac{dA}{2}\right)dx.\sin\theta$

$\qquad = w\left(A+\dfrac{dA}{2}\right)dz.$

As shown in Figure 171, this component opposes the flow. Also,

$$\text{Force on outlet opposing flow} = (p+dp)(A+dA).$$

The pressure on the element exerted by the pipe can be assumed to vary linearly between p and $(p+dp)$, i.e. can be written $(p+C.dp)$ where C is some constant and of little interest since the term containing it will be neglected in what follows. Since the pipe is conical there will be—along the axis and assisting the motion—an unbalanced force equal to the product of the mean of this pressure and the axial projection of the surface area, viz. $(p+C.dp)dA$. Since the force on the inlet section in the direction of motion is pA, the net force in this direction is given by

$$F = pA + (p + C.dp)\,dA - w\left(A + \frac{dA}{2}\right)dz - (p+dp)(A+dA)$$

$$= pA + p.dA + C.dp.dA - wA.dz - \frac{w}{2}.dA.dz - pA - p.dA - A.dp - dp.dA$$

$$= -wA.dz - A.dp \quad \text{(neglecting the products of differentials).}$$

Passing the inlet section:

$$\text{Volume/sec} = Au$$
$$\text{Weight/sec} = w(Au)$$
$$\therefore \quad \text{Mass/sec} = w(Au)\frac{1}{g}.$$

The change in the velocity of this mass is du so that

$$\text{Change in momentum/sec} = \left(\frac{wAu}{g}\right)du.$$

Since, from Newton's 2nd Law, the rate of change in momentum is equal to the applied force, we have

$$\left(\frac{wAu}{g}\right)du = -wA.dz - A.dp$$

or
$$w.dz + \frac{w}{g}u.du + dp = 0 \quad \text{(dividing by } A \text{ and transposing)}$$

$$\therefore \quad dz + \frac{1}{g}u.du + \frac{1}{w}.dp = 0$$

$$\int dz + \frac{1}{g}\int u.du + \frac{1}{w}\int dp = \text{a constant}$$

whence
$$\underline{z + \frac{1}{2}\cdot\frac{u^2}{g} + \frac{p}{w} = \text{a constant.}}$$

This equation for the loss-free steady flow of an incompressible fluid was first derived by the Swiss mathematician Daniell Bernoulli (1700–82). As already shown, the sum of the first and last terms is—relative to the chosen datum—the hydrostatic energy per unit weight, while the middle term is evidently the kinetic energy per unit weight. Thus the Bernoulli equation shows that the energy total is constant when there is no loss. This means that, for flow of an inviscid incompressible fluid between, say, points 1 and 2 in a system, we can write

$$\underline{z_1 + \frac{1}{2}\cdot\frac{u_1^2}{g} + \frac{p_1}{w} = z_2 + \frac{1}{2}\cdot\frac{u_2^2}{g} + \frac{p_2}{w}.}$$

It has been shown that the datum head, z, represents the height energy per unit weight, while the pressure head, p/w, represents the pressure energy in the same way. Hence the third quantity, $(\frac{1}{2})(u^2/g)$, which represents the kinetic energy, is also known as the *kinetic* (or velocity) *head*. (The units of energy/unit weight are J/N or metres, which are the units of head.) The constant in the Bernoulli equation is therefore the sum of three heads (or energies per unit weight) and this is denoted by H. This *total head*, H, is therefore the gross theoretical work which could be done by unit quantity of fluid. Note that if z is constant (as it is in a horizontal pipe), p will fall as u rises and vice versa.

EXAMPLE. The circular orifice of a fountain is 30 mm diameter, has its axis vertical and is fed from a lake the surface of which remains at 7·5 m above the plane of the orifice. Assume no loss and find:

(a) the initial velocity of the jet,
(b) the diameter of the jet at a height of 3 m.

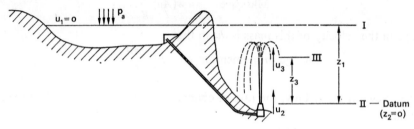

FIG. 172

Solution.

(a) If the plane of the orifice, Figure 172, is taken as datum then

$$z_2 = 0$$
$$z_1 = H = 7\cdot 5$$
$$p = p_1 = p_a \quad \text{(atmospheric)}$$
$$u_1 = 0 \quad \text{and} \quad u_2 \text{ is to be found}$$

since
$$z_2 + \frac{1}{2}\left(\frac{u_2^2}{g}\right) + \frac{p_2}{w} = z_1 + \frac{1}{2}\left(\frac{u_1^2}{g}\right) + \frac{p_1}{w}$$

∴
$$0 + \frac{1}{2}\left(\frac{u_2^2}{g}\right) = H + 0$$

i.e.
$$u_2 = \sqrt{(2gH)} \quad \text{theoretically}$$
$$= \sqrt{(2\times 9\cdot 81 \times 7\cdot 5)}$$
$$\simeq 12\cdot 1 \text{ m/s}.$$

(b) Taking the same datum, $p_3 = p_2 = p_a$ as before and $z_3 = 10$

so that
$$0 + \frac{1}{2}\left(\frac{12\cdot 1^2}{9\cdot 81}\right) = 3 + \frac{1}{2}\left(\frac{u_3^2}{9\cdot 81}\right)$$

$$12\cdot 1^2 - u_3^2 = 3(2\times 9\cdot 81)$$

whence
$$u_3^2 = 88\cdot 34$$

and
$$u_3 = 9\cdot 4 \text{ m/s}.$$

Assuming the jet to remain homogeneous (bubble-free) and circular,

$$A_3 u_3 = A_2 u_2 \quad \text{(continuity equation)}$$

i.e.
$$\left(\frac{\pi}{4} d_3^2\right) u_3 = \left(\frac{\pi}{4} d_2^2\right) u_2$$

where
$$d_2 = 0{\cdot}03 \text{ m.}$$

$$\therefore \quad d_3^2 = \left(\frac{u_2}{u_3}\right) d_2^2 = \left(\frac{12{\cdot}1}{9{\cdot}4}\right) 0{\cdot}03^2$$

from which
$$\underline{d_3 = 0{\cdot}034 \text{ m.} \quad (34 \text{ mm.})}$$

Viscosity

If a circular vessel containing liquid is given an axial rotation, an eddy-free motion (called *laminar flow*) is imparted to the contents, showing that the container exerts a shear force on the layer of liquid molecules in contact with it (the so-called boundary layer) and that this layer has a similar effect on its neighbour, the effect being transmitted throughout the fluid. As mentioned earlier, the ability to transmit shear is not possessed by ideal fluids so that existing (actual) fluids possess the property known as *viscosity*, this being defined by BS 188 : 1957 as "that property of a fluid which determines the resistance offered to a shear force under laminar flow conditions". Since work has to be done (energy expended) to maintain relative motion between layers, i.e. to overcome the shear resistance, it follows that heat is generated in the fluid, the amount being a function of the viscosity.

Dynamic Viscosity

Let a shaft, S, rotate as shown in Figure 173(a) and let it be separated from a stationary housing, H, by a film of fluid.

Since AB is small the flow may be assumed laminar and the velocity distribution may be assumed linear as shown in Figure 173(b), the boundary layer at A being stationary, while that at B has the shaft velocity, U. Under the shear stress on faces BC and AD the element $ABCD$ will change in shape to AB_1C_1D, i.e. the face AB will turn through some angle φ. If q is the shear stress in the film, then

$$\frac{d\varphi}{dt} = \text{a constant} \times q$$

or
$$q = \text{a constant} \times \frac{d\varphi}{dt}.$$

This constant is called the *dynamic viscosity* and is denoted by μ so that

$$\underline{q = \mu \cdot \frac{d\varphi}{dt}.}$$

The dynamic viscosity of a fluid is defined in BS 188 : 1957 as "the tangential force on unit area in either of two parallel planes unit distance apart when one area moves in its own plane at unit velocity relative to the other".

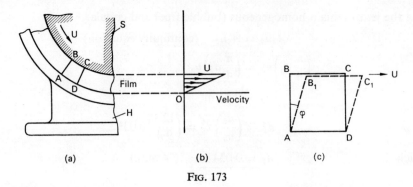

Fig. 173

A fluid which obeys the above law (i.e. deforms—or shears—at a rate which is proportional to the shear stress) is called a *Newtonian fluid*. Over their normal temperature range lubricating oils are in this category.

Transposing the above equation gives

$$\mu = q \left/ \frac{d\varphi}{dt} \right.$$

The dimensions of μ are therefore

$$\frac{\text{Shear stress}}{\text{Angular velocity}} = \frac{\text{Force}}{\text{Area}} \times \frac{\text{Time}}{\text{Radians}} \quad \text{(and radians are dimensionless)}$$

$$= \frac{\text{Mass} \times \text{Acceleration}}{(\text{Length})^2} \times \text{Time}$$

$$= \frac{M}{LT}.$$

The units of dynamic viscosity are therefore kg/m s.

When converted to g/cm s such (cgs) units are given the name *Poise* (after Poiseuille) and denoted by P. Unless the fluid is very viscous, the centipoise (0·01 P = 1·0 cP) is of a more convenient size. For freshly distilled water at 20°C, $\mu = 1\cdot0020$ cP.

The dimensions of μ may also be written

$$\frac{\text{Shear stress}}{\text{Angular velocity}} = \frac{N}{m^2} \times \frac{\text{Seconds}}{\text{Radian}}.$$

Alternative units are therefore Ns/m².

Kinematic Viscosity

This is defined as the ratio

$$\frac{\text{Dynamic viscosity}}{\text{Density}}$$

and is denoted by ν (nu) so that

$$\nu = \frac{\mu}{\varrho}.$$

Since density $= \dfrac{\text{Mass}}{\text{Volume}}$

it has dimensions M/L^3 so that the dimensions of kinematic viscosity are

$$\frac{M}{LT} \div \frac{M}{L^3} \quad \text{or} \quad \frac{L^2}{T}.$$

The units of kinematic viscosity are therefore m²/s.

When converted to cm²/s such (cgs) units are given the name *Stokes* (after Sir George Stokes, 1819–1903) and denoted by S. Unless the fluid is very viscous, the centistoke is of a more convenient size (1·0 cS = 0·01 S). For freshly distilled water at 20°C, v = 1·0038 cS.

Velocity Gradient

Let E and F, Figure 174(a), be points dy apart in adjacent layers of a fluid which separates a stationary shaft, A (e.g. a wagon axle) from a housing, B.

Let B have a velocity U, assume the flow to be laminar and let the velocities of points E and F be, respectively, u and $(u+du)$. In time dt the section AB will have turned through

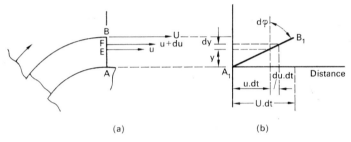

FIG. 174

a small angle $d\varphi$ to A_1B_1, Figure 174(b), while point F will have moved (relative to point E) through a distance $du.dt$, so that

$$d\varphi = \frac{du.dt}{dy} \quad \text{or} \quad \frac{d\varphi}{dt} = \frac{du}{dy}.$$

Since

$$q = \mu . \frac{d\varphi}{dt}$$

we have

$$q = \mu . \frac{du}{dy}.$$

Thus the shear stress is proportional to the rate of change in velocity as well as to the viscosity of the fluid and, for this reason, is greatest at the boundary.

Transposing the above equation gives

$$\mu = q \bigg/ \frac{du}{dy},$$

or

$$\text{Dynamic viscosity} = \frac{\text{Shear stress}}{\text{Velocity gradient}}.$$

(This is the Newtonian definition of viscosity.)

Reynolds Number

Laminar flow has been defined as the eddy-free or non-turbulent motion of one layer of a fluid relative to its neighbours. Such flow is also described as *viscous* or *streamline*, a streamline being defined as a continuous line in the fluid across which there is no flow, i.e. its direction is at all points that of the fluid velocity.

If the mean velocity of the liquid in a pipe is increased gradually from zero, a value is reached eventually at which the flow changes in character from laminar to turbulent, i.e. the streamlines become disturbed by eddies and ultimately—as the velocity rises—disrupted completely. Osborne Reynolds (1842–1912), who was the first to differentiate between the two types of flow and to point out (in 1883) that the laws governing fluid behaviour are different in the two cases, named this transition point the *higher critical velocity*. He also discovered that, when the velocity was reduced gradually from a value in the turbulent range, the flow reverted to laminar at a different, lower velocity, u_c, and he not only named this the *lower critical velocity* but showed that, for a pipe of diameter d, this value was obtainable from the equation

$$u_c = 2300\left(\frac{v}{d}\right)$$

where v = kinematic viscosity.

For a given value of v (i.e. for a given fluid) it is evident that the value of u_c—below which the flow *must* be laminar—is sensitive to the size of pipe. Thus if d is small (as in a capillary tube) u_c will be very high, so that practical velocities will be less than u_c and the flow will be laminar. Alternatively, if the quantity ud/v—which is called a *Reynolds Number* and denoted by R—is less than 2300, the flow must be laminar. The figure 2300 is called the *Critical Reynolds Number* and denoted by R_c.

The Reynolds Number can be shown to be the ratio of the forces of inertia and viscosity, a small value indicating that viscous forces predominate and vice versa.

EXAMPLE. At 15°C (59°F) the kinematic viscosity of a certain SAE 20 oil is 240 cS. Assume constant temperature and determine the lower critical velocity in a pipe of internal diameter 6 mm.

Solution.
$$v = 240 \text{ cS}$$
$$= 2.4 \text{ Stokes, i.e. cm}^2/\text{s},$$
$$= \frac{2.4}{100^2}$$
$$= 0.00024 \text{ m}^2/\text{s}$$
$$d = 0.006 \text{ m}$$
$$u_c = R_c\left(\frac{v}{d}\right)$$
$$= 2300\left(\frac{0.00024}{0.006}\right)$$
$$= 92 \text{ m/s}.$$

Since the kinematic viscosity of water at this temperature is about 1·0 cS, the lower critical velocity of water in this pipe would be 92/240 or about 0·38 m/s.

Steady Laminar Flow along a Uniform Pipe

Figure 175 shows a uniform pipe of internal radius r. When it is full there is no free surface and the only boundary layer is the wetted area of the pipe. Since this is stationary, the graph of velocity against radius has the form shown.

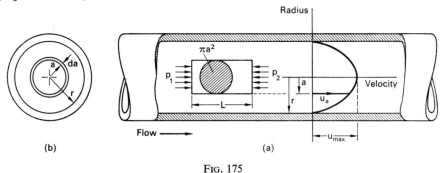

Fig. 175

Consider the cylindrical element of fluid of length L and radius a, co-axial with the pipe and moving at some uniform velocity, u, under a pressure difference (p_1-p_2). Then,

$$\text{Force producing motion} = (p_1-p_2)\pi a^2.$$

If q is the shear stress at radius a, i.e. on the cylindrical surface of the element,

$$\text{Force opposing motion} = q(2\pi aL)$$

$$= \mu(2\pi aL)\frac{du}{da}$$

since

$$q = \mu \cdot \frac{du}{da}.$$

Since u is constant at that radius, these forces must be equal and opposite so that

$$\mu(2\pi aL)\frac{du}{da} = -(p_1-p_2)\pi a^2$$

or

$$\mu \cdot du = -\frac{(p_1-p_2)}{2L} a \cdot da$$

i.e.

$$\mu \int_u^0 du = -\frac{p_1-p_2}{2L} \int_a^r a \cdot da.$$

$$\therefore \quad \mu(0-u) = -\frac{p_1-p_2}{2L} \cdot \frac{(r^2-a^2)}{2}$$

so that

$$u = \frac{p_1-p_2}{4\mu L}(r^2-a^2).$$

This equation shows that the velocity variation across any diameter is parabolic.

173

Now consider a hollow cylindrical element of radius a and thickness da, Figure 175(b), having the same length L and moving at the same velocity, u.

Section of element $= 2\pi a \cdot da$

$\therefore$ Elemental volume/sec $= (2\pi a \cdot da) \dfrac{p_1-p_2}{4\mu L}(r^2-a^2)$

$\qquad\qquad\qquad = \dfrac{\pi(p_1-p_2)}{2\mu L}(r^2 a - a^3)\,da$

$\therefore$ Total volume/sec, $Q = \dfrac{\pi(p_1-p_2)}{2\mu L}\displaystyle\int_0^r (r^2 a - a^3)\,da$

$\qquad\qquad\qquad = \dfrac{\pi(p_1-p_2)}{2\mu L}\left[r^2\dfrac{a^2}{2} - \dfrac{a^4}{4}\right]_0^r$

$\qquad\qquad\qquad = \dfrac{\pi(p_1-p_2)r^4}{2\mu L}\left(\dfrac{1}{2} - \dfrac{1}{4}\right)$

whence $\qquad\qquad Q = \dfrac{\pi r^4}{8\mu}\left(\dfrac{p_1-p_2}{L}\right).$

This is known as the Hagen–Poiseuille Formula.

At the pipe axis, $a = 0$, so that the maximum velocity is given by

$$u_{max} = \left(\dfrac{p_1-p_2}{4\mu L}\right)r^2.$$

If the velocity were uniform across the section and equal to u_{max}, the volume passing per second would be that of the cylinder $(\pi r^2)u_{max}$. However, the actual volume passing per second is that of the paraboloid (of base radius r and height u_{max}) enclosed by this cylinder and having half its volume. The mean velocity must be, therefore, $u_{max}/2$ or

$$u_{mean} = \left(\dfrac{p_1-p_2}{8\mu L}\right)r^2.$$

Hence, alternatively,

Total volume/sec, $\quad Q = \text{section} \times u_{mean}$

$\qquad\qquad\qquad = \pi r^2 \left(\dfrac{p_1-p_2}{8\mu L}\right)r^2$

or, as before, $\qquad\qquad Q = \dfrac{\pi r^4}{8\mu}\left(\dfrac{p_1-p_2}{L}\right).$

Transposing gives: $\qquad \dfrac{p_1-p_2}{L} = \left(\dfrac{8\mu}{\pi r^4}\right)Q.$

The quantity in the brackets is a constant for a given liquid in a given uniform pipe, while the left-hand side of the equation is the pressure drop per unit length due to viscous resistance within the fluid itself. Evidently this pressure drop increases in direct proportion to the rate of flow as well as to the dynamic viscosity.

EXAMPLE. At 15°C the kinematic viscosity and mass density of a certain SAE 20 oil are, respectively, 240 cS and 870 kg/m³. Determine the pressure drop due to viscous resistance per metre length of pipe of internal diameter 6 mm when the mean velocity is 1·5 m/s. If the pump has a mechanical efficiency of 0·6, find the input power required.

Solution. From the example on p. 172, it is known that the critical velocity for this liquid in this pipe at this temperature is 92 m/s. For a velocity of 1·5 m/s the flow is evidently laminar so that use can be made of the equation just derived.

Kinematic viscosity, $\quad \nu = \dfrac{240}{100}\left(\dfrac{1}{100^2}\right) \quad$ since $\quad$ 1 m = 100 cm,

$$= 0.00024 \text{ m}^2/\text{s}$$

Dynamic viscosity, $\quad \mu = \nu\varrho$

$$= 0.00024 \times 870 \quad \dfrac{\text{m}^2}{\text{s}}\dfrac{\text{kg}}{\text{m}^3}$$

$$= 0.209 \text{ kg/m s}$$

Pipe radius, $\quad r = \dfrac{0.006}{2} = 0.003 \text{ m}$

Volume/sec, $\quad Q = (\pi r^2)u$

$$= (\pi \times 0.003^2)1.5$$

$$= 42.4 \times 10^{-6} \text{ m}^3/\text{s}$$

Pressure drop, $\quad p_1 - p_2 = \left(\dfrac{8\mu}{\pi r^4}\right)QL \quad$ and $\quad L = 1.0 \text{ m}$

$$= \left[\dfrac{8 \times 0.209}{\pi(0.003)^4}\right]\dfrac{42.4}{10^6} \times 1.0 \left(\dfrac{\text{Ns}}{\text{m}^2}\right)\left(\dfrac{1}{\text{m}^4}\right)\left(\dfrac{\text{m}^3}{\text{s}}\right) \text{m}$$

$$= 279\ 000 \text{ Pa (N/m}^2\text{)}$$

Work done/sec = Mean velocity × Net force overcome

$$= u(p_1 - p_2)\pi r^2$$

$$= 1.5(279\ 000)\pi(0.003)^2 \quad \left(\dfrac{\text{m}}{\text{s}}\right)\left(\dfrac{\text{N}}{\text{m}^2}\right)\text{m}^2$$

$$= 11.8 \text{ Nm/s, i.e. J/s}$$

Required power $= \dfrac{11.8}{0.6}$

$$= \underline{19.7 \text{ W.}}$$

Conversion of Units of Dynamic Viscosity

Prior to the introduction of SI, there were no less than six different units available for the definition of dynamic viscosity. Since it is probable that the reader will encounter these, a table has been compiled of conversion factors deduced from the knowledge that

$$\text{Force} = \text{mass} \times \text{acceleration}$$
$$1{\cdot}0 \text{ lbf} = 1{\cdot}0 \text{ slug} \times 1{\cdot}0 \text{ ft/s}^2$$
$$1{\cdot}0 \text{ lbf} = 32{\cdot}2 \text{ lb} \times 1{\cdot}0 \text{ ft/s}^2$$

Hence
$$x \text{ lbf} \left(\frac{s}{ft^2}\right) = 32{\cdot}2x \text{ lb} \frac{ft}{s^2} \left(\frac{s}{ft^2}\right)$$

i.e.
$$x \text{ lbf s/ft}^2 = 32{\cdot}2x \text{ lb/ft s}.$$

Thus, to convert slug units (lbf s/ft²) to lb/ft s, one must multiply by 32·2, i.e. by g. (Conversely, to convert to slug units, one must divide by g.)

Similarly, in the cgs system, g = 981 cm/s² so that

$$x \text{ gf s/cm}^2 = 981x \text{ g/cm s} \quad \text{(or poise)}.$$

Thus, to convert gf/cm² to poise, one must multiply by 981.

Since 1·0 lb = 453·6 g and 1·0 ft = 30·48 cm, it follows that

$$x \frac{lb}{ft \ s} = x \left(\frac{453{\cdot}6}{30{\cdot}48}\right) \frac{1}{s} \frac{g}{cm \ s}$$

i.e.
$$x \text{ lb/ft s} = 14{\cdot}88x \text{ g/cm s} \quad \text{(or poise)}.$$

Thus, to convert lb/ft s to poise, one must multiply by 14·88. It follows that, to convert lbf s/ft² to poise, one must multiply by 14·88 × 32·2, i.e. by 479.

Since
$$\frac{kg}{m \ s} = \frac{g}{1000 \ cm \ s} \frac{100}{} = \left(\frac{g}{cm \ s}\right) \frac{1}{10} = \frac{poise}{10},$$

it follows that, to obtain SI units (kg/m s, or Ns/m²), the dynamic viscosity in poise must be divided by 10.

Alternatively, 1·0 SI unit = 10 poise = 1000 centipoise.

lb/ft s	lbf s/ft²	gf s/cm²	kg/m s (SI)	g/cm s (poise)	g/cm s × 10² (centipoise)
1·0	0·0311	0·0152	1·488	14·88	1488
32·2	1·0	0·488	47·9	479	47 900
65·6	2·049	1·0	98·1	981	98 100
0·672	0·02089	0·0102	1·0	10	1000
0·0672	0·002089	0·00102	0·1	1·0	100
0·000672	0·00002089	0·0000102	0·001	0·01	1·0

Finally,
$$\text{lbf s/in}^2 \text{ (Reyns)*} = \text{lbf s/ft}^2 \div 144 = \text{cP } (1{\cdot}45 \times 10^{-7}).$$

* After Reynolds.

Notes

1. No numerical value of viscosity has any significance unless the temperature is specified.
2. The denser of two liquids of identical dynamic viscosity will have the lesser kinematic value.

EXAMPLE. At a given temperature the dynamic viscosity of a certain lubricating oil is stated to be 0·025 lb/ft s. Convert this to (a) lbf s/ft², (b) gf s/cm², (c) centipoise, (d) kg/m s.

Solution. From the table:

$$0.025 \text{ lb/ft s} = 0.025 \times 0.0311 = 778 \times 10^{-6} \text{ lbf s/ft}^2$$
$$= 0.025 \times 0.0152 = 380 \times 10^{-6} \text{ gf s/cm}^2$$
$$= 0.025 \times 1488 = 37.2 \text{ cP}$$
$$= 0.025 \times 1.488 = 0.0372 \text{ kg/m s}.$$

Steady Turbulent Flow along a Uniform Pipe

It has been shown that steady flow (constant volume/sec) along a uniform pipe will be laminar if the Reynolds Number is less than 2300. The fall in pressure per unit length due to viscous resistance is then proportional to Q (and hence to the mean velocity, u) as well as to μ and to $1/d^4$. The velocity distribution is parabolic, Figure 176(a), the maximum being twice the mean.

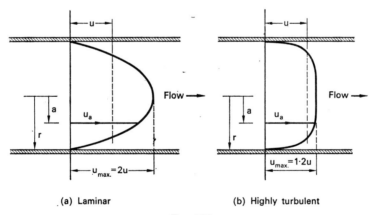

(a) Laminar (b) Highly turbulent

FIG. 176

Above the critical velocity, however, the streamlines—except in a thin layer close to the pipe surface—are disrupted by eddies, and cross-currents are superimposed on the main current. As a result of this turbulent mixing the fluid velocity is at no point constant in either direction or magnitude so that superimposed on the (unidirectional) shear stress associated with laminar flow is a continuously varying shear stress. The energy loss is greater therefore than under laminar conditions, the pressure drop per unit length being proportional to u^n (instead of to $u^{1·0}$), the value of n depending on the Reynolds Number and approaching 2·0 at high values of R. The effective velocity distribution under turbulent conditions is as shown in Figure 176(b), the ratio of maximum to mean values falling as R increases and approaching 1·2 at high values.

The foregoing assumes a smooth pipe. If the surface is sufficiently rough for the "high spots"—called *rugosities* (Latin: *ruga*, wrinkle)—to penetrate the residual laminar zone, the resistance to flow rises sharply since such penetrations are themselves the sources of eddy currents.

The Darcy Formula

Suppose the element of fluid, length L, radius r, Figure 177, to move at a mean velocity, u, as a result of a pressure difference, $p_1 - p_2$. Then

$$\text{Force producing motion} = (p_1 - p_2)\pi r^2.$$

If q is the shear stress at radius r (i.e. at the pipe surface) at unit mean velocity, then

Force opposing motion at unit velocity $= q(2\pi rL)$

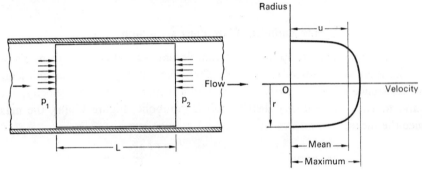

FIG. 177

If q increases as the square of u, then

Force opposing motion at velocity $u = qu^2(2\pi rL)$

Hence $(p_1 - p_2)\pi r^2 = qu^2(2\pi rL)$

i.e. $$p_1 - p_2 = \frac{2}{r} \cdot qLu^2$$

$$= \frac{4}{d}(2g)qL\left(\frac{u^2}{2g}\right)$$

so that $$\frac{p_1 - p_2}{w} = \frac{4}{d}\left(\frac{2gq}{w}\right)L\left(\frac{u^2}{2g}\right)$$

or $$\frac{p_1}{w} - \frac{p_2}{w} = \frac{4fL}{d}\left(\frac{u^2}{2g}\right)$$

where $$f = \frac{2gq}{w}.$$

Now $u^2/2g$ is the kinetic head per unit weight (Nm/N or m,) while the left-hand side of the equation is the fall in pressure head per unit weight. Thus the Darcy Formula as it is called (after the Frenchman H. P. G. Darcy, 1803–58) enables the fall in head to be expressed conveniently in terms of the kinetic head. This fall is known as the *head loss* and is denoted, usually, by h_f, so that

$$h_f = \frac{4fL}{d}\left(\frac{u^2}{2g}\right).$$

Note that, since $u^2/2g$ and $(p_1-p_2)/w$ have the same units (head in m) and L/d is a ratio, the so-called Darcy Coefficient, f, is just a number. Since, as already pointed out, the loss (i.e. q) varies as u^n where n is not constant, this coefficient is not itself a true constant. Early experiments conducted by the German, Heinrich Blasius, led to what is known as Blasius' Law, namely

$$f = 0.079\, R^{-0.25}.$$

This may be used with little error up to about $R = 100\,000$ and shows that, for a given pipe and fluid, the value of f falls gradually with rise in velocity.

Now, for laminar flow, the pressure drop in N/m² is given in terms of the flow in m³/s by

$$p_1 - p_2 = \left(\frac{8\mu L}{\pi r^4}\right)Q \quad \text{where} \quad Q = \pi r^2 u$$

$$= \left(\frac{8\mu L}{r^2}\right)u \quad \text{and} \quad r = d/2 \quad \text{while} \quad \mu = \nu\varrho$$

so that $\quad \dfrac{p_1}{w} - \dfrac{p_2}{w} = \left[\dfrac{32(\nu\varrho)L}{d^2 w}\right]u \quad$ and $\quad \dfrac{\varrho}{w} = \dfrac{1}{g}$

i.e. $\quad h_f = \left(\dfrac{4L}{d}\right)\left(\dfrac{16\nu}{d}\right)\left(\dfrac{u}{2g}\right) \quad$ (rearranging)

$$= \left(\frac{4L}{d}\right)16\left(\frac{\nu}{ud}\right)\left(\frac{u^2}{2g}\right) \left(\text{multiplying by } \frac{u}{u}\right)$$

$$= \frac{4L}{d}\left(\frac{16}{R}\right)\frac{u^2}{2g} \quad \text{since} \quad \frac{\nu}{ud} = \frac{1}{R}$$

so that $\quad h_f = \dfrac{4fL}{d}\left(\dfrac{u^2}{2g}\right) \quad$ where $\quad f = \dfrac{16}{R} \quad$ (or $16R^{-1}$).

Thus, if the kinematic viscosity is known, i.e. if the Reynolds Number can be calculated, the Darcy Formula may be used for laminar flow conditions by taking the Darcy Coefficient as $16/R$.

EXAMPLE. At 20°C the kinematic viscosity of water may be taken as 1·0038 cS. Estimate:

(a) the lower critical velocity along a copper pipe 12 mm internal diameter, assuming $R_c = 2300$;
(b) the volume passing per second at this critical velocity;
(c) the mean velocity if the time taken to fill a 20 litre can is 70 s;
(d) the value of the Reynolds Number at this velocity;
(e) the value of the Darcy Coefficient using the Blasius Law;
(f) the head loss on a length of 6 m.

Solution.

(a) $v = 1\cdot0038 \times 10^{-6}$ m²/s, $d = 0\cdot012$ m

$\therefore$ Pipe section, $\quad A = \dfrac{\pi}{4}(0\cdot012)^2 = 113 \times 10^{-6}$ m²

$$u_c = R\left(\dfrac{v}{d}\right) = 2300\left(\dfrac{1\cdot0038}{0\cdot012 \times 10^6}\right) = \underline{0\cdot192 \text{ m/s.}}$$

(b) Volume/s at $u_c = Au_c$

$$= \dfrac{113 \times 0\cdot192}{10^6}$$

$$= \underline{21\cdot7 \times 10^{-6} \text{ m}^3\text{/s.}}$$

(This corresponds to about 1·5 litres per minute.)

(c) Volume/s $= \dfrac{1}{70}\left(\dfrac{20}{1000}\right) = 286 \times 10^{-6}$ m³/s

$\therefore \quad u = \dfrac{286 \times 10^{-6}}{113 \times 10^{-6}} = \underline{2\cdot53 \text{ m/s.}}$

(d) Corresponding Reynolds Number, $\quad R = \dfrac{ud}{v}$

$$= \dfrac{2\cdot53 \times 0\cdot012 \times 10^6}{1\cdot0038}$$

$$= \underline{30\,200.}$$

(e) From Blasius' Law, $\quad f = 0\cdot079 R^{-0\cdot25}$

$$= \dfrac{0\cdot079}{30\,200^{+0\cdot25}}$$

$$= \dfrac{0\cdot079}{13\cdot17}$$

$$= \underline{0\cdot006.}$$

(f) Head loss, $\quad h_f = \dfrac{4fL}{d}\left(\dfrac{u^2}{2g}\right) \quad$ where $\quad L = 6$ m

$$= \dfrac{4 \times 0\cdot006 \times 6}{0\cdot012}\left(\dfrac{2\cdot53^2}{2 \times 9\cdot81}\right)$$

$$= \underline{3\cdot91 \text{ m.}}$$

A quicker way of arriving at this result is to consult the Guide published in 1970 by the Institution of Heating and Ventilating Engineers.

Steady Flow through a Sudden Increase in Section

An increase in pipe diameter as shown in Figure 178 will disturb the flow pattern and cause eddy currents in the so-called "dead water" between the main stream and the pipe. The consequent energy dissipation is responsible for a fall in head or "shock loss".

Suppose the pipe to be horizontal and the pressure on the annular surface (A_2-A_1) to be p. If there is no radial acceleration of the fluid in the plane of the change in section, then $p \simeq p_1$. (This has been borne out by experiment.) Since $u_1 > u_2$ the momentum of an

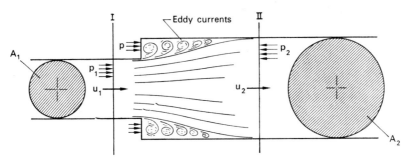

Fig. 178

element of fluid is reduced and this requires a decelerating force. Considering the pressures on the fluid between the sections:

Force producing deceleration $= p_2 A_2 - p_1 A_1 - p_1(A_2 - A_1)$
$= p_2 A_2 - p_1 A_1 - p_1 A_2 + p_1 A_1$
$= (p_2 - p_1) A_2.$

(Note that $p_2 > p_1$ because $u_2 < u_1$.)

Mass passing per second $= \dfrac{wQ}{g}$ where $Q = A_2 u_2$

$= \dfrac{w A_2 u_2}{g}$

Change in velocity $= u_1 - u_2$ (reduction)

Rate of change in momentum $= \dfrac{w A_2 u_2}{g}(u_1 - u_2)$

Equating: $(p_2 - p_1) A_2 = \dfrac{w A_2 u_2}{g}(u_1 - u_2)$

i.e. $\dfrac{1}{w}(p_2 - p_1) = \dfrac{u_2(u_1 - u_2)}{g}$ (neglecting the small force due to friction).

Loss = change in total energy

or $\quad h = H_1 - H_2$

$$= \left(z_1 + \frac{1}{2} \cdot \frac{u_1^2}{g} + \frac{p_1}{w}\right) - \left(z_2 + \frac{1}{2} \cdot \frac{u_2^2}{g} + \frac{p_2}{w}\right) \quad \text{and} \quad z_2 = z_1$$

$$= \frac{1}{2g}(u_1^2 - u_2^2) + \frac{1}{w}(p_1 - p_2)$$

$$= \frac{1}{2g}(u_1^2 - u_2^2) - \frac{1}{w}(p_2 - p_1)$$

$$= \frac{u_1^2 - u_2^2}{2g} - \frac{u_2(u_1 - u_2)}{g} \quad \text{(substituting)}$$

$$= \frac{u_1^2 - u_2^2 - 2u_2(u_1 - u_2)}{2g}$$

$$= \frac{1}{2g}(u_1^2 - u_2^2 - 2u_1 u_2 + u_2^2 + u_2^2)$$

so that $\quad h = \dfrac{1}{2g}(u_1^2 - u_2^2).$

When the pipe ends at a large tank, u_2 is approximately zero so that putting $u_1 = u$ we obtain

$$h = \frac{u^2}{2g} = \text{kinetic head.}$$

The loss may be reduced by substituting a conical section (called a *diffuser*) for the sudden change, so reducing the volume of the "dead" region. It has been shown experimentally that the loss is a minimum when the semi-angle of the cone is about 5° as shown in Figure 179.

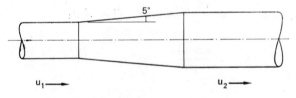

Fig. 179

Steady Flow through a Sudden Reduction in Section

Figure 180(a) shows a uniform pipe containing liquid under pressure and having its right-hand end closed by a plate containing a relatively small concentric hole having the section shown in Figure 180(b). Such a hole is called an *orifice* and in its vicinity the liquid accelerates towards the axis before issuing as a jet, the diameter of which is less than that of the orifice owing to the curvature of the streamlines. If d is the orifice diameter, the minimum diameter of the jet occurs at about $d/2$ from the plane of the orifice and is known as

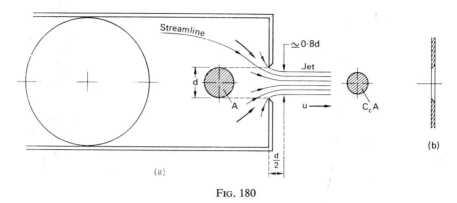

Fig. 180

the *vena contracta*. At this point the pressure in the fluid has fallen to atmospheric, the fluid has maximum velocity (i.e. the acceleration is complete) and the sides of the jet are parallel.

Suppose the total head of liquid relative to the pipe axis to be H and let the velocity of the jet be u. Then, since the energy of the jet is wholly kinetic,

$$\frac{1}{2} \cdot \frac{u^2}{g} = H \quad \text{(neglecting loss)}$$

i.e. $\quad u = \sqrt{(2gH)} \quad$ theoretically.

The actual velocity of the jet is less than this because of friction so that the above value must be multiplied by a correction factor. Thus

Actual velocity of jet $= C_v \cdot \sqrt{(2gH)}$

where $\qquad C_v = $ *coefficient of velocity*.

In general the value of C_v is found (experimentally) to lie between 0·94 and 0·98.

The difference between the section of the orifice and that of the vena contracta must be allowed for by the introduction of a second correction factor. Thus, if A is the orifice section,

Actual section of jet $= C_c A$

where $\qquad C_c = $ *coefficient of contraction*.

For a sharp-edged orifice the value of C_c has been found (experimentally) to depend on the relative sizes of pipe and orifice, the variation being as shown in Figure 181.
This shows that when the pipe is very large relative to the orifice, the value of C_c approaches 0·6. It follows that the least possible jet diameter for a sharp-edged orifice of diameter d is $\sqrt{(0\cdot6)}d$ or $0\cdot78d$ approx.

Now, Actual flow $=$ Actual section $\times$ Actual velocity

$$= C_c A \times C_v \cdot \sqrt{(2gH)}$$

so that $\qquad Q = C_d \cdot A \cdot \sqrt{(2gH)}$

where $\qquad C_d = C_c C_v$.

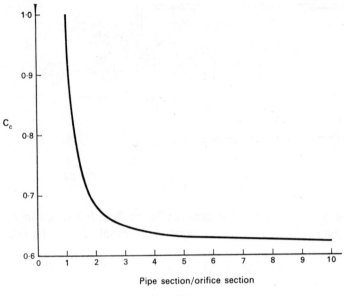

Fig. 181

The quantity C_d is called the *coefficient of discharge* and may be defined as the ratio of the actual flow (found experimentally) to that which would occur at ideal velocity without contraction. Since the value of C_v is almost unity, it is evident that the loss at an orifice is due almost entirely to the reduction in section of the jet between the plane of the orifice and the vena contracta.

If the edge of the orifice is not sharp, i.e. is given a radius, the value of C_d is increased, becoming nearly unity for a so-called "bell-mouth". The value of C_d thus lies between the extremes shown in Figure 182.

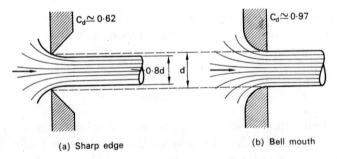

Fig. 182

Note that the acquisition of momentum by the fluid as it issues from the orifice results in the exertion of a reaction (on the pipe in this case) equal to the product of mass/sec and velocity.

Now suppose a uniform pipe of section A_2 to have a restriction of reduced section A, normal to the direction of flow as shown in Figure 183.

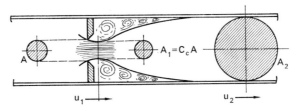

Fig. 183

The section of the vena contracta will be $C_c A$ and this may be denoted by A_1. Then, if the corresponding velocity is u_1,

$$\text{Volume/sec, } Q = A_1 u_1 = C_c A u_1 = A_2 u_2$$

so that
$$u_1 = \frac{1}{C_c}\left(\frac{A_2}{A}\right) u_2$$

or Velocity at vena contracta $= \dfrac{1}{C_c}\left[\dfrac{\text{Pipe section}}{\text{Orifice section}}\right] \times$ Velocity along pipe.

Since the fluid section increases suddenly from A_1 to A_2 there will be a loss of head given by

$$h = \frac{1}{2g}(u_1 - u_2)^2$$

$$= \left[\frac{1}{C_c}\left(\frac{A_2}{A}\right) u_2 - u_2\right]^2 \frac{1}{2g} \quad \text{(substituting for } u_1\text{)}$$

$$= \left[\frac{1}{C_c}\left(\frac{A_2}{A}\right) - 1\right]^2 \frac{u_2^2}{2g}$$

or $\quad h = k\left(\dfrac{u_2^2}{2g}\right) \quad$ where $\quad k = \left[\dfrac{1}{C_c}\left(\dfrac{A_2}{A}\right) - 1\right]^2$.

Thus the loss can be expressed in terms of the kinetic head.
If $A_2 = A$, the system reduces to that shown in Figure 184.

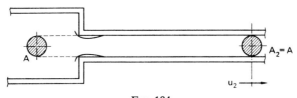

Fig. 184

Putting $A_2/A = 1\cdot 0$ in the expression for the constant, we obtain

$$k = \left(\frac{1}{C_c} - 1\right)^2.$$

If the reduction is considerable, say 75 per cent (or 4 to 1), then $C_c \simeq 0.62$ so that

$$k = \left(\frac{1}{0.6}-1\right)^2 \simeq 0.45.$$

It is common to be on the safe side and to assume that $k = 0.5$, i.e. that

$$h = 0.5\left(\frac{u_2^2}{2g}\right).$$

However, a better solution is to determine the value of k from the graph shown in Figure 185 which is based on experimental results.

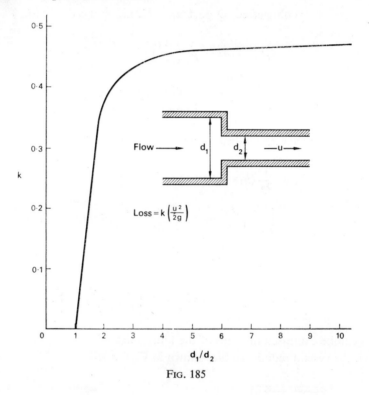

FIG. 185

EXAMPLE. A plate containing an orifice 50 mm diameter is mounted transversely in a horizontal pipeline 100 mm diameter. Find, for a water velocity of 0·75 m/s, the pressure drop between a tapping just upstream of the plate and

(a) a second tapping opposite the vena contracta,
(b) a third tapping downstream of the vena contracta.

Solution. Referring to Figure 186,

Velocity at vena contracta, $u_2 = \dfrac{1}{C_c}\left(\dfrac{\text{Pipe section}}{\text{Orifice section}}\right) \times$ Velocity along pipe

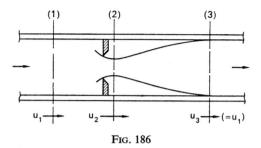

Fig. 186

Now, (Pipe section)/(Orifice section) = $100^2/50^2 = 4$, so that, from Figure 181, $C_c = 0.64$.

Hence $$u_2 = \frac{4 \times 0.75}{0.64} = 4.69 \text{ m/s}.$$

Neglecting the loss between tappings (1) and (2),

$$z_1 + \frac{1}{2} \cdot \frac{u_1^2}{g} + \frac{p_1}{w} = z_2 + \frac{1}{2} \cdot \frac{u_2^2}{g} + \frac{p_2}{w}$$

and $z_2 = z_1$ (since pipe is horizontal)

so that
$$\frac{p_1 - p_2}{w} = \frac{u_2^2 - u_1^2}{2g}$$

$$= \frac{4.69^2 - 0.75^2}{2 \times 9.81}$$

$$= 1.095 \text{ m}.$$

This is the pressure drop due to the increase in kinetic head between sections (1) and (2). Some of this will be recovered due to the reduction in kinetic head between sections (2) and (3), the difference being the shock loss resulting from the expansion. For this loss:

$$k = \left[\frac{1}{C_c}\left(\frac{\text{Pipe section}}{\text{Orifice section}}\right) - 1\right]^2$$

$$= \left[\frac{1}{0.64}(4) - 1\right]^2$$

$$= 27.5.$$

Between tappings (2) and (3) the loss is given by

$$h = k\left(\frac{u_3^2}{2g}\right) \quad \text{and} \quad u_3 = u_1 = 0.75 \text{ m/s}$$

$$= \frac{27.5 \times 0.75^2}{2 \times 9.81}$$

$$= \underline{0.79 \text{ m}}.$$

Alternatively, using the Bernoulli Equation for sections (1) and (3) and accounting for the loss,

$$\left[z_1 + \frac{1}{2} \cdot \frac{u_1^2}{g} + \frac{p_1}{w}\right] = \left[z_3 + \frac{1}{2} \cdot \frac{u_3^2}{g} + \frac{p_3}{w}\right] + \text{Loss} \quad \text{and} \quad z_3 = z_1, u_3 = u_1$$

so that $\quad \dfrac{p_1 - p_2}{w} = \text{Loss} = k\left(\dfrac{u_3^2}{2g}\right)\quad$ as above.

EXAMPLES 7

1. The greatest and least depths of immersion of a circular drain cover 1·22 m diameter are 1·53 and 0·61 m respectively. Find the position and magnitude of the resultant force on the upper side taking the specific weight of water as 9800 N/m³ (12 200 N at a depth of 1·12 m).
2. The four sloping sides of a square swimming pool 1·83 m deep are plane and trapezoidal in shape, the upper and lower edges of each being respectively 4·58 and 3·05 m in length. Determine the magnitude and position of the resultant thrust on one side taking $w = 9800$ N/m³ (63·3 kN at a depth of 1·18m).
3. Find the head of water required to maintain a flow of 70 m³/h through an orifice 64 mm diameter given that the coefficient of discharge is 0·79 (3 m approx.).
4. A jet issues from a circular orifice in the vertical side of a tank. If, at a distance x from the vena contracta (measured along the orifice axis), the fluid has fallen through a vertical distance y, show that the velocity at the vena contracta is given by $u = (gx^2/2y)^{0.5}$. If the orifice is at a depth H below the surface, show that the coefficient of velocity is given by $C_v = (x^2/4Hy)^{0.5}$.
5. The axis of a circular orifice in the side of a canal is to be at a depth of 7·6 m. Take $C_d = 0.6$ and determine the least diameter required for the discharge to be 285 m³/h (205 mm).
6. Water is pumped along a pipe 64 mm diameter to a fountain 490 m from the pump and 6·1 m above it. The pump head is 85 m while the Darcy Coefficient for the pipe is 0·0045. If the orifice has a coefficient of velocity of 0·94 and a diameter of 20 mm, find
    (a) the head loss in the delivery pipe (7·1 m);
    (b) the initial velocity of the jet (35 m/s).
7. (a) State the Newtonian definition of viscosity.
    (b) Discuss the statement: "No numerical value of viscosity has any significance unless the temperature is specified."
    (c) Explain why it is possible for two liquids of identical dynamic viscosity to have differing kinematic values.
8. At atmospheric pressure and 0°C the dynamic viscosity of water is 1·793 cP. Express this value in
    (a) lbf s/ft² (37.5×10^{-6});
    (b) Reyns (0.26×10^{-6});
    (c) gf s/cm² (18.3×10^{-6}).
9. A fuel oil having a dynamic viscosity of 52.7×10^{-4} kg/m s at 10°C is pumped along a pipeline 100 mm diameter at the rate of 455 litres/min. If the specific weight at this temperature is 9120 N/m³, find the Reynolds Number (16 600).
    If the pipeline is 1220 m long find the head loss assuming a Darcy Coefficient of 0·0058 (12·2 m).
    If the tank inlet is 15·25 m above the pump, estimate the pump input power required assuming it to have an efficiency of 0·6 (3·13 kW).
10. A tank discharges under gravity into a second tank, the difference in fluid level being 4·875 m. The connecting pipe is 520 m long and its diameter changes suddenly from 75 mm to 50 mm after the first 43 m. Assume a Darcy Coefficient of 0·005 for both pipe sections and, for all changes in section, assume a contraction coefficient of 0·58. Estimate:
    (a) the velocity of discharge into the second tank (3·6 m/s);
    (b) the rate of discharge (7·27 litres/s).

8

COMBUSTION AND HEAT TRANSFER

CONCEPTS AND DEFINITIONS

System

For the purpose of studying the relation between work done, temperature change and heat flow, a *system* may be defined as any engineering assembly containing a quantity of working fluid and separated from its environment by a continuous *boundary*, real or imaginary. The boundary may be elastic—e.g. a bicycle tube—or it may be nominally rigid, e.g. a bomb calorimeter.

Process

This is the name given to the consecutive steps leading to a modification to the *state* or condition of a system. If a process does not alter the quantity of working fluid or other matter within the boundary of a system, the system is said to be *closed* and the process is described as *non-flow*. The boundary is then crossed by work and heat only, e.g. a fixed quantity (mass) of, say, air being compressed in a cylinder. A *flow process* is one which permits matter, as well as work and heat, to cross the system boundary. Such a system is said to be *open*, e.g. the discharge from a rocket nozzle.

A change in the state of a closed system can be brought about either by displacing the boundary by force—i.e. by doing work or letting work be done—or by bringing the boundary into contact with the boundary of a system at a different temperature, i.e. by permitting heat to flow across it. Each of these actions is a process.

Fluid

This may be a gas or mixture of gases (e.g. the contents of an exhaust pipe) or a liquid, or a mixture of gases and evaporated liquid—e.g. air/fuel mixture leaving a carburettor. The different forms of the working fluid—solid, liquid, vapour—are called *phases*. A closed system is defined completely when its boundary and the composition and mass of the fluid it contains are known.

Properties

The thermodynamic properties of a system are those characteristics which are useful in describing its condition or *thermodynamic state* and are six in number, viz.

Pressure	(force/unit area of boundary)	Changes observable by the senses.
Specific volume	(volume of unit mass)	
Temperature	(to some suitable scale)	
Internal energy	see p. 195	Changes not observable by the senses.
Enthalpy	see p. 197	
Entropy	see p. 191	

Experiment has shown that, if any two of these properties can be varied independently o each other in a given closed system, their instantaneous corresponding values determine the thermodynamic state of unit mass. This state then fixes the values of the other four properties and makes possible their calculation. Hence, at any particular state, each property has a single numerical value irrespective of the process by which the fluid came to be in that state. The degree of variation in the properties is related to the quantities of heat and work which cross the system boundary. If neither heat nor work crosses the boundary, there is no variation in any property and the system is said to be in *thermal equilibrium*.

Cycle

If a system undergoes a series of processes and is restored to its original state at the end of the final process, it is said to have passed through a *cycle*.

Temperature

If two systems suffer no observable change when their boundaries are brought into contact, they are said to be at the same *temperature*. It follows that, if two systems are each at the temperature of a third system, they are at the same temperature, i.e. they are in thermal equilibrium. If the third system is a thermometer, the temperatures of any two other systems may be compared. All that is required of the third system (thermometer) is that it should possess some property which changes appreciably with change in temperature (preferably in linear fashion) and that such changes should be easily observable. In the case of the common mercury thermometer, notice is taken of the change in length of a column of mercury in a glass tube. (The variation of such length with temperature is approximately linear.)

Work and Heat

Work is the transfer of mechanical energy—i.e. it is associated with the motion of a force—and is said, conventionally, to be negative when it is done *on* a system by all or part of its environment. Examples are the compression of air in a cylinder by a plunger (called *displacement work* since the boundary is moved) and the rotation of an impeller against

COMBUSTION AND HEAT TRANSFER

fluid resistance by an applied torque (called *stirring* or *shear work*). Conversely, a work output is said to be positive.

Heat, like work, is a transient quantity and may be defined as the transfer of energy across a boundary in association with a difference in temperature. A heat input to a system is said, conventionally, to be positive. The conventions are shown in Figure 187(a) and (b), the boundaries being indicated by dotted lines.

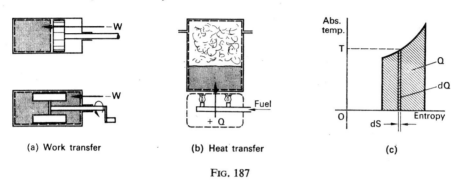

(a) Work transfer   (b) Heat transfer   (c)

FIG. 187

Entropy

It is useful to be able to draw a graph, Figure 187(c), using absolute temperature as ordinate, under which the area represents "heat", i.e.

$$dQ = T.dS$$

where, say, S is the independent variable.

Then
$$Q = \int T.dS$$

or
$$dS = \frac{1}{T}.dQ.$$

The variable invented for this purpose has been given the name *entropy*.

PROPERTIES AND BEHAVIOUR OF GASES

Boyle's Law

The statement that compression of a quantity of gas at constant temperature (i.e. isothermally) results in a hyperbolic pressure rise was made—as a result of experiment—by Robert Boyle (1627–91) and is known as Boyle's Law. Such compression may be represented by either the hyperbola of Figure 188 or the equation

$$pV = \text{constant}.$$

The most useful form of this equation is

$$p_2 V_2 = p_1 V_1$$

so that, after an isothermal reduction in volume, the new pressure of a quantity of gas is,

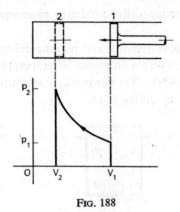

Fig. 188

according to Boyle's Law, given by

$$p_2 = \left(\frac{V_1}{V_2}\right) p_1.$$

The quantity V_1/V_2 is known as the *compression ratio* and evidently is greater than unity.

Since the value of the constant depends upon the temperature of compression (i.e. is a function of it) we can also write mathematically

$$pV = f(T).$$

More recent (and accurate) work has shown that gas behaviour deviates from this simple law. However, for analytical purposes it is sometimes convenient to assume that gases do so behave and to define as *ideal* or *perfect* a (hypothetical) gas which obeys Boyle's Law. Some gases do in fact comply with the law (very nearly) over a limited range of low pressure.

Charles' Law

The statement that the reduction in volume of a quantity of ideal gas at constant pressure is proportional to the temperature drop was made by Jacques Charles (1746–1823) and is known as Charles' Law. As a result of experiment he deduced that, at constant pressure, the actual volume reduction per degree Celsius[†] fall is $V/273$ where V is the volume at 0°C. He then reasoned that, if the temperature could be reduced to -273°C without change in phase (i.e. without liquefaction occurring) the volume would become zero. This temperature therefore has been given the name *Absolute Zero* and the temperature scale starting from this point (which is artificial and obtained by extrapolation—see Fig. 190) is called the *Kelvin Scale* (°K) after William Thompson (1824–1907) later Lord Kelvin.

The other scale starting from absolute zero (and now superseded) but using Fahrenheit degrees, is called the *Rankine Scale* (°R) after W. J. M. Rankine (1820–72). The four scales are shown for comparison in Figure 189.

[†] The name Celsius replaces Centigrade by an international decision made in 1948.

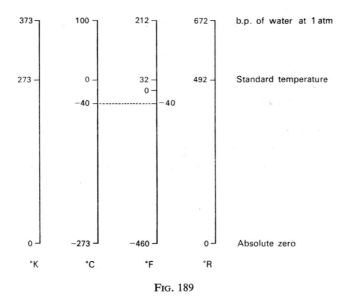

Fig. 189

It is worth remarking that, although temperatures as low as 0·005°K have been attained under laboratory conditions, absolute zero will not be reached until a perfect "insulator", i.e. non-transmitter of heat, has been developed. This is improbable.

Standard Temperature and Pressure

A standard atmosphere (1 atm) is defined as 101 325 N/m² and corresponds to the pressure due to a head of approximately 760 mm mercury at 273°K. For comparative purposes gas properties are always quoted on this basis which is now known as *standard temperature and pressure* or s.t.p. This has superseded n.t.p. as used formerly.

Characteristic Constant

If the absolute temperature is denoted by T, then Charles' Law may be represented by the equation

$$\frac{V}{T} = \text{constant}.$$

The most useful form of this equation is

$$\frac{V_2}{T_2} = \frac{V_1}{T_1}$$

so that, after a temperature drop at constant pressure, the new volume of a quantity of gas is, according to Charles' Law, given by

$$V_2 = \left(\frac{T_2}{T_1}\right) V_1.$$

Now the equation for a quantity of ideal gas may, if the quantity has unit mass, be written

$$pv = f(T)$$

where v is the specific volume,

or

$$pv = RT$$

where R is a constant the value of which is different for each ideal gas.
R is called therefore the *Characteristic Constant* and, if the product pv is plotted against T over a practicable range with a real gas, will be the slope of the straight line obtained. Figure 190 shows such a graph in which the part ab has been extrapolated to the theoretical absolute zero.

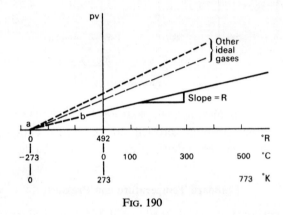

Fig. 190

Since $v = V/m$, where V is the volume at s.t.p. of m kg of gas, we have

$$pV = RmT.$$

It has been found by experiment that 1 kg air occupies 0·773 m³ at 273°K and 101 325 N/m² so that, for air at s.t.p., substitution in the above equation gives

$$101\ 325 \times 0\cdot 773 = R(1\cdot 0 \times 273)$$

whence
$$R = 287 \text{ Nm/kg deg K (or J/kg deg K)}.$$

Again, since pv/T is constant, we have also the useful relation called the *Combined Law* (because it is in fact a combination of the laws of Boyle and Charles) namely

$$\frac{p_2 v_2}{T_2} = \frac{p_1 v_1}{T_1}$$

or, multiplying both sides by m,

$$\frac{p_2 V_2}{T_2} = \frac{p_1 V_1}{T_1}, \quad \text{since} \quad V = mv.$$

COMBUSTION AND HEAT TRANSFER

The 1st Law of Thermodynamics

J. P. Joule (1818–89) found experimentally (in 1843) that to raise the temperature of a fixed quantity of water by a fixed amount it was necessary to expend a fixed quantity of mechanical energy. The 1st Law formalises his discovery and states that there is a fixed rate of exchange between thermal and mechanical energy; it is thus an expression of the principle of conservation of energy.

An alternative statement of the 1st Law is as follows: "The net work received by the surroundings of a closed system, when the system experiences a cycle, is proportional to the net heat given up by the surroundings."

(Although there exists no formal proof of this relation, no one has ever succeeded in disproving it or any deduction made from it.)

The basic unit of energy (thermal and mechanical) in the SI is the newton metre, Nm, also called the joule, J.

The Non-flow Energy Equation

In a non-flow process the boundary is, by definition, crossed by work or heat only, the quantity of fluid remaining constant. Also, since there is no change in position, the potential and kinetic energies must each remain constant. Now energy can change only from one form to another (i.e. it is "conserved") so that, if the quantity of thermal energy entering the fluid across the boundary (positive according to convention) is greater than the quantity of mechanical energy leaving (expended by) the fluid (negative according to convention), the difference must remain in the fluid. This difference increases the vibration and random motion of the molecules (i.e. raises the temperature) and represents an increase in *internal energy*.

(The converse evidently holds good, an expenditure of mechanical energy in excess of the heat input resulting in a temperature drop.)

Hence,

$$\text{Heat input} - \text{Work output} = \text{Change in internal energy}$$

or

$$Q - W = E.$$

If the process is a complete cycle, i.e. if the final state is identical to the initial state, then

$$Q - W = 0.$$

The above are known as "non-flow energy equations".

Specific Heat Capacity

Originally, the ratio of the heat required to raise the temperature of unit mass of substance through one degree to the heat required to perform the operation on unit mass of water was known as the *specific heat* of the substance. The heat required to raise unit mass of water through one degree was taken as the basic unit of heat and this was unsatisfactory because different authorities defined heat units for different positions in the temperature scale. Specific heats therefore varied also.

MECHANICAL TECHNOLOGY FOR HIGHER ENGINEERING TECHNICIANS

By international agreement the joule is now defined as the basic energy (heat) unit so that the use of water as a criterion is avoided and the ratio mentioned above is eliminated. Substances are now said to have a *specific heat capacity*, that for water being 4186·8 J/kg deg K.

Unlike solids and liquids, a gas has an infinite number of specific heat capacities, each of which corresponds to a particular process.

The Steady Flow Energy Equation

Suppose 1 kg fluid of specific volume v_1 and having internal energy e_1 to cross the boundary of an open system at uniform velocity u_1 at a point at a height z_1 above some chosen datum against a pressure p_1 as shown in Figure 191 and suppose that, after having received a quantity of heat Q and done a quantity of work W, the fluid emerges with velocity u_2 against a pressure p_2 at height z_2 and has then internal energy and specific volume e_2 and v_2 respectively.

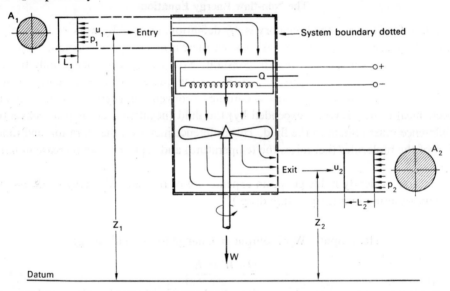

FIG. 191

If the inlet pipe section is A_1 and 1 kg fluid occupies a length L_1 then,

Work done in getting the fluid
into the system against pressure $p_1 = (p_1 L_1) A_1$

or Flow work done *on* system $= p_1 v_1.$

Energy at entry = sum of internal, kinetic and height energies

$$= e_1 + \frac{1}{2}\frac{u_1^2}{g} + z_1$$

Energy at exit $= e_2 + \dfrac{1}{2}\dfrac{u_2^2}{g} + z_2.$

196

COMBUSTION AND HEAT TRANSFER

If the outlet pipe section is A_2 and 1 kg fluid occupies a length L_2 then,

Work done in getting the fluid out
of the system against pressure $p_2 = (p_2 A_2) L_2$
or Flow work done *by* system $= p_2 v_2$.

Since the velocities are uniform, i.e. the flow is steady, the energy leaving the system must be equal to that entering it, i.e.

$$\begin{bmatrix}\text{Flow work done}\\ \text{on system}\end{bmatrix} + \begin{bmatrix}\text{Energy at}\\ \text{entry}\end{bmatrix} + Q = W + \begin{bmatrix}\text{Energy at}\\ \text{exit}\end{bmatrix} + \begin{bmatrix}\text{Flow work done}\\ \text{by system}\end{bmatrix}$$

or,
$$p_1 v_1 + e_1 + \frac{u_1^2}{2g} + z_1 + Q = W + e_2 + \frac{u_2^2}{2g} + z_2 + p_2 v_2.$$

This is called the *steady flow energy equation*. (The changes in datum energy are usually negligible.)

It is usual to refer to the sum of flow work and internal energy as the *enthalpy* of the fluid and to denote this by h. Thus, in general,

$$h = pv + e.$$

Note that, since p, v and e are thermodynamic properties of the fluid, it follows that enthalpy is such a property also. Its units are evidently those of internal energy. Neglecting the datum energy terms, the steady flow energy equation can now be written

$$h_1 + \frac{u_1^2}{2g} + Q = h_2 + \frac{u_2^2}{2g} + W.$$

The 2nd Law of Thermodynamics

This was first propounded by R. J. E. Clausius (1822–88) who stated that

Heat will flow unaided only to a cooler body.

The law states, therefore, that heat can be extracted *from* a cooler body *only by expending mechanical energy*. This is a fundamental limitation on the two-way interchangeability of thermal and mechanical energy and is exemplified by the need of a refrigerator for an energy input of some kind. The law may be said to express the principle that total conversion of a quantity of heat into work is impossible. Thus no machine devised by man can (or ever will be able to) convert *all* of a given quantity of heat into mechanical energy—although the converse is possible. In fact, if T_1 and T_2 are, respectively, the absolute temperatures of heat reception and rejection by a system, the theoretical maximum convertible proportion of a heat input, Q, is

$$\left(\frac{T_1 - T_2}{T_2}\right) Q.$$

Since there is a practical (lower) limit to T_2, the converted fraction (i.e. the efficiency) can be increased only by raising T_1. This is the underlying reason for the ever-rising operating temperatures of "heat engines" and other thermal devices.

Reversibility

If a temperature difference exists between two points in a fluid, heat is transferred (necessarily, unavoidably and without external help) from the point at the higher to that at the lower temperature (2nd Law of Thermodynamics). This happens *in fact* and, since the process cannot be reversed without the doing of work, it is called *irreversible*. All heat transfer processes are irreversible.

Again, if a quantity of fluid expands in any practical way, some of the output work is done against friction (if only on account of turbulence in the fluid) so that this kind of process is irreversible too. It can be stated that *all thermodynamic processes encountered in engineering are inherently irreversible*. Alternatively, a process is irreversible if the energy changes cannot be reversed and if the consecutive states of the fluid during the process cannot be retraced in the reverse sequence.

It is, however, convenient to imagine that turbulence (and hence internal friction) can be eliminated by carrying out, say, an expansion at an infinitely low speed so that the fluid passes through a series of equilibrium states. Also that mechanical loss can be eliminated and that heat can be transferred in association with an infinitely small temperature difference. In other words it is convenient to imagine that processes can be carried out under ideal conditions, i.e. can be made reversible. Ideal efficiencies can then be deduced on this basis and theoretical efficiencies under proposed conditions can be compared with them.

Heat Transfer at Constant Volume

Suppose the boundary not to be displaced (no displacement work done) and suppose no stirring work to be done either. If the specific heat capacity at constant volume is denoted by c_v then the heat input required to raise the temperature of m kg of gas from T_1 to T_2 will be given by
$$Q = mc_v(T_2 - T_1).$$
But
$$Q - W = E$$
and
$$W = 0$$
so that
$$\underline{E = mc_v(T_2 - T_1).}$$

Thus the only effect of the heat supplied is to change (increase) the internal energy of the gas. This equation in fact holds good whatever the process since it is really a statement of Joule's Law, viz. that the change in internal energy is proportional only to temperature change.

Heat Transfer at Constant Pressure

Suppose, as the temperature of m kg of gas rises from T_1 to T_2, that no stirring work is done as above but that the volume now increases in such a way as to maintain the pressure constant. Since the boundary is displaced (enlarged) work will be done at the expense of some of the heat supplied, i.e. more heat is absorbed than in above. (Alternatively, not all he heat supplied remains in the gas, i.e. the increase in internal energy is less.) If the specificf

heat capacity at constant pressure is denoted by c_p then this (greater) heat input will be given by
$$Q = mc_p(T_2 - T_1).$$

The ratio c_p/c_v is denoted by γ (gamma) and is evidently greater than unity. Since the values of c_p and c_v depend on the temperatures of heat transfer, the value of γ will be constant only if c_p and c_v vary in the same way.

Constant Pressure Process

Suppose the ideal (frictionless) piston shown in Figure 192(a) to be freely supported by V_1 m³ of ideal gas, i.e. suppose it to maintain constant pressure.

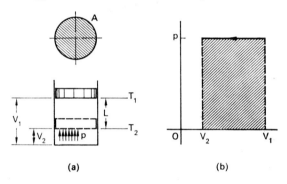

FIG. 192

If heat is extracted, the volume will fall to V_2 and the piston will travel downward a distance L. Hence

Work done on gas $= -(pA \times L)$ (conventionally negative)

i.e. $\qquad W = -p(V_1 - V_2)$ since $AL =$ swept volume,

$\qquad =$ shaded area under graph, Figure 192(b).

But $p_2V_2 = RmT_2$ and $p_1V_1 = RmT_1$, so that $-p(V_1-V_2) = -Rm(T_1-T_2)$.
Hence, $\qquad W = -Rm(T_1 - T_2).$

Evidently there is a fall in temperature while the heat is being extracted (i.e. $T_1 > T_2$) so that, while the work is being done,

$$\text{Heat loss, } Q = -mc_p(T_1 - T_2)$$

The change (reduction) in internal energy is also given by

$$E = -mc_v(T_1 - T_2).$$

Hence, using the energy equation,
$$Q - W = E$$
$$-mc_p(T_1 - T_2) + Rm(T_1 - T_2) = -mc_v(T_1 - T_2)$$
i.e. $\qquad -c_p + R = -c_v$
or $\qquad R = c_p - c_v.$

Thus R is equal to the difference between the two values of specific heat capacity.

Constant Temperature Process

Suppose a quantity of ideal gas, V_1, to be compressed without rise in temperature (i.e. isothermally) to some lesser volume, V_2. (The compression would have to take place infinitely slowly to allow the heat time to transfer across the gas boundary.)

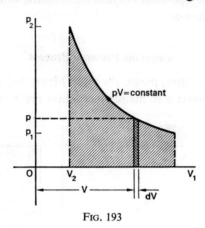

Fig. 193

Referring to Figure 193,

Work done = area under curve (conventionally negative)

i.e.
$$W = -\int_{V_2}^{V_1} p \, dV$$

where
$$p = p_1 V_1 \left(\frac{1}{V}\right)$$

∴
$$W = -p_1 V_1 \int_{V_2}^{V_1} \frac{1}{V} \, dV$$

so that
$$W = -p_1 V_1 \log_e \left(\frac{V_1}{V_2}\right).$$

Since the temperature is constant, all this work crosses the gas boundary as heat so that $E = 0$. Hence $Q = -W$, i.e.

$$Q = p_1 V_1 \log_e \left(\frac{V_1}{V_2}\right).$$

Polytropic Process *(General Case)*

In this case the reduction in volume is accompanied by a rise in temperature, i.e. $T_2 > T_1$. There is, therefore, an increase in internal energy. Referring to Figure 194 we have

$$W = -\int_{V_2}^{V_1} p \, dV$$

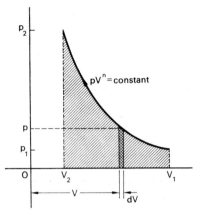

Fig. 194

where
$$p = p_2 V_2^n \left(\frac{1}{V^n}\right).$$

Hence
$$W = -p_2 V_2^n \int_{V_2}^{V_1} \frac{1}{V^n} dV$$

$$= -\frac{p_1 V_1 - p_2 V_2}{1-n} \quad \text{(after some algebra)}$$

or
$$W = -\frac{p_2 V_2 - p_1 V_1}{n-1}.$$

Putting $p_2 V_2 = RmT_2$ and $p_1 V_1 = RmT_1$ we obtain
$$p_2 V_2 - p_1 V_1 = Rm(T_2 - T_1)$$

Hence,
$$W = -\frac{Rm(T_2 - T_1)}{n-1}.$$

Increase in internal energy, $E = mc_v(T_2 - T_1)$ so that, from the energy equation, $Q - W = E$ we obtain,
$$Q = E + W$$

or
$$Q = mc_v(T_2 - T_1) - \frac{Rm(T_2 - T_1)}{n-1}.$$

Since not all the work done is absorbed by the fluid, i.e. since $W > E$, it follows that Q is negative and represents a heat loss. Note that the value of n depends on the process.

Adiabatic Process *(Particular Case)*

An adiabatic process is, by definition, one during which no heat is transferred across the boundary. Thus, in the energy equation, $Q = 0$. Assuming polytropic compression of a quantity of ideal fluid we have
$$0 = mc_v(T_2 - T_1) - \frac{Rm(T_2 - T_1)}{n-1}$$

i.e.
$$\frac{Rm(T_2-T_1)}{n-1} = mc_v(T_2-T_1)$$

whence
$$R = c_v(n-1)$$
but
$$R = c_p - c_v$$
so that
$$c_v(n-1) = c_p - c_v$$
or
$$nc_v - c_v = c_p - c_v$$
whence
$$n = \frac{c_p}{c_v} = \gamma.$$

Hence, for adiabatic processes, pV^γ = constant. The work input required for adiabatic compression is therefore given by

$$W = \frac{p_2V_2 - p_1V_1}{\gamma - 1} = \frac{Rm(T_2 - T_1)}{\gamma - 1} \quad \text{thermal units.}$$

Note that γ is a property of the gas. Unlike the index, n, it does not depend upon the particular process of compression.

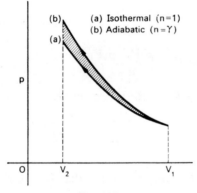

(a) Isothermal ($n=1$)
(b) Adiabatic ($n=\gamma$)

FIG. 195

Referring to Figure 195 it is evident that more work (represented by the shaded area) is needed to reduce a gas volume (by a given amount) adiabatically than isothermally. Since there is no heat loss, the final temperature and pressure are higher than in the previous case.

Atomic Structure

The atom consists of a relatively large, positively charged nucleus which is at the centre of the orbits of one or more (relatively small) negatively charged electrons, the whole being electrically neutral. Due to the opposite sign of the electric charges on nucleus and electron, the large nucleus attracts the electron to it (and vice versa) and it is this attraction which provides the centripetal force needed to keep the electron on its path. The number of orbiting electrons is known as the *atomic number* and is different for each element. With few exceptions, the so-called *atomic weight* increases with the number of electrons.

The potential energy of an electron is the work which would have to be done in moving it from infinity to its distance (radius) from the nucleus, and this work can be shown to equal $-mu^2$ where m is the electron mass and u is the velocity corresponding to its radius. The kinetic energy of an electron is $+\frac{1}{2}(mu^2)$ so that the net energy of an electron in orbit is $-\frac{1}{2}(mu^2)$ and is constant. It follows that if an electron changes its orbit for another at a different radius, there will be a change in the net energy of the atom to which it belongs, and, as a result, either release or absorption of energy.

Molecular Weight

The smallest quantity of a substance which can exist alone is called a *molecule*. The molecules of oxygen, hydrogen, nitrogen and carbon monoxide contain two atoms and are given the name *diatomic*. Each substance, whether an element or a compound, is denoted by a chemical symbol comprising one or more letters, the number of atoms per molecule being denoted by a figure subscript. The number of molecules under consideration is denoted by a figure prefix, e.g. $5O_2$ represents five molecules of oxygen each consisting of two atoms.

The so-called *molecular weight*,[†] M, depends upon the substance, that of the oxygen molecule being taken as 32 exactly and used as a reference for those of other gases. As already mentioned, the oxygen molecule contains two atoms so that its atomic weight is 16. The atomic weights of several other substances are shown in the following table.

Name of substance	Chemical symbol	Atomic weight
Oxygen	O_2	16
Hydrogen	H_2	1
Nitrogen	N_2	14
Carbon	C	12
Sulphur	S	32

The Clausius Kinetic Theory

Except at very high pressure, the volume of the molecules in a gas is very small relative to the volume occupied by the gas so that their spacing is large and the gas is highly compressible. According to the Clausius Theory, the molecules move within the gas boundary in random fashion and with a velocity which increases with rise in temperature so that their kinetic energy, i.e. the internal energy of the gas, also increases with temperature. Further, any molecule must approach other molecules continually, in succession, sufficiently closely for inter-molecular forces to become significant, the result being consecutive changes in molecular direction. Evidently the molecules themselves must be perfectly elastic since otherwise there would be an energy loss at each "collision" and the molecules would, eventually, settle on the bottom of the container. (This does not happen.) Further, the change in momentum of molecules rebounding from the sides of the container is the cause of "pressure". Hence, for a given volume, pressure rises with temperature and vice versa.

† Molecular weight *ratio* would be more appropriate.

The Universal Gas Constant

The following table gives experimental values of several constants for the commoner (non-ideal) gases encountered in engineering, the molecular weight of oxygen being taken as 32 exactly. (For engineering calculations it is permissible to assume that all values of M are whole numbers.)

The values of ϱ (specific mass, or density) are those at s.t.p. and have been used to calculate the values of R from the relation

$$R = \frac{pv}{T} = \frac{p}{T\varrho} = \frac{101\,325}{273\varrho} = \frac{371\cdot5}{\varrho} \text{ J/kg deg K.}$$

(The figures in the table refer to pure nitrogen. The nitrogen left by the removal of oxygen from air contains about 1·8 per cent by weight of other gases, mainly argon. For such a mixture the value of R is about 295 J/kg deg K.)

Gas	Symbol	M	ϱ kg/m³	c_p	c_v	$\frac{c_p}{c_v} = \gamma$	$R = \frac{p}{T\varrho}$ J/kg °K	RM
Oxygen	O_2	32	1·429	915	655	1·396	260	8320
Nitrogen	N_2	28·016	1·250	1035	738	1·4	297	8325
Hydrogen	H_2	2·016	0·089	14 300	10 130	1·41	4170	8400
Air	—	28·97	1·293	1005	718	1·4	287	8325
Carbon monoxide	CO	28·01	1·250	1040	743	1·398	297	8320
Carbon dioxide	CO_2	44·01	1·977	844	656	1·285	188	8270
Sulphur dioxide	SO_2	64·06	2·926	643	516	1·245	127	8130
Methane	CH_4	16·04	0·717	2230	1712	1·3	518	8310
Ethane	C_2H_6	30·07	1·356	1765	1491	1·18	274	8230
Propane	C_3H_8	44·09	2·020	1690	1506	1·123	184	8120
Ethylene	C_2H_4	28·05	1·260	1564	1269	1·23	295	8280
Acetylene	C_2H_2	26·04	1·175	1710	1394	1·225	316	8245

Inspection of the table reveals that the product RM is almost constant at 8310 and that the ratio M/ϱ is almost constant at 22·41. It is therefore reasonable to assume that, in general,

$$M = 22\cdot41\varrho.$$

Since the units of ϱ are kg/m³, it follows that the mass in kg of 22·41 m³ *of any gas* will be equal numerically to the molecular weight, i.e. for any gas

$$\text{Mass of } 22\cdot41 \text{ m}^3 = M \text{ kg.}$$

This quantity of gas is called therefore *one kg-mole* or *kilogramme molecular volume*. (This is abbreviated usually to either "mole" or "mol".) Alternatively, we may write

$$\text{Volume of } M \text{ kg at s.t.p.} = 22\cdot41 \text{ m}^3.$$

It follows that

1 mole of O_2 (i.e. 22·41 m³) at s.t.p. has a mass of 32 kg
1 mole of H_2 at s.t.p. has a mass of 2·016 kg
1 mole of CO_2 at s.t.p. has a mass of 44·01 kg

Now, for m kg of gas occupying V m³ at T °K and p N/m² we have
$$pV = RmT, \quad \text{where} \quad R = \text{characteristic constant}.$$
Hence, for M kg of gas, i.e. for 1 mole of gas, occupying v_0 m³,
$$pv_0 = RMT. \quad (v_0 \text{ is called the "molar volume".})$$
If the product RM (which is nearly constant at 8310 and, in fact, has an average value of 8314·3) is denoted by R_0 we obtain
$$pv_0 = R_0 T.$$
It follows that, for n moles of gas occupying V m³,
$$pV = nR_0 T, \quad \text{where} \quad R_0 = 8314 \text{ J/kg deg K}.$$

This is known as the *universal gas equation*, the quantity R_0 being called the *universal gas constant*.

The variations in M/ϱ and RM, small though they are, show that real gases do not obey exactly the law $pV = R_0 T$. However, their behaviour becomes more nearly perfect as the pressure is reduced while, on the other hand, the deviation from the law increases both as the pressure rises and as the temperature of liquefaction is approached.

Molar Heat

If c_v and c_p are respectively the specific heat capacities of a gas at constant volume and constant pressure, we have, at constant volume,
Heat to raise 1 mole through 1 deg $= c_v M = k_v$ say.
And at constant pressure
Heat to raise 1 mole through 1 deg $= c_p M = k_p$ say.

Since $\quad\quad\quad\quad\quad\quad\quad c_p = c_v + R,$
∴ $\quad\quad\quad\quad\quad\quad\quad c_p M = c_v M + RM \quad \text{and} \quad RM = R_0$
so that $\quad\quad\quad\quad\quad\quad k_p = k_v + R_0$
i.e. $\quad\quad\quad\quad\quad\quad\quad \underline{k_p - k_v = 8314 \text{ J/kg deg K}}.$

The quantities k_p and k_v are known as the *molar heats* of a gas.

Chemical Thermodynamics

Heat is obtained for engineering applications by burning (usually) a mixture of fuel and air in some kind of furnace or combustion chamber. The gaseous products of combustion are then either permitted to expand under controlled conditions—so transforming a proportion of their heat content (i.e. internal energy) into mechanical work—or are made to transfer part of their heat content to another fluid via a *heat exchanger*.

* This was suspected by Avogadro (1776–1856) who, in 1811, deduced that "different ideal gases at a given temperature and pressure have the same number of molecules per unit volume", i.e. that M is proportional to ϱ. The number of molecules per mole of (any) gas is $27·3 \times 10^{25}$ and this is known as "Avogadro's Number".

Combustion itself may be described as the rapid chemical combination of a substance with oxygen (oxidation) involving a rearrangement of the atoms and a release of thermal energy as their electrons are redistributed. The oxygen is usually supplied in the form of air which, for engineering purposes, may be assumed to contain 21 per cent O_2 by volume, the remainder being N_2. It follows that, in 100 m³ air there are 21 m³ O_2 and 79 m³ N_2. Alternatively, 100 moles of air contain 21 and 79 moles respectively of oxygen and nitrogen. Since the ratio $79/21 = 3.76$ we can say that

$$\text{Volume of } N_2 = 3.76 \times \text{Volume of } O_2$$

Since the ratio $100/21 = 4.76$ we can also so that

$$\text{Volume of air} = 4.76 \times \text{Volume of } O_2.$$

Similarly, the gravimetric (by weight) percentages are 23 and 77 so that, alternatively, since $77/23 = 3.35$,

$$\text{Weight of } N_2 = 3.35 \times \text{Weight of } O_2,$$

$$\text{Weight of air} = 4.35 \times \text{Weight of } O_2.$$

Also,
$$\underline{\text{Mass of air} = 4.35 \times \text{mass of oxygen.}}$$

The chemistry of the relations between the constituents of the air/fuel mixture, or *reactants*, and the composition of the combustion products is known as *stoichiometry* pronounced "sto-ee-keometry" to rhyme with "geometry". In addition to oxygen and its accompanying nitrogen (which is not changed by the reaction although its presence reduces the rate of combustion and transfers some of the energy released) the most important reactants are carbon, hydrogen and sulphur. The last-named has been included because, although only small quantities of it are found in solid and liquid fuels, the products of its combustion are corrosive. Appreciable sulphur content may lead therefore to excessive expenditure on maintenance of plant.

When a fuel contains more than one reactant, the mass of oxygen required per kilogram is the sum of the quantities required for each constituent, less any oxygen already present in the fuel. The mass of air required is then this figure multiplied by 100/23, i.e. by 4.35.

The mass of air supplied per kilogram of fuel is known as the *air/fuel ratio* and may be more or less than the theoretical amount.

The Combustion of Hydrogen

This may be represented by the following *chemical* equation which shows the reactants on the left-hand side and the products on the right-hand side:

$$2\,H_2 + O_2 = 2\,H_2O.$$

In words: Two molecules of H_2 combine with one molecule of O_2 to form two molecules of steam. This equation complies with the requirement that, since mass is not destroyed in a chemical reaction, there must be the same mass of each element among the products as in the fuel originally, i.e. the four hydrogen atoms on the left-hand side (two in each molecule of hydrogen) must reappear on the right-hand side. This they do, since there are two in

COMBUSTION AND HEAT TRANSFER

each molecule of steam. There exists, therefore, what is known as *hydrogen balance*. Similarly, there must, in this case, be oxygen balance.

Inspection of the equation reveals that there has been a reduction in volume, because all molecules—although of different "weights"—are of the same size. It can be deduced, therefore, that 2 m³ H_2 require 1 m³ O_2 for combustion but produce only 2 m³ steam. Alternatively, moles may be written for cubic metres.

If atomic weights are substituted in the equation it will read

$$2(1 \times 2) + (16 \times 2) = 2[(1 \times 2) + 16]$$

or $\qquad\qquad 4 \;+\; 32 \;=\; 36$

i.e. $\qquad\qquad \underline{1 \;+\; 8 \;=\; 9}$ (Mass equation)

In words: One kilogram of H_2 requires eight kilograms of O_2 to form nine kilograms of steam.

The heat resulting from the reaction has been found (experimentally) to be 121·4 MJ and this is called the *Lower Calorific Value*, or L.C.V. If the steam is cooled until it condenses, then the latent heat of evaporation will be recovered. This amounts to 2·4 MJ/kg so that the 9 kg steam when condensing will reject a further 21·6 MJ. Addition of this to the L.C.V. gives a total of 143 MJ/kg and this is known as the *Higher Calorific Value* (H.C.V.) of hydrogen. It follows that all fuels containing H_2 have two calorific values.

Evidently the *air* required for the combustion of 1 kg H_2 is $(8 \times 4·35)$ or 34·8 kg.

The Combustion of Carbon

One chemical equation is: $2C + O_2 = 2CO$. This means that two molecules of the element carbon (which is monatomic and solid) will, if the temperature is high enough, combine with one molecule of oxygen (which is diatomic and gaseous at the "right" temperature) to form two molecules of the new diatomic compound carbon monoxide. This is also a gas and toxic. Since the carbon volume is negligible relatively, the products have twice the volume of the reactants. Substituting weights we obtain

$$(2 \times 12) + (16 \times 2) = 2(12 + 16)$$

or $\qquad\qquad 24 + 32 = 56$

i.e. $\qquad\qquad \underline{1 + 1\tfrac{1}{3} = 2\tfrac{1}{3}}$

Thus, when 1 kg carbon and $\tfrac{4}{3}$ kg oxygen are brought together at the right temperature they will react to form $\tfrac{7}{3}$ kg carbon monoxide. In so doing they will emit 10·25 MJ.

Since carbon monoxide will combine with oxygen to form carbon dioxide (i.e. it will itself burn) the reaction described above is not "complete". The equation for the complete combustion of carbon is

$$C + O_2 = CO_2.$$

Substituting weights we obtain

$$12 + (16 \times 2) = [12 + (16 \times 2)]$$

or $\qquad\qquad 12 + 32 = 44$

i.e. $\qquad\qquad \underline{1 + 2\tfrac{2}{3} = 3\tfrac{2}{3}}$

Thus, if the oxygen supply is doubled, no carbon monoxide is (theoretically) formed and the combustion is "complete". The left-hand side of the equation is said to be the *stoichiometric* air/fuel ratio. The heat emitted during combustion is 34 MJ in contrast to the mere 10·25 MJ obtained from the partial combustion first considered and this disparity emphasises the need to ensure adequate air supply.

From the foregoing it is evident that, if the oxygen supplied per kilogram of carbon lies between $\frac{4}{3}$ and $\frac{8}{3}$ kg, the products will be a mixture of CO and CO_2.

The Combustion of Carbon Monoxide

The chemical equation is: $2CO + O_2 = 2CO_2$. Substituting weights we obtain

$$2(12+16)+(16\times2) = 2[12+(16\times2)]$$

or
$$56 + 32 = 88$$

i.e.
$$1 + \tfrac{4}{7} = \tfrac{11}{7}.$$

Now, 1 kg carbon produces 10·25 MJ when burnt to $\frac{7}{3}$ kg CO and 34 MJ when burnt to $\frac{11}{3}$ kg CO_2. It follows that $\frac{7}{3}$ kg CO produce $(34-10\cdot25)$ or 23·75 MJ when burnt to CO_2. The calorific value of carbon monoxide is therefore $23\cdot75 \div \frac{7}{3}$ or 10·2 MJ/kg.

Coal gas (also known as "town" gas) contains 5–15 per cent CO, blast furnace gas 25–30 per cent CO and so-called "producer" gas about the same proportion. The last-named is made by blowing steam through smouldering (i.e. incompletely burnt) wood, coke or coal. The incomplete combustion "produces" CO and its heat decomposes the steam into O_2 and H_2, the O_2 going to form more CO. In addition to the 25 per cent CO, the mixture given off contains 10–15 per cent H_2.

The Combustion of Sulphur

The chemical equation is:

$$S + O_2 = SO_2.$$

Substituting weights we obtain

$$32 + (16\times2) = 32 + (16\times2)$$

i.e.
$$1 + 1 = 2.$$

Thus equal weights of sulphur and oxygen react together (emitting 9·3 MJ/kg sulphur) to produce the sum of their weights in sulphur dioxide. This is colourless, heavier than air and smells of decayed fish! It is also soluble in water so that, if present in the products of combustion together with condensed steam, it will form sulphurous acid:

$$SO_2 + H_2O = H_2SO_3$$

or
$$1 + \tfrac{9}{32} = 1\tfrac{9}{23} \quad \text{(substituting weights).}$$

The Hydrocarbon Family

The many combinations of carbon and hydrogen, whether liquid or gaseous, are known collectively as *hydrocarbons* and form an exception to the rule that molecules generally consist of relatively few atoms. The last five gases in the table on p. 204 are some of the less complex, the methane molecule (the "marsh gas" of old) being the simplest of all in composition. It has one carbon atom linked to four hydrogen atoms. North Sea gas contains at least 90 per cent methane.

The addition of further hydrogen and carbon atoms (in multiples of one and two respectively) to the methane molecule gives a succession of hydrocarbons known as the *Paraffin Series*, the general formula being C_nH_{2n+2}. Alternative methods of representing the first three in this series are shown in Figure 196.

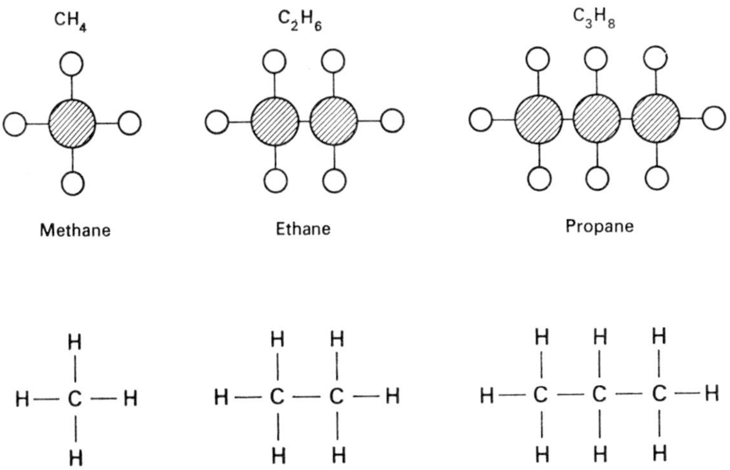

FIG. 196

Evidently the series can be extended indefinitely. Note that, since a molecule is three-dimensional, the above are not accurate pictures of the shapes.

Now the power of atoms to combine or link with other atoms in various ways to form molecules is called *valency*. From the diagram of the methane molecule, Figure 196, the valencies of carbon and hydrogen are, respectively, 4 and 1. It follows that, if the full valency of each atom is filled by links, each to a separate atom, the compound has no urge to combine further, i.e. it is inherently stable. On this basis all members of the paraffin series are stable and are said to be *saturated*.

On the other hand, if the number of effective links is less than the valency, there will exist between two carbon atoms what is called a *double bond*. The extra bonds or links—denoted by dotted lines in Figure 197—are a source of weakness and render such molecules relatively unstable.

The subtraction of two hydrogen atoms from each member of the paraffin[†] series gives (Fig. 197(b)) what is known as the *Olefin Series* (having the general formula C_nH_{2n}) while a second similar subtraction gives the *Acetylene Series* (Fig. 197(c)) having the general formula C_nH_{2n-2}. In the following table, which shows these three series, the atmospheric boiling point of each hydrocarbon has been included to show which are normally gaseous.

† Derived from Latin *par-affinus* showing no desire to combine.

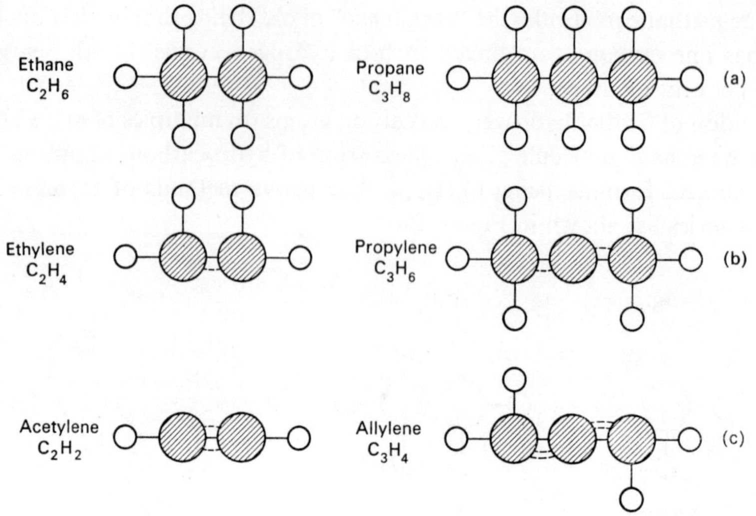

Ethane C$_2$H$_6$
Propane C$_3$H$_8$ (a)

Ethylene C$_2$H$_4$
Propylene C$_3$H$_6$ (b)

Acetylene C$_2$H$_2$
Allylene C$_3$H$_4$ (c)

FIG. 197

Paraffin series			Olefin series			Acetylene series		
C$_n$H$_{2n+2}$		b.p. °C	C$_n$H$_{2n}$		b.p. °C	C$_n$H$_{2n-2}$		b.p. °C
Methane	CH$_4$	−161·5						
Ethane	C$_2$H$_6$	−88·6	Ethylene	C$_2$H$_4$	−103·7	Acetylene	C$_2$H$_2$	−84
Propane	C$_3$H$_8$	−42·1	Propylene	C$_3$H$_6$	−47·7	Allylene	C$_3$H$_4$	−24
Butane	C$_4$H$_{10}$	−0·5	Butylene	C$_4$H$_8$	−0·5	Crotonylene	C$_4$H$_6$	−27·5
Pentane	C$_5$H$_{12}$	36·1	Amylene	C$_5$H$_{10}$	37·1	Valylene	C$_5$H$_8$	47·2
Hexane	C$_6$H$_{14}$	68·7	Hexylene	C$_6$H$_{12}$	67·8	Hexoylene	C$_6$H$_{10}$	79·5

Both propane and butane are highly volatile, are liquefied by compression and are sold in "bottles". Butane‡ is added to petrol to give the volatility necessary for the good starting of petrol motors under low temperature conditions. Further members of the paraffin series (heptane, octane, etc.) are liquids of successively higher boiling point and density, until at about C$_{22}$H$_{25}$ the compounds are solid at room temperature and are of the nature of paraffin wax.

Referring to Figure 198(a), which represents a molecule of "straight-chain" butane, it can be seen that, without altering the composition, the atoms could be rearranged as shown in Figure 198(b). To distinguish between these two possibilities, straight-chain hydrocarbons carry the prefix "n" (for normal) while the branched chain variants—which are called *isomers*—carry the prefix "iso". It is pointed out that, although the chemical compositions are identical, the substances are slightly different, e.g. n-pentane has a specific gravity of 0·621 and boils at 36°C while for iso-pentane the figures are 0·619 and 28°C respectively.

Both normal and iso-paraffins are extremely reluctant to take part in chemical reactions.

Referring to Figure 198(b) it is seen that the central carbon atom is linked to one hydrogen atom and to three so-called *methyl groups*. Iso-butane can be regarded therefore as

‡ Flame temperature in air 1880°C (propane 1900°C).

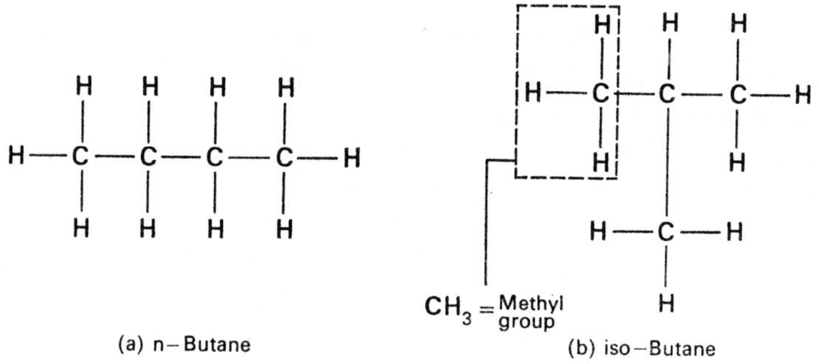

FIG. 198 (a) n-Butane (b) iso-Butane CH$_3$ = Methyl group

methane with three of its hydrogen atoms replaced by methyl groups. This gives rise to the alternative (and more precise) description, *tri-methyl methane*.

In addition to the paraffins, olefins and acetylenes, there exist two further groups of hydrocarbons, viz. the *naphthenes* and the *aromatics*. The members of the first of these groups have the same composition and general formula as the corresponding members of the olefin series (C_nH_{2n}), but differ from them in molecular arrangement. This arrangement, shown in Figure 199, gives rise to the term *saturated ring compound* and the prefix "cyclo". The group CH_2 having two available links is called the *methylene group* and the solid members of this series are asphaltic in nature.

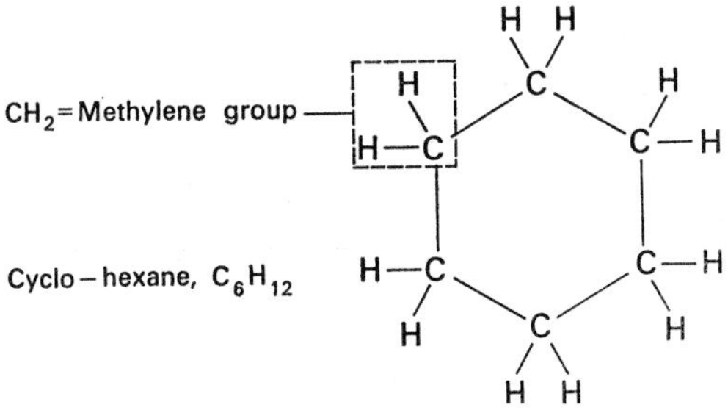

FIG. 199 CH$_2$ = Methylene group — Cyclo-hexane, C_6H_{12}

The aromatics—so called because of their strong smell—have the general formula C_nH_{2n-6} and are evidently unsaturated. Their molecular structure, which also is of the basic ring type, is typified by that of *benzene* (C_6H_6), the molecular arrangement of which is shown in Figure 200, the dotted lines representing double bonds:

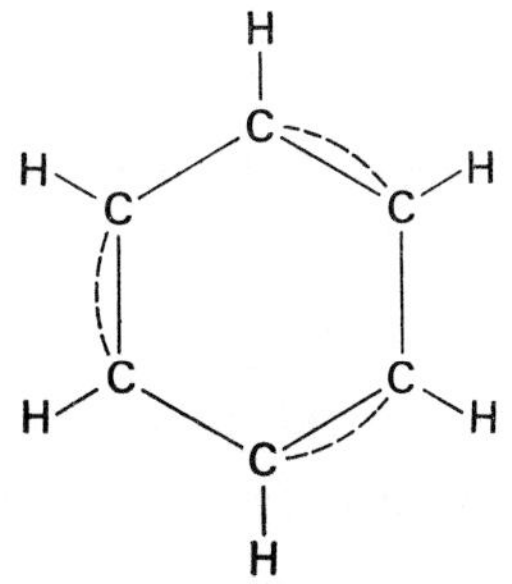

FIG. 200

Benzene is the chief constituent of what is known as "benzole", the other constituent aromatics being *toluene* (C_7H_8) and xylene (C_8H_{10}). All three are obtained by distilling coal tar, a by-product of the carbonisation of coal.

Figure 201 shows, for some of the paraffin series, the variations in composition and boiling point. For example, the percentage by weight of hydrogen in heptane (C_7H_{16}) is

$$\left[\frac{(1\times 16)}{(12\times 7)+(1\times 16)}\right]100 \quad \text{or} \quad \underline{16 \text{ per cent exactly.}}$$

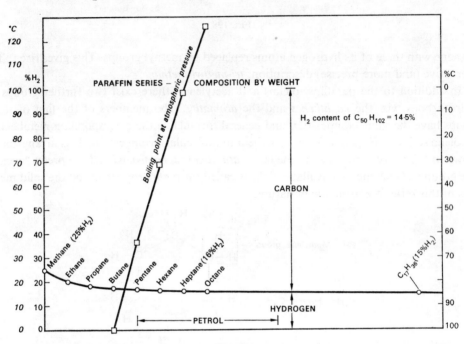

Fig. 201

The Combustion of Acetylene

Acetylene does not occur naturally and one way of making it is as follows. Calcium oxide (quicklime) and coke are heated together in an electric furnace to produce calcium carbide and carbon monoxide:

$$CaO + 3C = CaC_2 + CO.$$

The calcium carbide is then made to react with water to produce acetylene and calcium hydroxide:

$$CaC_2 + 2H_2O = C_2H_2 + CaO_2H_2.$$

Acetylene, which is a gas at normal temperature, combines with oxygen to produce the oxy-acetylene flame (3200°C), the equation being

$$2C_2H_2 + 5O_2 = 4CO_2 + 2H_2O.$$

COMBUSTION AND HEAT TRANSFER

Thus, 2·5 volumes of O_2 are required per volume of C_2H_2 so that (2·5×4·76) or 11·9 volumes of *air* are required, i.e.

$$\text{Volumetric air/fuel ratio} = 11\cdot 9.$$

Substitution of weights in the equation gives

$$2[(12\times 2)+(1\times 2)]+5(16\times 2) = 4[12+(16\times 2)]+2[(1\times 2)+16]$$

or
$$1+\frac{40}{13} = \frac{44}{13}+\frac{9}{13}$$

so that 40/13 or 3·08 kg of O_2 are required per kg of C_2H_2, i.e. (3·08×4·35) or 13·4 kg of *air* are required. Hence

$$\text{Gravimetric air/fuel ratio} = 13\cdot 4.$$

In general, the greater the H_2 content of a hydrocarbon the higher the stoichiometric air/fuel ratio.

Derivation of Chemical Equations

If the coefficients of a presumed chemical equation are unknown, they can be deduced from a consideration of the principle of conservation of mass. The symbol for acetylene is C_2H_2 so that combustion implies the combination of O_2 with C (i.e. the production of CO_2) and of O_2 with H_2, i.e. the production of H_2O. Taking the coefficient of the fuel as 1·0, i.e. assuming one molecule (or one mole), the equation will be as shown, the unknown coefficients being denoted by x, y and z. By equating the total number of atoms of each constituent in the reactants to the corresponding total in the products we obtain:

$$C_2H_2 + x\, O_2 = y\, CO_2 + z\, H_2O.$$

For carbon	2	$= y$		$\therefore y = 2$
For hydrogen	2	$=$	$2z$	$\therefore z = 1$
For oxygen		$2x = 2y$	$+\ z$	$\therefore x = \frac{5}{2}$
Hence		$C_2H_2 + \frac{5}{2}O_2 = 2\, CO_2 + H_2O$		
or		$2C_2H_2 + 5\, O_2 = 4\, CO_2 + 2\, H_2O$ (as already used).		

Equations for other hydrocarbons may be deduced in the same way.

Influence of Air/Fuel Ratio on the Combustion of Hexane

Hitherto only the combustion of chemically correct mixtures has been considered, the CO_2 and steam produced being associated, in the exhaust, with the unavoidable nitrogen.

If insufficient O_2 is supplied (so-called "rich mixture") the H_2, having a greater "affinity" for O_2 than has carbon, may be assumed to form steam in the usual way. It follows that not all the carbon is converted to CO_2, some forming CO. The exhaust will therefore (theoretically) contain H_2O, CO_2, CO and N_2. (If the mixture is *very* rich, some carbon goes without O_2 entirely and appears as black smoke, i.e. as very fine solid particles.)

On the other hand, if too much O_2 is supplied (so-called "weak mixture") the excess O_2

will be mixed with the other products. The exhaust will therefore (theoretically) contain H_2O, CO_2, O_2 and N_2. Further, the imperfect mixing of the reactants in practice results in the simultaneous presence in the exhaust of both CO and O_2 even when the mixture is "correct". Thus it is evident that the quantity and composition of the products depend, not only on the air/fuel ratio but on the method of carburation and on the design of the combustion chamber and associated passages.

If the percentage composition by weight of the fuel is known, the theoretical composition of the products can be calculated for any air/fuel ratio. Suppose 1 kg heptane (C_7H_{16}) to be supplied with, say, 14 kg air. The equation for chemically complete ("correct") combustion is

$$C_7H_{16} + 11\,O_2 = 7\,CO_2 + 8\,H_2O$$

so that the stoichiometric air/fuel ratio is

$$\frac{\text{Mass of } O_2 \times 4\cdot 35}{\text{Mass of fuel}}, \quad \text{i.e.} \quad \frac{(16\times 2)11\times 4\cdot 35}{(12\times 7)+16} \quad \text{or} \quad 15\cdot 3.$$

Since less than the required air is supplied, the mixture is rich. Carbon monoxide must be expected therefore in the exhaust. Representing the unknown volume coefficients by x, y and z, the equation for the actual combustion will be

$$C_7H_{16} + 11\left(\frac{14\cdot 0}{15\cdot 3}\right)O_2 = x\,CO_2 + y\,CO + z\,H_2O$$

or $\qquad C_7H_{16} + 10\cdot 07\,O_2 \qquad = x\,CO_2 + y\,CO + z\,H_2O.$

Hence $\qquad 7 \qquad\qquad\qquad\qquad = x \qquad + y \qquad\qquad$ (for C)

and $\qquad 16 \qquad\qquad\qquad\qquad = \qquad\qquad\qquad 2z \qquad$ (for H_2)

and $\qquad\qquad 20\cdot 14 \qquad\qquad = 2x \quad + \quad y+ \quad z \quad$ (for O_2).

Thus $\qquad\qquad\qquad\qquad x+y = 7, \qquad$ i.e. $\quad y = 7-x$

and $\qquad\qquad\qquad\qquad\quad 2z = 16, \qquad$ i.e. $\quad z = 8$

and $\qquad\qquad\qquad\qquad 2x+y+z = 20\cdot 14$

so that $\qquad\qquad\qquad\qquad\quad x = 5\cdot 14 \quad$ and $\quad y = 1\cdot 86.$

The actual combustion is represented by the equation

$$C_7H_{16} + 10\cdot 07\,O_2 = 5\cdot 14\,CO_2 + 1\cdot 86\,CO + 8\,H_2O.$$

There will also be present in the exhaust ($10\cdot 07 \times 3\cdot 35$) or 33·8 volumes of nitrogen. If the steam is condensed, i.e. if the products are what is known as "dry", the theoretical percentage composition by volume will be

$$CO_2 = \left(\frac{5\cdot 14}{5\cdot 14 + 1\cdot 86 + 33\cdot 8}\right)100 = 12\cdot 6 \text{ per cent}$$

$$CO = \left(\frac{1\cdot 86}{40\cdot 8}\right)100 \qquad\qquad = 4\cdot 6 \text{ per cent}$$

$$N_2 = \left(\frac{33\cdot 8}{40\cdot 8}\right)100 \qquad\qquad = 82\cdot 8 \text{ per cent}.$$

Now suppose 1 kg heptane to be supplied with, say, 18 kg air. This is more than the stoichiometric requirement—i.e. the mixture is weak—so that oxygen will appear in the exhaust. The equation can be written

$$C_7H_{16} + 11\left(\frac{18}{15\cdot3}\right)O_2 = x\,CO_2 + y\,O_2 + z\,H_2O.$$

Solving the three simultaneous equations we obtain

$$C_7H_{16} + 12\cdot95\,O_2 = 7\,CO_2 + 1\cdot95\,O_2 + 8\,H_2O.$$

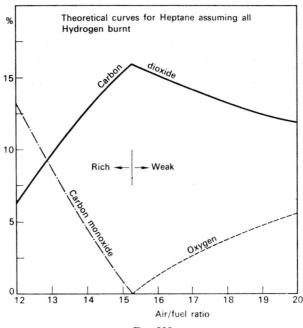

FIG. 202

For the dry products, which include ($12\cdot95 \times 3\cdot35$) or $43\cdot35$ volumes of nitrogen, we have (in theory) by volume:

$$CO_2 = \left(\frac{7}{7+1\cdot95+43\cdot35}\right)100 = 13\cdot4 \text{ per cent}$$

$$O_2 = \left(\frac{1\cdot95}{52\cdot2}\right)100 = 3\cdot74 \text{ per cent}$$

$$N_2 = \left(\frac{43\cdot35}{52\cdot2}\right)100 = 83 \text{ per cent.}$$

The six values just obtained are shown below in tabular form together with corresponding figures for air/fuel ratios of 12 and 20. Figure 202 presents this information graphically.

215

C_7H_{17}	Theoretical percentage composition			
Ratio	O_2	CO_2	CO	N_2
12	—	6.29	13.2	80.5
14	—	12.60	4.6	82.8
15.3	—	15.95	—	83.8
18	3.74	13.40	—	83.0
20	5.62	11.95	—	82.2

Note that the volumetric nitrogen content is a maximum, theoretically, when the mixture is "correct".

Pump Petrol*

A given hydrocarbon has a given boiling point at a given pressure and will boil away at this temperature in the same way as water. Petrol as sold for use in motor vehicles consists of a range (or "blend") of hydrocarbons, each having its own boiling point so that, when heated, the constituent having the lowest boiling point (i.e. the most *volatile*) is the first

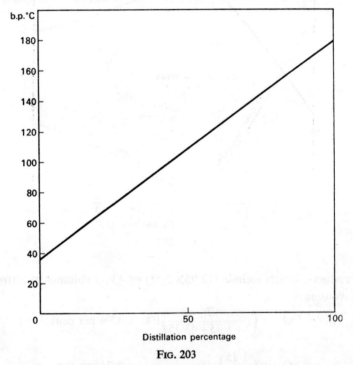

FIG. 203

to vaporise. In a typical fuel this constituent is pentane (C_5H_{12}) which boils at 36°C while the least volatile element vaporises at about 180°C. Figure 203 shows a typical distillation curve.

The stoichiometric air/fuel ratio is a mean of the ratios of all the constituent hydrocarbons

* The word "petrol" was first used on 28.9.1876 by Eugen Langen in a letter to Adolf Schmidt of Liège, Benzin (petrol) was first determined chemically by Faraday and named after Professor Benzin of Berlin University.

and this figure, for any blend of pump petrol, lies between 14·9 and 14·5, an average figure being 14·7.

Now, 1 kg C needs 2·67 kg O_2 for complete combustion and 1 kg H_2 needs 8 kg O_2. If, therefore, 1 kg pump petrol contains, say, y kg H_2 and $(1-y)$ kg C we have

$$[(1-y)2\cdot67+(y\times8)]4\cdot35 = 14.7$$

whence
$$y = 0\cdot133 \text{ kg.}$$

An average pump petrol contains therefore 13·3 per cent hydrogen by weight. If this fuel is assumed to be a single boiling point hydrocarbon, C_nH_x say, then

$$13\cdot3 = \left[\frac{(1\times x)}{(12\times n)+(1\times x)}\right]100$$

or
$$0\cdot133\,(12n+x) = x \quad \text{whence} \quad \underline{x = 1\cdot84n.}$$

The equivalent hydrocarbon can thus be described as

$$\underline{C_nH_{1\cdot84n}} \quad \text{or, very nearly,} \quad \underline{C_6H_{11}}.$$

The theoretical variations in the combustion products of this "equivalent hydrocarbon" are represented by the dotted lines of Figure 204 and are based on the assumptions that all the hydrogen is burnt first and that all the remaining oxygen then reacts with the carbon to form CO_2 mixed with either CO or O_2.

Since, in practice, part of the hydrogen also finds no oxygen when the mixture is rich while part of the remaining oxygen (but less than 1 per cent) finds neither carbon nor hydro

C_6H_{11}	Theoretical % composition by volume			
A/F ratio	O_2	CO_2	CO	N_2
12	—	9·3	10·8	80·0
13·5	—	13·8	4·4	81·8
14·7	—	17·0	—	83·0
17	3·3	14·5	—	82·1
19	5·5	12·9	—	81·6

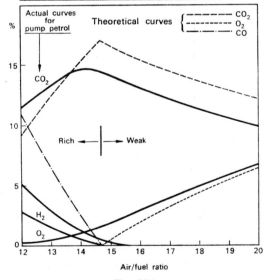

Fig. 204

gen, the exhaust of a petrol motor supplied with a rich mixture contains steam, H_2, CO_2, CO, O_2 and N_2.

Assuming the steam to be condensed, the actual percentages of these gases—as obtained by analysis, N_2 excepted—are represented by the full lines of Figure 204. The presence of hydrogen (together with less than the expected amount of carbon monoxide) indicates that, where less than the stoichiometric air is supplied, a more reasonable hypothesis would be that the carbon is first burnt to carbon monoxide and that the remaining oxygen is then shared between this and the hydrogen.

Relation between Air/Fuel Ratio and Power Developed

If the speed of a petrol motor is maintained constant at full throttle opening, the power output at different values of air/fuel ratio can be determined with the aid of a dynamometer. The plotting of specific fuel consumption (s.f.c.) against brake mean effective pressure (b.m.e.p.) will result in a curve having, typically, the form shown in Figure 205.

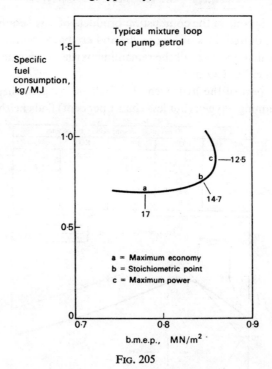

Fig. 205

This is called a *mixture loop*. Point "a" represents maximum economy in operation and corresponds to an air/fuel ratio of about 17. Point "b" represents stoichiometric operation and point "c" represents maximum power output, this corresponding to a ratio of about 12·5.

To ensure that the temperature of the exhaust valve head does not exceed about 700°C —i.e. to prevent valve "burning"—when a motor is run at full throttle on pump petrol, it is advisable to use an air/fuel ratio not exceeding 13·5.

Detonation in the Spark-ignited Petrol Motor

Combustion proper does not begin at the instant at which the arc bridges the spark gap but after a *delay period* of the order of 0·002 sec. The angle turned through by the crank during this period depends on (and increases with) the crank speed, e.g. at 5000 rev/min the angle is $360\left(\frac{5000}{60}\right)0\cdot002$ or 60 deg. For this reason the timing of the discharge from the ignition coil is adjusted—by a centrifugally operated mechanism—to occur proportionately further in advance of the outer (top) dead centre as the speed rises. After combustion proper has begun, the initial velocity of the flame front (radially outwards from the spark) is low. As combustion proceeds, thermal energy is transferred by radiation and turbulent mixing to the unburnt part of the charge, the temperature and pressure of which are rising rapidly.

When combustion is almost complete, the temperature and pressure of the remaining charge—known as the "end gas"—may attain certain critical values with the result that ignition will occur suddenly at many points remote from the flame front. Such self-ignition, although not detonation proper, is accompanied by a violent rise in pressure which is heard as a "knocking" or "pinking" sound. The period during which the end gas is exposed to radiation from the advancing flame front depends upon the shape and size of the combustion chamber and on the position in it of the point of ignition. It follows that, provided valves of reasonable size can be accommodated, the smaller the cylinders the better, particularly as, for a given capacity, the cyclic fluctuation in torque is reduced as the number of cylinders is increased. It follows also that, ideally, the combustion chamber should be hemispherical in form and that the spark gap should be as near as is practicable to its centre.

Alternatively, the rising pressure in advance of the flame front may, under certain conditions, cause a sudden and violent increase in the *rate of combustion* (and hence in the pressure) of the end gas. The pressure rise—in the form of a *shock* or *detonation* wave—and the flame front then traverse the end gas together. As the wave reaches the gas boundary and is reflected back and forth it is heard as a metallic ringing noise of somewhat higher frequency than that due to self-ignition.

The stationary layer of gas on the boundary (i.e. on the surfaces of the combustion chamber) is disturbed by the excessive turbulence which accompanies detonation so that there is a rise in the rate at which heat is transferred from the gas to its surroundings and hence an increased power loss.

The conditions in the end gas just described lead to the formation of certain organic peroxides and it is these which induce detonation. Members of the straight-chain paraffin series—which form such peroxides with relative ease—are therefore prone to detonation. On the other hand the branched chain molecular structure of hydrocarbons such as iso-octane and methanol are better able (for reasons not yet fully understood) to resist pre-combustion reactions of this nature.

To decide the degree of so-called "knock resistance" of a particular petrol, the performance of a sample of it is compared with that of samples of "good" and "poor" reference fuels. For this purpose the sample is ignited in the combustion chamber of a cylinder provided with variable compression ratio which is part of what is known as the CFR[†] motor.

[†] Co-ordinating Fuel Research committee, sponsored in the U.S.A. by a consortium of interested organisations.

The "poor" reference fuel is n-heptane while the "good" reference fuel is an iso-octane known as tri-methyl pentane. The molecular arrangements of these two hydrocarbons are as shown in Figure 206.

The mixture of these two which detonates at the same compression ratio as the test sample is known as the *Octane Number*. For example, if such a mixture consisted of 90 per cent iso-octane and 10 per cent n-heptane, the sample would be rated "90 Octane".* It is appropriate here to point out that, since there is no connection between Octane Number and calorific value, economic considerations demand the use of the cheapest fuel having the requisite "anti-knock" properties.

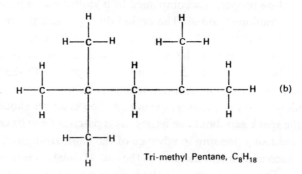

FIG. 206

Typical Properties of Pump Fuel

The following figures are representative of "100 Octane" pump petrol which can be used unmodified with compression ratios of up to 10:

Lower calorific value	44·2 MJ/kg
Stoichiometric air/fuel ratio	14·7
Specific gravity	0·75
Percentage carbon by weight	86·7
Flash point	51°C
Atmospheric boiling range	35–185°C
*Latent heat of vaporisation	0·326 MJ/kg

It is seen that, in theory, the thermal energy released per kilogram of air consumed is

$$\frac{44 \cdot 2}{14 \cdot 7} \quad \text{or} \quad 3 \cdot 01 \text{ MJ}.$$

---

* Thanks to modern additives, fuels are available which have "anti-knock" performances superior to that of tri-methyl pentane and hence octane numbers greater than 100.

Special Additions to Petrol

The petrol used in racing cars is diluted, usually, by adding methyl or ethyl alcohol, some properties of which are given in the following table:

Properties	Methyl alcohol ("Methanol")	Ethyl alcohol ("Ethanol")
Chemical formula	CH_3OH	C_2H_5OH
LCV, MJ/kg	20·00	26·80
* Latent heat, MJ/kg	1·095	0·865
Octane number	100+	100+
Specific gravity	0·794	0·789
Flash point	0°C (32°F)	14°C (57°F)
Atmospheric b.p.	65°C (149°F)	78°C (172°F)
Volumetric air/fuel ratio	7·15	14·28
Gravimetric air/fuel ratio	6·52	9·08

* Change in specific enthalpy during evaporation.

The equation for the combustion of methyl alcohol is

$$2\,CH_3OH + 3\,O_2 = 2\,CO_2 + 4\,H_2O$$

so that $\frac{3}{2} \times 4{\cdot}76$ volumes of air are required per volume of vapour, i.e. the volumetric air/fuel ratio is 7·15. The corresponding gravimetric ratio is 6·52. The addition of alcohol to petrol reduces therefore the required air/fuel ratio, i.e. makes it necessary to enrich the mixture and so to consume more fuel per kilogram of air. This means that, since the rate of vaporisation is increased, more heat is extracted from the inlet tract. This added cooling effect is enhanced by the relatively high latent heat of vaporisation, viz. 1·095 MJ/kg as compared to 0·326 MJ/kg for "normal" petrol. In turn the cooling effect increases the density of the incoming charge, the overall result more than compensating for the relatively poor calorific value and giving a net increase in mean effective pressure.

The equation for the combustion of ethyl alcohol is

$$C_2H_5OH + 3\,O_2 = 2\,CO_2 + 3\,H_2O$$

from which may be deduced an approximate gravimetric air/fuel ratio of 9.1. Thus, in theory, the thermal energy released per kilogram of air consumed is

$$\frac{26{\cdot}8}{9{\cdot}08} \quad \text{or} \quad \underline{2{\cdot}95\ \text{MJ}}.$$

For methyl alcohol the corresponding figure is

$$\frac{20{\cdot}0}{6{\cdot}52} \quad \text{or} \quad \underline{3{\cdot}07\ \text{MJ}}.$$

The second of these values is greater than that for "normal" petrol.

Vaporisation

Although the forces of attraction (cohesive forces) between the molecules of a liquid are greater than those between the molecules of a gas (but less than those between the molecules of a solid) the molecules still have random motion within the liquid boundary. The mean molecular velocity —and hence kinetic energy—depends upon the temperature of the liquid and rises with it. When a molecule near a free surface has sufficient kinetic energy and its velocity is in the right direction, it is able to penetrate the surface, i.e. to escape from the liquid. Such escaping molecules form a *vapour* above the surface, their escape being called *vaporisation* (or *evaporation*). Further vaporisation is encouraged if the vapour already formed is carried away by a current of air, but if the liquid is in a container, the escaped molecules remain above the liquid surface and exert on it a *vapour pressure.** If the vapour is still able to absorb molecules when all the liquid is evaporated, it is said to be *unsaturated*. If sufficient additional liquid is introduced, molecules will continue to escape until a state is reached in which the number of molecules escaping per second is equal to the number per second which penetrate the surface in the opposite direction. The vapour is then said to be *saturated* and the state is referred to as one of *dynamic equilibrium*. The force per unit liquid surface is then called the *saturation vapour pressure* or s.v.p.

If the temperature is raised to a new value, the kinetic energy of the molecules—and the rate at which they escape—increases so that the density of the vapour (i.e. the number of molecules per cc) also increases. This continues until, at the new temperature, a new state of equilibrium corresponding to a new s.v.p. is attained.

When quoted in tables of properties, the term "vapour pressure" is used to define the pressure in a specified volume containing a specified quantity of liquid at a specific temperature, e.g. the stated vapour pressure of hexane at 0°C is 50 mm Hg or about 6900 N/m^2.

If the water vapour present in, say, a roomful of air is expressed as a percentage of the amount required to saturate that air at the same temperature, the result is known as the *relative humidity*. When the relative humidity is low, say 45 per cent, the air feels dry since sweat evaporates (vaporises) readily. If the temperature of a room is reduced at constant pressure, the air will eventually become saturated and liquid will form on the boundaries, i.e. vapour will condense. The temperature at which this happens is known as the *dew point*.

Some Remarks on Vehicle Carburation

The essentials of a simple open-choke spray carburettor are shown in Figure 207.

The air/fuel ratio, i.e. the quality of the charge, is determined by the design of the carburettor which varies it (about an initial setting) to suit a variety of operating conditions. A secondary function of the carburettor is to break up the liquid fuel into a mist of fine particles so that it may be vaporised more easily during its subsequent passage along the inlet tract. The fuel mist is produced by placing the fuel jet at the throat of a venturi so that the greater (atmospheric) pressure on the liquid in the float chamber causes liquid to issue into the air stream.

---

* Some vapour pressure exists even in the absence of a container.

The rate of admission of the air/fuel mixture (and hence the weight of the charge) is varied by means of a "throttle" valve located in the inlet tract between the carburettor and the inlet valve, as shown in Figure 207. When the throttle is wide open, the pressure drop across it is zero and the point of lowest pressure is at the venturi throat, but as soon as the throttle is closed to a point where the mixture velocity is higher than that at the throat, the throttle itself becomes the point of lowest pressure. When the throttle is in the so-called "idling" position, the throat pressure is practically atmospheric (101 kPa), while—in a typical present-day petrol motor—the pressure on the cylinder side of the throttle is of the order of 300 mm Hg or 40 kPa. (This may be expressed as (760—300) or 460 mm Hg "vacuum", i.e. about 60 kPa below atmospheric.)

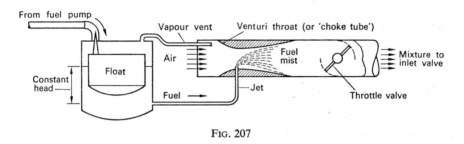

FIG. 207

In the case of the more volatile part of the fuel mist, vaporisation results from the absorption of heat from the inlet tract and throttle assembly, while the least volatile part may not be completely vaporised until it has been compressed into the combustion chamber itself. It is only necessary to heat the induction tract to the point where vaporisation is complete by the end of the compression stroke since further heating reduces the density of the charge and hence the volumetric efficiency.

When, from the idling position, the throttle is opened suddenly, the pressure on the cylinder side of it rises sharply (from about 40 kPa) to near atmospheric. This causes fuel to condense out (precipitate) on the sides of the inlet tract and so weakens the mixture just at the wrong moment. For this reason open-choke carburettors incorporate a syringe-type pump (coupled to the throttle lever) which supplies a separate orifice in the throat and enriches the mixture as the throttle is opened. When the throttle is closed, the plunger is returned by a spring and the cylinder of the so-called "accelerator pump" is recharged via a non-return valve.

If the throttle opening is increased at constant crank speed, the mixture velocity rises, the throat pressure falls and the charge density increases. Since the "delay period" shortens with increase in density, combustion begins earlier so that ignition could occur later. Automatic retardation of the ignition point in such circumstances is effected by a diaphragm ("vacuum retard") which makes use of the reduction in vacuum associated with the throttle movement.

EXAMPLE. When running at full throttle on a pump fuel approximating to C_6H_{11} a six cylinder petrol motor develops 80 kW at 4500 rev/min, this corresponding to a specific fuel consumption of 0·08 kg/MJ. If there are three inlet pipes each 35 mm diameter, calculate the air inlet velocity at 20°C assuming the air/fuel ratio to be chemically correct. (By weight (i.e. mass) the fuel has a composition of 86·7 per cent and carbon 13·3 per cent hydrogen.)

Solution.

$$\text{Fuel consumed/s} = \left(\frac{80\times 1000}{10^6}\right)0\cdot 08 = 0\cdot 0064 \text{ kg}$$

$$\therefore \text{O}_2 \text{ required/s} = 0\cdot 0064\,[(0\cdot 867\times 2\cdot 67)+(0\cdot 133\times 8)]$$
$$= 0\cdot 0216 \text{ kg}$$

At s.t.p., specific volume $= \dfrac{22\cdot 41}{M}$ m³/kg

$$\therefore \text{Volume of O}_2/\text{s} = \left(\frac{22\cdot 41}{32}\right)0\cdot 0216$$
$$= 0\cdot 0152 \text{ m}^3 \text{ at s.t.p.}$$
$$= 0\cdot 0152\left(\frac{293}{273}\right)$$
$$= 0\cdot 0163 \text{ m}^3 \text{ at 20 °C and 101 325 N/m}^2.$$

Corresponding air volume/s $= 0\cdot 0163\times 4\cdot 76$
$$= 0\cdot 0775 \text{ m}^3 \text{ for 3 carburettors}$$
$$= 0\cdot 0258 \text{ m}^3 \text{ per carburettor}$$

Inlet pipe section $= \dfrac{\pi}{4}(0\cdot 035)^2 = 0\cdot 00096 \text{ m}^2$

Air velocity $= \dfrac{0\cdot 0258}{0\cdot 00096} = 27 \text{ m/s approx.}$

EXAMPLE. The composition by mass (gravimetric analysis) of pump petrol approximates to 86·7 per cent carbon, 13·3 per cent hydrogen and the specific gravity may be taken as 0·75. Assume that specific volume at s.t.p. is 22·41/M and determine for 1 litre the theoretical values of

(a) the mass of air required for combustion;
(b) the volume of this air at s.t.p.;
(c) the mass of the products of combustion;
(d) the percentage composition by mass of the products;
(e) the volume of the products at 120°C and atmospheric pressure;
(f) the mean velocity of these products along an exhaust pipe 50 mm diameter if the fuel is consumed in a motor delivering 80 kW at a specific fuel consumption of 0·08 kg/MJ.

Solution.

(a) Mass of 1 litre fuel $= 0\cdot 75$ kg
O_2 required/litre $= [(0\cdot 0867\times 2\cdot 67)+(0\cdot 133\times 8)]0\cdot 75$
$= 2\cdot 53$ kg

COMBUSTION AND HEAT TRANSFER

Air required/litre $= 2\cdot53 \times 4\cdot35$
$= 11\cdot0$ kg.

(b) Volume of oxygen $= 2\cdot53\left(\dfrac{22\cdot41}{32}\right) = 1\cdot77$ m³ at s.t.p.

Volume of air $= 1\cdot77 \times 4\cdot76 = 8\cdot42$ m³ at s.t.p.

(c) Mass of steam produced $= (0\cdot133 \times 0\cdot75)9 = 0\cdot897$ kg

Mass of CO_2 produced $= (0\cdot0867 \times 0\cdot75)3\cdot67 = 2\cdot385$ kg

Mass of N_2 supplied $= 0\cdot77 \times$ air supplied
$= 0\cdot77 \times 11\cdot0$
$= 8\cdot47$ kg

∴ Mass of products $= 0\cdot897 + 2\cdot385 + 8\cdot47 = 11\cdot75$ kg.

(*Check*: $11\cdot0$ kg air $+ 0\cdot75$ kg fuel.)

(d) Percentage mass of steam $= \left(\dfrac{0\cdot897}{11\cdot75}\right)100 = 7\cdot63$

Percentage mass of $CO_2 = \left(\dfrac{2\cdot39}{11\cdot75}\right)100 = 20\cdot35$

Percentage mass of $N_2 = \left(\dfrac{8\cdot47}{11\cdot75}\right)100 = 72\cdot0$.

(e) At s.t.p., Volume of steam $= 0\cdot897\left(\dfrac{22\cdot41}{18}\right) = 1\cdot12$ m³ (if not condensed)

Volume of $CO_2 = 2\cdot39\left(\dfrac{22\cdot41}{44}\right) = 1\cdot22$ m³

Volume of $N_2 = 8\cdot47\left(\dfrac{22\cdot41}{28}\right) = 6\cdot78$ m³

∴ Volume of products $= 1\cdot12 + 1\cdot22 + 6\cdot78 = 9\cdot12$ m³ at 0°C

Volume at 120°C $= 9\cdot12\left(\dfrac{120 + 273}{273}\right) = 12\cdot5$ m³.

(f) Consumption/min $=$ joules/min $\times$ kg/J
$= (80 \times 1000)60\left(\dfrac{0\cdot08}{10^6}\right)$
$= 0\cdot384$ kg

Time to consume 0·75 kg $= \dfrac{0\cdot75}{0\cdot384} = 1\cdot95$ min $= 117$ s

Volume/s $= Au$ where $A =$ pipe section, $u =$ mean gas velocity

∴ $\dfrac{12\cdot5}{117} = \dfrac{\pi}{4}(0\cdot050^2)u$

i.e. $u = \dfrac{12\cdot5 \times 4}{117\pi(0\cdot050^2)}$

$= 5\cdot45$ m/s.

EXAMPLE. A volumetric analysis of the products of combustion from a furnace burning anthracite showed 16 per cent CO_2, 81 per cent N_2 and 1 per cent CO. If the dry fuel contained 90 per cent carbon and 3 per cent hydrogen by weight, calculate the stoichiometric air/fuel ratio. Estimate also the actual air/fuel ratio and the percentage excess air supplied.

Solution.

Weight of O_2 required $= (0\cdot90\times2\cdot67)+(0\cdot03\times8)$
$= 2\cdot64$ kg

Weight of air required $= 2\cdot64\times4\cdot35$
$= 11\cdot5$ kg.

This is the stoichiometric air.

Suppose the actual amount of air supplied per kg fuel to be W. Then the weight of N_2 supplied is $0\cdot77W$ and the ratio by weight of carbon to nitrogen in the reactants is

$$\frac{0\cdot9}{0\cdot77W}.$$

Since the molecular weights of carbon and nitrogen are, respectively, 12 and 28, the ratio by weight of carbon to nitrogen in the products is

$$\frac{12(0\cdot16+0\cdot01)}{28(0\cdot81)} \quad \text{or} \quad 0\cdot09.$$

Equating these ratios we obtain

$$0\cdot09 = \frac{0\cdot9}{0\cdot77W}$$

i.e. $$W = \frac{0\cdot9}{0\cdot09\times0\cdot77} = \underline{13 \text{ kg.}}$$

This is the actual air supplied.

Percentage excess $= \left(\dfrac{13\cdot0-11\cdot5}{11\cdot5}\right)100$

$= \underline{13.}$

HEAT TRANSFER

General

Thermal energy is required not only to be obtained from and converted to other forms of energy, but to be *transferred*, i.e. conveyed, from one point to another. Alternatively, it may be necessary to minimise the rate of heat transfer. A knowledge of fundamentals is necessary therefore to the designer of, for example, systems for promoting heat transfer from (cooling) transformer oil, compressed air, used steam and so on; for promoting heat transfer to (heating) liquid fuel (to reduce viscosity), solid metals (for melting and heat treatment); and for discouraging heat transfer ("insulating") in a particular direction in the case of furnaces, cold storage rooms and so on. Since heat can only be transferred in a direction of reducing temperature, i.e. down a temperature gradient, it follows that a temperature difference is essential to each of the three basic mechanisms of heat transfer mentioned below.

Conduction

This depends upon that property of the medium which facilitates the passage of (i.e. conducts) thermal energy through it, relative motion between adjacent layers not being necessary. Conduction is exemplified by the flow of energy along a welding rod from the hotter end. This results from the transfer (by impact) of kinetic energy from the hotter (faster moving) molecules to adjacent, cooler molecules.

Convection

This may be defined as the transfer of energy between points in the same medium as a result of relative motion between adjacent layers. Convection is exemplified by the carriage of thermal energy up a chimney by the rising products of combustion. Note that, strictly speaking, the energy transferred in this way is regarded as heat only during its flow across the boundary, i.e. into and out of the system.

Radiation

This is the transmission of energy ("radiant" heat) by means of electromagnetic waves which are of the same nature as radio and light waves and appear to require no medium for their propagation. If a body is not in a state of thermal equilibrium it will either radiate energy or absorb heat radiated from other bodies. The earth is so heated by solar energy.

In engineering situations, thermal energy is usually transferred by two or more of the above methods simultaneously.

Conduction through a Uniform Sheet

Heat has been defined as the transfer of energy across a boundary in association with a difference in temperature. For a given material the rate of transfer by conduction per unit area in the direction of temperature fall (J/m² h usually) is found, experimentally, to be proportional to the temperature gradient (degK/m usually) so that, per unit area,

Rate of flow = a constant × temperature gradient.

Suppose a steady heat flow, Q, to cross a slice of area A and thickness dL as shown in Figure 208 and suppose that, during the transfer, the temperature falls by dT. Then, since the slope is negative and a heat input is, conventionally, positive,

$$Q = -kA \frac{dT}{dL}.$$

This equation is due to the French physicist J. B. Biot, 1774–1862.

MECHANICAL TECHNOLOGY FOR HIGHER ENGINEERING TECHNICIANS

The constant, k, is known as the *thermal conductivity* and its value depends upon the material. If one second is taken as the unit of time it follows that, since

$$k = -\frac{Q}{A}\frac{dT}{dL},$$

its units are Jm/m²s°K or, since J/s = W, Wm/m² deg K. (The temptation to reduce this to W/m deg K should be resisted.)

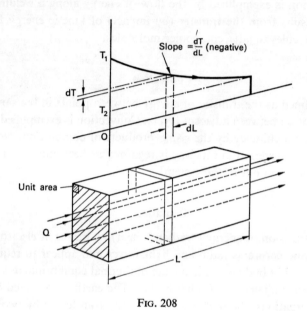

Fig. 208

Thus
$$k = \frac{\text{Power}}{\text{Unit section of path}} \bigg/ \frac{\text{Temperature difference}}{\text{Unit length of path}}$$

or
$$k = \frac{W}{m^2} \bigg/ \frac{\text{deg K}}{m}.$$

From the Biot equation it is evident that, when k is constant, the slope of the temperature graph is constant and given by $(T_1-T_2)/L$. Then

$$Q = +kA\left(\frac{T_1-T_2}{L}\right)$$

where T_1 and T_2 are *surface* temperatures.

In fact for most materials k varies with temperature although, if the temperature range is not too great, a mean value may be assumed with little error. Mean values of k for a few representative materials are given in the following table.

Material	Btu in/ft² h °F	Wm/m² deg K
Copper	2700	389
Aluminium	1570	226
Carbon steel	350	50·5
Porcelain	10	1·44
Firebrick	8·5	1·22
Common brickwork	8·0	1·15
Window glass	7·3	1·05
Concrete	5·0	0·72
Coke breeze slab	4·5	0·65
Water	4·3	0·62
Asbestos cement	2·5	0·36
Plaster	2·0	0·29
Snow	2·0	0·29
Glass fibre sheet	1·7	0·24
Asbestos	1·4	0·20
Refractory brick	1·3	0·19
Polyvinylchloride (PVC)	1·25	0·18
Plaster board	1·2	0·17
Polystyrene sheet	1·2	0·17
Natural rubber	1·1	0·16
Timber	1·0	0·14
Hardboard	0·6	0·086
Thatch	0·5	0·072
Cork	0·4	0·058
Wool and felt	0·3	0·043
Cellular polyurethane	0·27	0·039
Cellular polystyrene	0·25	0·036
Air	0·18	0·026

Note:

$$1\left(\frac{\text{Btu in}}{\text{ft}^2 \text{ h °F}}\right) = \frac{1055 \times 0 \cdot 0254 \times 9}{0 \cdot 305^2 \times 3600 \times 5} = 0 \cdot 1442 \left(\frac{\text{Jm}}{\text{m}^2 \text{ s °K}}\right)$$

$$= 0 \cdot 1442 \text{ Wm/m}^2 \text{ deg K}.$$

(If, for convenience, the length of the path is expressed in mm, the values of k given in the table may be changed to W mm/m² deg K by multiplying by 1000.)

Transposing the Biot equation gives

$$\frac{L}{k} = \frac{A}{Q}(T_1 - T_2).$$

If k has units Wm/m² deg K, this ratio has units m² deg K/W. It is called the *thermal resistance* of a material.

It is appropriate here to point out that the temperature of the boundary of a solid is not that of the fluid exerting pressure on it. The inert (stagnant) boundary layer of fluid has a temperature gradient across it and so has the inevitable film of deposit-cum-oxide.

Now suppose energy to be transmitted unidirectionally at a constant rate by conduction through several, say three, sheets of different material in series, as shown in Figure 209.

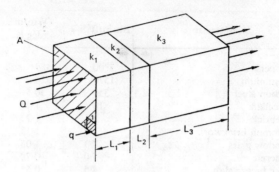

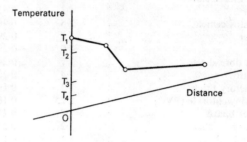

Fig. 209

Since the rate of transfer is the same through each slice,

$$Q = k_1 A\left(\frac{T_1-T_2}{L_1}\right) = k_2 A\left(\frac{T_2-T_3}{L_2}\right) = k_3 A\left(\frac{T_3-T_4}{L_3}\right)$$

Hence $\quad (T_1-T_2)+(T_2-T_3)+(T_3-T_4) = \dfrac{Q}{A}\left(\dfrac{L_1}{k_1}+\dfrac{L_2}{k_2}+\dfrac{L_3}{k_3}\right)$

i.e. $\quad T_1-T_4 = q\left(\dfrac{L_1}{k_1}+\dfrac{L_2}{k_2}+\dfrac{L_3}{k_3}\right) \quad$ putting $\quad \dfrac{Q}{A} = q.$

Or, Total temperature drop = q(sum of thermal resistances). Alternatively, in unit time,

$$\text{Energy transfer/unit area} = \frac{\text{Temperature difference causing transfer}}{\text{Sum of thermal resistances}}.$$

(Note the similarity to an electrical system obeying Ohm's Law.)

The reciprocal of the sum of the thermal resistances between two external surfaces is called the *thermal conductance* and denoted by C. It has units W/m² deg K.

Thus $\quad\quad\quad\quad\quad\quad C = \dfrac{1}{\Sigma(L/K)} = \dfrac{1}{R} \quad$ say,

so that, denoting the energy transfer per unit area in unit time by q we have

$$q = C \text{ (temperature difference between external surfaces).}$$

EXAMPLE. The temperature at the inner surface of a furnace lined with firebrick 250 mm thick is 780°C. The brick casing is 115 mm thick and separated from the lining by an

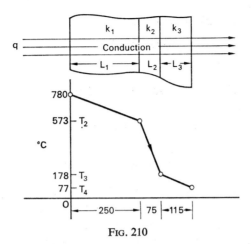

Fig. 210

"insulating" layer 75 mm thick. If the conductivities in Wm/m² deg K for lining, insulating and casing materials are, respectively, 1·21, 0·19 and 1·14, and the rate of energy release inside the furnace is 1·0 kW per square metre of surface, estimate the temperature at the outer surface and at each interface.

Solution. Referring to Figure 210, the thermal resistances are

$$R_1 = \frac{L_1}{k_1} = \frac{0{\cdot}250}{1{\cdot}21} = 0{\cdot}207$$

$$R_2 = \frac{L_2}{k_2} = \frac{0{\cdot}075}{0{\cdot}19} = 0{\cdot}395$$

$$R_3 = \frac{L_3}{k_3} = \frac{0{\cdot}115}{1{\cdot}14} = 0{\cdot}101.$$

If good contact between the layers is assumed, then

Sum of thermal resistances, $R = R_1 + R_2 + R_3 = 0{\cdot}703$ m² deg K/W.

Total temperature drop, $T_1 - T_4 = qR$
$$= 1000 \times 0{\cdot}703$$
$$= 703 \text{ deg C.}$$

Temperature at outer surface, $T_4 = 780 - 703 = \underline{77°\text{C}}.$

Similarly, $\quad T_1 - T_2 = qR_1$
$$= 1000 \times 0{\cdot}207$$
$$= 207 \text{ deg C.}$$

Temperature at first interface, $T_2 = 780 - 207 = \underline{573°\text{C}}.$

Similarly, $\quad T_2 - T_3 = qR_2$
$$= 1000 \times 0{\cdot}395$$
$$= 395 \text{ deg C.}$$

Temperature at second interface, $T_3 = 573 - 395 = \underline{178°\text{C}}.$

As a check:
$$T_3 - T_4 = qR_3$$
$$= 1000 \times 0.101$$
$$= 101 \text{ deg C.}$$

Temperature at outer surface, $T_4 = 178 - 101 = \underline{77°C}$.

Note that the steepest temperature gradient occurs in the material having the lowest value of k.

Heat Transfer through a Boundary Layer

A solid of uniform thickness often separates fluids. These may be gaseous or liquid, the same or different. If the fluid temperatures are not equal there will be a transfer of energy through the solid in the manner already described. Figure 211 shows the simple case of a hot gas (fluid A) separated from the atmosphere (fluid B) by a furnace boundary, S.

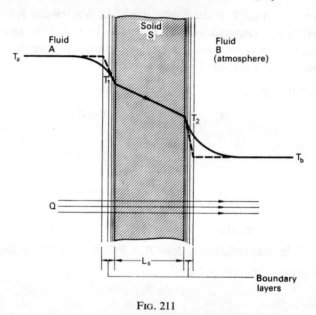

Fig. 211

If the temperature of the atmosphere is plotted as the wall is approached it will rise as shown from, say, T_b (at some distance from the wall) to T_2 at the wall itself, the temperature gradient becoming progressively steeper as the stationary (stagnant) boundary layer is crossed. The heat transfer through this boundary layer is by conduction at the wall itself and thereafter by a combination of conduction and convection, the latter increasing with distance from the wall. These two effects are not only inseparable but have superimposed on them that due to radiation. The phenomena are repeated in the reverse order during heat transfer through the other boundary layer, the temperature falling most sharply at the surface itself.

If the "resistance" offered by the first boundary layer to heat flow is denoted by R_{s1} then the temperature difference required to maintain a steady flow of energy from fluid A

to solid S is given by

$$T_a - T_1 = \frac{Q}{A} \cdot R_{s1}.$$

The reciprocal of R_{s1} is called the *surface coefficient* and its units are W/m² deg K. The second surface (solid to atmosphere) will have a similar coefficient, R_{s2}, the total resistance of the solid and its two boundary layers being given by

$$R = R_{s1} + \frac{L_s}{k_s} + R_{s2}.$$

If the fluids are separated by several different layers in series then

$$R = R_{s1} + \Sigma\left(\frac{L}{k}\right) + R_{s2} \quad \text{(m² deg K/W)}.$$

Heat Losses from Structures

The thermal conductance, C, is defined in terms of *surface* temperature which is not simple to measure so that where buildings are concerned it is customary to compute heat transfer per unit area per unit difference in external and internal *air* temperature, measurement of which *is* simple. This quantity—which takes surface resistances into account—is known as the *thermal transmittance* (or "U-value") and is so defined in BS 874 : 1956. Coefficients for different types of structure may be found either experimentally or—in the case of a series of layers—be computed from an estimate of the various resistances.

Thus, when an internal surface of resistance R_{s1} is separated from an external surface of resistance R_{s2} by a series of layers of total resistance $\Sigma(L/k)$ we have, *when conditions are steady*,

Transmittance coefficient, $\quad U = \dfrac{1}{R_{s1} + \Sigma(L/k) + R_{s2}}.$

Since no account is taken of the length of the path, the units of U are the same as those of C, namely W/m² deg K. Taking surface resistances into account, the total loss in a given time per unit external area under steady conditions is then given by

$$\text{Loss} = U \times \text{Mean air temperature difference} \times \text{seconds}.$$

Since conditions are rarely steady, the figure so arrived at is only an estimate.

[*Note.* The resistance of a corrugated surface is, approximately, 0·8 of the value for the corresponding plane surface.]

EXAMPLE. A cavity wall 6 m long 3 m high consists of two rows of bricks each 115 mm thick separated by an air space the thermal resistance of which may be taken as 0·176 m² deg K/W. The thermal conductivity of the brickwork is 1·15 Wm/m² deg K and the values of R_{s1} and R_{s2} may be taken respectively as 0·124 and 0·076 m² deg K/W.
Find:
(a) the U-value for this wall,
(b) the heat loss per hour when the internal and ambient temperatures are respectively 25° and 10°C.

Solution.

For brickwork, $$R_b = \frac{L}{k} = \frac{2 \times 0.115}{1.15}$$
$$= 0.2 \text{ m}^2 \text{ deg K/W}$$

For air space, $R_a = 0.176 \text{ m}^2 \text{ deg K/W}$

For internal surface, $R_{s1} = 0.124 \text{ m}^2 \text{ deg K/W}$

For external surface, $R_{s2} = 0.076 \text{ m}^2 \text{ deg K/W}$

∴ Total resistance, $R = 0.2 + 0.176 + 0.124 + 0.076$
$$= 0.576 \text{ m}^2 \text{ deg K/W}$$

Transmittance coefficient, $$U = \frac{1}{0.576}$$
$$= \underline{1.74 \text{ W/m}^2 \text{ deg K.}}$$

Area of wall, $A = 6 \times 3 = 18 \text{ m}^2$.

∴ Heat loss/second $= UA(T_1 - T_2)$
$$= 1.74 \times 18(25 - 10)$$
$$= 470 \text{ J.}$$

Heat loss/hour $= 470 \times 3600 = 1.7 \text{ MJ.}$

EXAMPLE. Two sheets of asbestos compound sandwich between them a heating element having an output of 3000 W/m², the sheet details being as follows:

Length of heat path	$L_1 = 15$ mm	$L_2 = 75$ mm
Thermal conductivity	$k_1 = 0.35$ Wm/m² deg K	$k_2 = 0.07$ Wm/m² deg K
Surface coefficient	$U_1 = 20$ W/m² deg K	$U_2 = 10$ W/m² deg K

Determine, for an ambient temperature of 20°C

(a) the temperature of the element,
(b) the percentage of the output transmitted by each sheet,
(c) the surface temperature of each sheet.

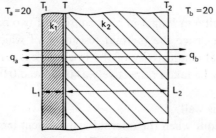

FIG. 212

Solution.

For side A:
$$R_1 = \frac{L_1}{k_1} = \frac{0.015}{0.350} = 0.0429 \text{ m}^2 \text{ deg K/W}$$

$$R_{s1} = \frac{1}{U_1} = \frac{1}{20} = 0.05 \text{ m}^2 \text{ deg K/W}$$

$$\therefore \Sigma R_a = 0.0429 + 0.05$$
$$= 0.0929 \text{ m}^2 \text{ deg K/W}$$

$$\therefore q_a = \frac{T - T_a}{\Sigma R_a}$$

$$= \frac{T - 20}{0.0929} \text{ W/m}^2.$$

For side B:
$$R_2 = \frac{L_2}{k_2} = \frac{0.075}{0.070} = 1.07 \text{ m}^2 \text{ deg K/W}$$

$$R_{s2} = \frac{1}{U_2} = \frac{1}{10} = 0.01 \text{ m}^2 \text{ deg K/W}$$

$$\therefore \Sigma R_b = 1.07 + 0.01$$
$$= 1.17 \text{ m}^2 \text{ deg K/W}$$

$$\therefore q_b = \frac{T - T_b}{\Sigma R_b}$$

$$= \frac{T - 20}{1.17} \text{ W/m}^2.$$

But,
$$3000 = q_a + q_b = (T - 20)\left(\frac{1}{0.0929} + \frac{1}{1.17}\right)$$
$$= (T - 20)(10.76 + 0.855)$$

so that
$$T - 20 = \frac{3000}{11.61} = 258$$

i.e.
$$\underline{T = 278 \text{ °C}.}$$

Hence
$$q_a = \frac{278 - 20}{0.0929} = 2780 \text{ W/m}^2$$

Percentage of output
$$= \left(\frac{2780}{3000}\right) 100 = 92.7$$

and
$$q_b = \frac{278 - 20}{1.17} = 220 \text{ W/m}^2$$

Percentage of output
$$= 100 - 92.7 = 7.3.$$

Since
$$T - T_1 = q_a R_1$$
$$\therefore 278 - T_1 = 2780 \times 0.0429 = 118$$
so that
$$\underline{T_1 = 278 - 118 = 160 \text{°C}.}$$

Similarly
$$T - T_2 = q_b R_2$$
$$\therefore 278 - T_2 = 220 \times 1.07 = 235$$
so that
$$\underline{T_2 = 278 - 235 = 43 \text{°C}.}$$

Conduction through a Uniform Pipe

Figure 213 represents the section of a uniform pipe containing a fluid at a temperature higher than atmospheric. If the internal and external temperatures are uniform and constant there will be a steady flow of energy in a radial direction as shown. The temperature at any intermediate surface of, say, radius r will be constant also.

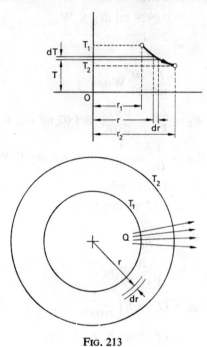

FIG. 213

Since $Q = -kA \cdot \dfrac{dT}{dr}$ and A increases with r, it follows that, if k is constant, the value of dT/dr falls with increase in r. Hence the graph of temperature against radius has the form shown, the slope dT/dr being negative.

Now, Area of flow path, $A = 2\pi rL$

where $\qquad\qquad\qquad\qquad L = $ length of pipe.

$\therefore$ Heat transfer, $\qquad\qquad Q = -k(2\pi rL)\dfrac{dT}{dr}$

or, transposing, $\qquad\qquad Q\left(\dfrac{1}{r}\right)dr = -2\pi kL \cdot dT.$

Supposing the *surface* temperatures to be T_1 and T_2 at, respectively, r_1 and r_2 we have

$$Q \int_{r_1}^{r_2} \left(\frac{1}{r}\right) dr = -2\pi k L \int_{T_1}^{T_2} dT$$

236

so that
$$Q \log_e \left(\frac{r_2}{r_1}\right) = -2\pi kL(T_2-T_1) = 2\pi kL(T_1-T_2)$$

whence
$$Q = \frac{2\pi kL(T_1-T_2)}{\log_e (r_2/r_1)} \quad \text{J/s or W}.$$

Heat Transfer through a Lagged Pipe

Suppose the pipe of Figure 213 to be insulated by a layer of material of thermal conductivity k_i and outer radius r_3 as shown in Figure 214 and let T_1, T_2 and T_3 be the surface temperatures. Then, if the thermal conductivity of the pipe material is k_p we have, using the expression just derived and assuming steady conditions, through the pipe itself:

$$Q = \frac{2\pi k_p L(T_1-T_2)}{\log_e (r_2/r_1)}, \quad \text{i.e.} \quad T_1-T_2 = \frac{Q \log_e (r_2/r_1)}{2\pi k_p L}$$

and through the insulation:

$$Q = \frac{2\pi k_i L(T_2-T_3)}{\log_e (r_3/r_2)}, \quad \text{i.e.} \quad T_2-T_3 = \frac{Q \log_e (r_3/r_2)}{2\pi k_i L}.$$

Now the boundary layers at the internal and external surfaces are similar to those at plane

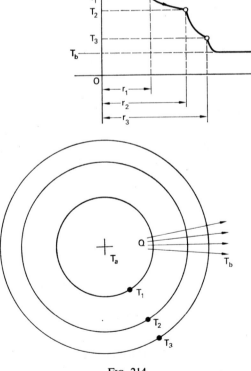

Fig. 214

surfaces and offer "resistances" to heat transfer, R_{s1} and R_{s2} say, which can be taken into account in the same way by the introduction of surface coefficients $U_{s1}\ (=1/R_{s1})$ and $U_{s2}\ (=1/R_{s2})$. Thus, if the fluid in the pipe is at temperature T_a and the fluid external to the pipe is at T_b we have, as with a plane surface,

$$T_a - T_1 = \left(\frac{Q}{A}\right) R_{s1} = \left(\frac{Q}{2\pi r_1 L}\right) R_{s1}$$

where $\quad R_{s1} = \dfrac{1}{U_{s1}}.$

and $\quad T_3 - T_b = \left(\dfrac{Q}{A}\right) R_{s2} = \left(\dfrac{Q}{2\pi r_3 L}\right) R_{s2}$

where $\quad R_{s2} = \dfrac{1}{U_{s2}}.$

Hence, $(T_a - T_1) + (T_1 - T_2) + (T_2 - T_3) + (T_3 - T_b)$

$$= \left(\frac{Q}{2\pi r_1 L}\right) R_{s1} + \frac{Q \log_e (r_2/r_1)}{2\pi k_p L} + \frac{Q \log_e (r_3/r_2)}{2\pi k_i L} + \left(\frac{Q}{2\pi r_3 L}\right) R_{s2}$$

i.e. $\quad T_a - T_b = \dfrac{Q}{2\pi L} \left[\dfrac{R_{s1}}{r_1} + \dfrac{\log_e (r_2/r_1)}{k_p} + \dfrac{\log_e (r_3/r_2)}{k_i} + \dfrac{R_{s2}}{r_3} \right].$

Hence, $\quad Q = \dfrac{2\pi L \text{ (Temperature difference causing transfer)}}{\left[\dfrac{R_{s1}}{r_1} + \dfrac{\log_e (r_2/r_1)}{k_p} + \dfrac{\log_e (r_3/r_2)}{k_i} + \dfrac{R_{s2}}{r_3} \right]}.$

Now, for plane sheets in series,

$$Q = \frac{\text{Temperature difference causing transfer}}{\text{Sum of thermal resistances}}$$

so that for cylinders in series, if the temperature difference is multiplied by $2\pi L$, each of the quantities in the large bracket may be regarded as the resistance offered by that section of the heat path (and denoted by R_1, R_2, R_3 and R_4) while the sum of these quantities may be regarded as the effective resistance to heat transfer between the fluids.

Denoting this sum by ΣR we obtain

$$Q = \frac{2\pi L(T_a - T_b)}{\Sigma R} \quad W.$$

External surface coefficients vary between 5·7 W/m² deg K for bright metallic surfaces and 10 W/m² deg K for the dullest matt surface. Reference should be made to BS 1588 :1963.

EXAMPLE. Steam at 205°C is carried 6 m from a boiler to a heat exchanger by a pipe 4 mm thick, 26 mm internal diameter and made of steel having a thermal conductivity of 50 Wm/m² deg K. The pipe has an internal surface coefficient of 6800 W/m² deg K and is lagged to an overall diameter of 120 mm with asbestos felt having a thermal conductivity of 0·08 Wm/m² deg K and an external surface coefficient of 10 W/m² deg K. Estimate the rate of heat loss from the pipe when the temperature of the boiler house is 25°C.

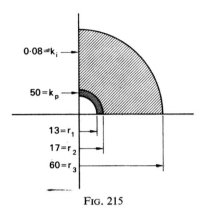

Fig. 215

Solution.

$$R_{s1} = \frac{1}{6800} \text{ m}^2 \text{ deg K/W}, \quad r_1 = 0.013 \text{ m} \quad \text{and} \quad \frac{r_2}{r_1} = \frac{17}{13} = 1.31$$

$$R_{s2} = \frac{1}{10} \text{ m}^2 \text{ deg K/W}, \quad r_3 = 0.060 \text{ m} \quad \text{and} \quad \frac{r_3}{r_2} = \frac{60}{17} = 3.53.$$

For inner surface, $R_1 = \dfrac{R_{s1}}{r_1} = \dfrac{1}{6800 \times 0.013} = 0.0113$ m deg K/W

For pipe, $R_2 = \dfrac{\log_e (r_2/r_1)}{k_p} = \dfrac{\log_e 1.31}{50} = 0.0054$ m deg K/W

For insulation, $R_3 = \dfrac{\log_e (r_3/r_2)}{k_i} = \dfrac{\log_e 3.53}{0.08} = 15.75$ m deg K/W

For outer surface, $R_4 = \dfrac{R_{s2}}{r_3} = \dfrac{1}{10 \times 0.060} = 1.67$ m deg K/W

Hence $\Sigma R = 0.0113 + 0.0054 + 15.75 + 1.67$

$= \underline{17.44 \text{ m deg K/W}}$ (or m² deg K/Wm).

(Notice that R_1 and R_2 could be neglected with little error.)
Temperature difference causing energy transfer $= 205 - 25$
$= 180$ deg K.

Hence, Heat loss, $Q = \dfrac{2\pi L \times 180}{\Sigma R}$ where $L = 6$ m,

$= \dfrac{2\pi \times 6 \times 180}{17.44}$

$= \underline{389 \text{ W}}.$

Note that, for each material, Q depends on the *ratio* of the radii and *not* on the difference in radius. The higher this ratio, the lower the rate of energy transfer.

EXAMPLES 8

1. Show that the chemical equation for the combustion of liquid butylene (C_4H_8) is
$$C_4H_8 + 6O_2 = 4CO_2 + 4H_2O.$$
Hence determine the stoichiometric air/fuel ratio for this hydrocarbon (14.9).

2. Find the stoichiometric air/fuel ratio for a sample of coal the gravimetric analysis of which is 0·80 carbon, 0·12 hydrogen, remainder incombustible (13·3). Determine the volumetric analysis of the theoretical products of combustion assuming the steam to be condensed (CO_2 15·5 per cent, N_2 84·5 per cent).

3. A producer gas consists (by volume) of the following mixture:

H_2	CH_4	C_4H_8	CO	CO_2	O_2	Total
0·52	0·23	0·05	0·12	0·03	0·05	1·000 m³

Calculate the stoichiometric volume of air which would have to be supplied at the same temperature (4·9 m³).

4. Convert the volumetric analysis of the mixture in question 3 to a gravimetric analysis by multiplying the volume of each constituent (per m³ mixture) and dividing each result by the sum of such results (H_2 7·6, CH_4 26·6, C_4H_8 20·3, CO 24·4, CO_2 9·6, O_2 11·6%).

5. Calculate the gravimetric air/fuel ratio required for the producer gas in question 3 using the gravimetric analysis as obtained in question 4 (10·4).

6. Show that the combustion equation for ethylene (C_2H_4) is
$$C_2H_4 + 3O_2 = 2CO_2 + 2H_2O.$$
Hence find the stoichiometric air assuming the fuel vaporised (5·2 m³).

7. (a) Discuss the influence of each of the following fuel characteristics on the performance of the spark-ignited petrol motor:
    1. Calorific value.
    2. Boiling point range.
    3. Octane number.
   (b) Repeat for the following design characteristics:
    1. Diameter of carburettor venturi.
    2. Length of inlet tract.
    3. Diameter of inlet valve.
    4. Shape of combustion chamber.
    5. Location of ignition point.
    6. Timing of spark.
    7. Compression ratio.
    8. Profile of valve operating cam.
   (c) Describe and explain the effect on performance of:
    1. Air temperature.
    2. Altitude.

8. A domestic heating system uses a liquid hydrocarbon fuel containing 16 per cent H_2. Calculate the percentage of CO_2 in the combustion products assuming stoichiometric air and condensed steam (14 per cent).

9. If the fuel in question 8 were supplied with air/fuel in the ratio 14/1, calculate the percentage of CO_2 in the combustion products assuming condensed steam and no hydrogen (11·6 per cent).

10. Explain how the CO_2 content of combustion products may be used as a guide to combustion efficiency.

11. The percentage analysis by volume at 70°F of a by-product gas mixture was estimated to be as follows:

CH_4	CO	H_2	CO_2	O_2	N_2
2	21	15	5	3	Remainder.

If the air supplied is to be at the same temperature and 40 per cent in excess of the stoichiometric, calculate the volume required per cubic metre of mixture (1·23 m³).

12. A sample of coal gas revealed the following composition by volume:
$$CH_4\ 39·5\%,\ CO\ 7·5\%,\ \text{remainder incombustible}.$$
An analysis of the combustion products showed 10% O_2 by volume. Estimate:
(a) the volume of air supplied/volume of gas (9·1 m³);
(b) the reduction in volume at combustion (10·9%).

13. The analysis of the dry products of combustion of a hydrocarbon fuel containing by weight 85·5 per cent carbon and 14·5 per cent hydrogen showed 14 per cent carbon dioxide. Estimate the value of the air/fuel ratio used given that oxygen was present (15·7).
14. Estimate the hourly rate of heat transfer through a simple brick wall 230 mm thick for which $k = 1·15$ Wm/m² deg K, given that the surface temperatures are 20°C and 0°C (363 kJ/m²).
15. Water at 88°C is contained in an unlagged tank, the material of which is 12·7 mm thick, has internal and external surface coefficients of 2730 and 10·2 W/m² deg K respectively and a thermal conductivity of 50 Wm/m² deg K. If the ambient temperature is 10°C estimate:
    (a) the hourly heat loss per square metre of surface (2·84 MJ);
    (b) the external temperature of the tank (87°C).
16. A heating element having an output of 2690 W/m² is sandwiched between two sheets of asbestos compound the details being as follows:
    Length of heat path: $L_1 = 12·7$, $L_2 = 63·5$ mm,
    Thermal conductivity: $k_1 = 0·274$, $k_2 = 0·088$ Wm/m² deg K,
    Surface coefficient: $U_1 = 20·0$, $U_2 = 8·5$ W/m² deg K.
    If the ambient temperature is 21°C, estimate
    (a) the temperature of the element (260°C);
    (b) the percentage of the output transferred by the thinner sheet (89·5);
    (c) the surface temperatures (143° and 54°C).
17. At a given load the I^2R loss per metre of single copper cable 6 mm diameter is 7 W. If the cable is insulated to a depth of 4 mm with PVC having a thermal conductivity of 0·18 Wm/m² deg K and a surface coefficient of 10 000 W/m² deg K, estimate the temperature of the surface of the metal when the ambient temperature is 20°C (36°C).

9

TRIBOENGINEERING

Introductory

Relative motion between two surfaces transmitting a force normal to them is resisted by what is called the force of *friction* acting in the plane of the surfaces in a direction opposite to that of the motion. Thus part of the energy input is absorbed in overcoming *frictional resistance*, i.e. it is converted into heat. This heat is not only wasted but brings about an unwelcome rise in temperature. To keep the energy loss and temperature rise within acceptable limits one of two courses may be adopted.

Either (a) load-bearing surfaces can be selected for their low frictional resistance (assuming a satisfactory rate of wear and compatibility with environment),

or (b) the two surfaces can be separated completely.

There are two methods of separating the surfaces:

   (1) by introducing a film of lubricant;
   (2) by interposing rolling elements.

If the motion is oscillatory there is a third possibility, namely, the separation of the surfaces by a flexible element (such as rubber) bonded to each.

Dry Friction

Simple experiments confirm the approximate validity of the so-called Coulomb "laws" of friction (C. A. Coulomb, 1736–1806) namely that, for smooth clean surfaces, the frictional force is

   (1) proportional to the transverse force between the surfaces,
   (2) independent of the (nominal) area of contact,
   (3) greater at the initiation of relative motion than subsequently.

However, the operative words in the foregoing are "smooth" and "clean". The variations from the "flat" of an element of the smoothest engineering surface (mirror finish) are measured in micro-metres so that, if elements of two such clean surfaces are brought lightly together, contact will be made initially at only a few points, i.e. where "high spots" coincide. At such points the elastic limit in compression will certainly be exceeded and plastic deformation will occur. This will continue until the actual (as distinct from the nominal) area of contact is capable of supporting the load. This area will be, approximately, proportional to the load.

Since the contact between the outermost atoms in each "high spot" is as intimate as that between adjoining atoms of the same metal, the opposing high spots are effectively welded together. Such welds are known as "junctions" and, if the two surfaces are given relative motion after pressing together, these junctions fail in shear. The shear force (which is necessary to start the motion) is proportional, theoretically, to the junction area and hence to the normal force and this relation is confirmed by experiment.

Since the junctions need not necessarily fail in the plane of contact, the surfaces might be expected to deteriorate, i.e. to become less smooth as the motion progresses until, eventually, seizure occurs. That this does not happen in practice is due to the fact that surfaces in engineering are seldom "clean" under atmospheric conditions; inevitably an oxide layer is present and this is accompanied by dust, dampness and grease from handling.

Boundary Lubrication

Such a contaminating mixture (of oxide, cutting fluid, dust, etc.) brings about partial separation of the surfaces and is in effect a lubricant (although it may be as little as a few molecules in thickness) and causes a remarkable drop in frictional resistance. The state is described as one of "boundary lubrication" and the theoretical treatment of friction in screws, belts, clutches and so on which now follows assumes this state to exist.

Coefficient of Friction

The first "law" of Coulomb friction states that the frictional force is proportional to the transverse load carried by the surfaces, i.e.

$$F = a \text{ constant} \times P$$

where P = transverse load and F = resistance to motion due to friction,

or
$$F/P = a \text{ constant}.$$

This constant is denoted by μ and called the *coefficient of friction*, so that

$$\underline{F = \mu P.}$$

Motion up an Inclined Plane

Consider a vertical load W to rest on a plane inclined θ to the horizontal as shown in Figure 216(a) and let uniform motion up the plane result from the application of a horizontal force F.

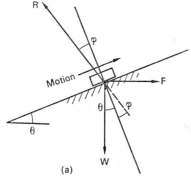
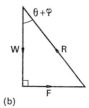

(a) (b)

Fig. 216

The reaction, R, of the plane is equal and opposite to the resultant of F and W and makes an angle $\varphi(= \tan^{-1} \mu)$ with the normal to the plane. From the force triangle

$$F = W \tan(\varphi + \theta).$$

It can be shown in a similar way that the horizontal force required to maintain uniform motion *down* the plane is

$$F = W \tan(\varphi - \theta).$$

Screw Friction

For a single-start screw the axial distance between corresponding points on adjacent threads is called the *pitch* and denoted by p. If the thread is square in section it may be treated as an inclined plane wrapped helically round a cylinder as shown in Figure 217.

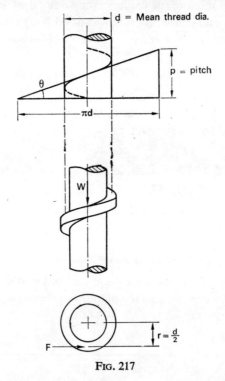

FIG. 217

Rotation of the screw against an axial load (applied, say, to a nut) is equivalent to making such load move up a plane inclined at an angle $\theta = \tan^{-1}(p/\pi d)$. Since, as already shown, the force F required at mean thread *radius* is $W \tan(\varphi + \theta)$, we have

Torque to produce rotation $= Fr$

i.e.
$$T = W \tan(\varphi + \theta) \frac{d}{2}.$$

If this torque is supplied by means of a force P applied at radius R (using, for example,

a spanner) then
$$P = \frac{T}{R}.$$

For an ideal screw $\mu = 0$, i.e. $\tan \varphi = 0$, so that the corresponding effort required at mean thread radius is $W \tan \theta$. Hence, for any practical screw

$$\text{Efficiency} = \frac{\text{Ideal effort}}{\text{Actual effort}} = \frac{W \tan \theta}{W \tan (\varphi + \theta)}$$

or
$$\eta = \frac{\tan \theta}{\tan (\varphi + \theta)}.$$

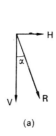
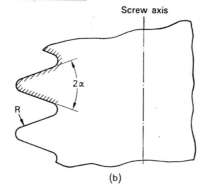

(a)

(b)

Fig. 218

If the section of the helix is not square but of vee form as shown in Figure 218(b), the normal force on the thread, R, may be resolved into a radial component $H(= R \sin \alpha)$ and an axial component $V(= R \cos \alpha)$ which is evidently equal to the axial load, i.e.

$$W = R \cos \alpha,$$

so that Normal force
$$R = \frac{W}{\cos \alpha}$$

instead of W as with a square threaded screw. Since the frictional force is proportional to $W/(\cos \alpha)$ instead of to W it follows that $\mu/(\cos \alpha)$ can be substituted for μ in the equations already derived for the force at mean thread radius.

Thus, for a square thread
$$F = W \tan (\varphi + \theta)$$
$$= W \left(\frac{\tan \varphi + \tan \theta}{1 - \tan \varphi \tan \theta} \right)$$
$$= W \left(\frac{\mu + \tan \theta}{1 - \mu \tan \theta} \right)$$

and for a vee thread
$$F = W \left[\frac{\dfrac{\mu}{\cos \alpha} + \tan \theta}{1 - \dfrac{\mu \tan \theta}{\cos \alpha}} \right].$$

EXAMPLE. A spanner having an effective length of 610 mm is used to tighten a nut of 10 mm pitch and 44 mm mean thread diameter, the thread being of square section and the nut face having a mean radius of 33 mm. Estimate the force required normal to the free end of the spanner to induce a tension in the bolt of 16 kN. Assume $\mu = 0.125$ throughout.

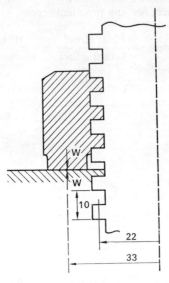

FIG. 219

Solution. Referring to Figure 219,

$$\tan \theta = \frac{p}{\pi d} = \frac{10}{44\pi} = 0.0723$$

$$\tan \varphi = \mu = 0.125$$

$$\text{Force at mean thread radius} = 16\,000\left(\frac{0.125 + 0.0723}{1 - (0.125 \times 0.0723)}\right)$$

i.e. $\qquad F = 3190$ N.

$$\text{Resisting torque due to thread friction} = 3190\left(\frac{0.044}{2}\right)$$

$$= 70 \text{ Nm.}$$

Frictional force at nut face $= \mu \times$ axial load
$= 0.125 \times 16\,000$
$= 2000$ N.

If this is assumed to act at the mean radius of 0·033 m (see next paragraph) then
Resisting torque due to face friction $= 2000 \times 0.033$
$= 66$ Nm.

Total resisting torque $= 70 + 66 = 136$ Nm.

Necessary force at 0·61 m radius $= \dfrac{136}{0.61} = 223$ N.

FRICTION BETWEEN TWO ANNULAR SURFACES

When a rotating shaft is loaded axially, as is the case when its axis is vertical or when it carries a helical gear or turbine blading, some form of annular bearing (Fig. 220) must be provided to prevent axial movement. The frictional torque at the bearing surfaces (which must be overcome before rotation can occur) must be kept to an acceptably low value.

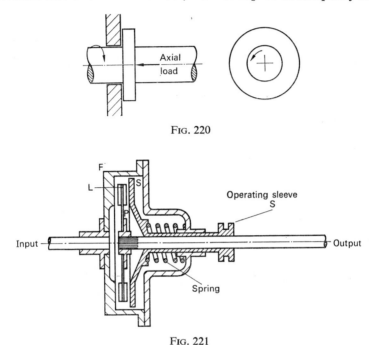

Fig. 220

Fig. 221

The friction rings or linings (L) on a flat clutchplate (P) are also annular in form and are shown in Figure 221, the plate being splined to the output shaft and free to move axially. To engage the clutch (i.e. to transmit torque from input to output) the sleeve (S) is permitted to slide to the left under the action of the engaging spring until the plate is gripped between the flywheel (F) and the sleeve face. When the engaging force reaches the value which makes the frictional torque on the plate greater than the resisting torque, the output shaft will begin to rotate, the input and output speeds being equal when slipping ceases.

Although the frictional torque must be as low as possible in a thrust bearing and as high as possible in a clutch, the theory is the same for each.

Assumptions Made

If the surfaces are coaxial and the force between them is normal to their plane, then the pressure on them will be uniform when they are new. However, since points at different radius have different velocity there will be some variation (very small) in the value of μ. This will apply particularly in the case of a flat pivot where the velocity at the centre is zero. Thus the initial wear will not be uniform but will increase with radius and be greatest at the outside. This variation in the rate of wear alters the pressure distribution, increasing the intensity towards the centre. This in turn alters the rate of wear so that the variations in wear and pressure are continuous. Ultimately the pressure may become uniform again though the surfaces will not necessarily regain their original profile.

In view of the foregoing it is usual to assume either

(1) that the pressure is uniform, or
(2) that the rate of wear is uniform.

The first assumption gives a slightly higher value of frictional torque and may be used in the estimation of power loss in a bearing. Clutch design is based, usually, on the second assumption.

Uniform Pressure

Assuming the surfaces to be new and referring to Figure 222:

Area of element	$= 2\pi r dr$
Normal force on element	$= 2\pi r dr \times p$
Frictional force on element	$= 2\pi r dr p \times \mu$
Frictional torque per element	$= 2\pi r dr p \mu \times r$
	$= 2\pi \mu p r^2 dr$ (assuming μ constant).

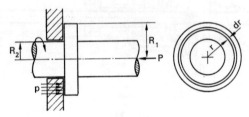

FIG. 222

The sum of all such elemental torques is the total torque, i.e.

$$T = \int_{R_2}^{R_1} 2\pi \mu p r^2 \, dr$$

$$= 2\pi \mu p \left(\frac{R_1^3 - R_2^3}{3} \right)$$

where

$$p = \frac{P}{\pi(R_1^2 - R_2^2)}$$

$$\therefore T = \frac{2}{3} \mu P \left(\frac{R_1^3 - R_2^3}{R_1^2 - R_2^2} \right).$$

When $R_2 = 0$ the outer radius may be denoted by R so that, for a flat pivot as shown in Figure 223,

$$T = \tfrac{2}{3} \mu P R.$$

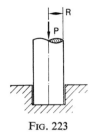

Fig. 223

In the case of a clutch the first expression derived gives the friction torque for one pair of engaged surfaces so that if (as in Fig. 221) both sides of a single plate are effective, this value must be doubled. In the case of a multi-plate clutch having n pairs of engaged surfaces the transmissible torque per pair must be multiplied by n.

Uniform Wear

Wear may be assumed to be proportional both to the frictional force and to the velocity of rubbing. The frictional force at any radius is proportional to the pressure there and the

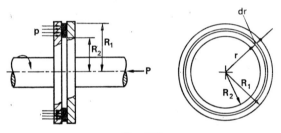

Fig. 224

velocity of rubbing is proportional to that radius so that the wear may be assumed proportional to the product of pressure and radius. (However, when the surfaces are new, the pressure is uniform so that, initially, the wear is proportional to the radius only.) Assuming then that $pr = k$ (where k is some constant) and referring to Figure 224:

Area of element $\qquad = 2\pi r\, dr$
Normal force on element $\qquad = 2\pi r\, dr \times p$
$\qquad = 2\pi k\, dr.$

Total force, i.e. axial load, $\qquad P = \int_{R_2}^{R_1} 2\pi k\, dr$

or $\qquad P = 2\pi k(R_1 - R_2)$

so that $\qquad k = \dfrac{P}{2\pi(R_1 - R_2)}.$

Frictional force on element $\qquad = $ Normal force $\times \mu$
$\qquad = 2\pi k\, dr \times \mu$
Frictional torque per element $\qquad = 2\pi k\, d\mu \times r$
$\qquad = 2\pi k\mu r\, dr.$

The sum of all such elemental torques is the total torque, i.e.

$$T = \int_{R_2}^{R_1} 2\pi k \mu r \, dr$$

$$= 2\pi k \mu \frac{(R_1^2 - R_2^2)}{2}$$

where
$$k = \frac{P}{2\pi(R_1 - R_2)}$$

so that
$$T = \frac{1}{2} \mu P \frac{(R_1 + R_2)(R_1 - R_2)}{(R_1 - R_2)}$$

∴
$$\underline{T = \frac{1}{2} \mu P (R_1 + R_2).}$$

When $R_2 = 0$, the outer radius may be denoted by R so that, for a flat pivot or other circular friction surface,

$$T = \tfrac{1}{2} \mu P R.$$

EXAMPLE. The output shaft of a reduction gear rotates uniformly at 200 rev/min and drives, via a single-plate friction clutch, a machine the moving parts of which have a moment of inertia of 30 kg m². If the clutch spring can exert an engaging force of 400 N estimate the minimum time required for the machine to attain full speed from rest, assuming uniform acceleration. The friction ring diameters are 230 mm and 380 mm, both sides are effective and $\mu = 0.25$. Estimate also the heat generated during the engagement period.

Solution. Assuming uniform wear:

$$T = 2[\tfrac{1}{2}\mu P(R_1 + R_2)]$$
$$= (0.25 \times 400)(0.190 + 0.115)$$
$$= 30.5 \text{ Nm.}$$

Angular acceleration,
$$\alpha = \frac{T}{I} = \frac{30.5}{30} = 1.015 \text{ rad/s}^2.$$

$$\omega_1 = 0 \quad \text{and} \quad \omega_2 = \frac{2\pi \times 200}{60} = 20.9 \text{ rad/s} = \omega_1 + \alpha t$$

∴ Time to attain full speed,
$$t = \frac{20.9}{1.015} = 20.6 \text{ s.}$$

Angle turned through by input during engagement,

$$\theta = \omega_2 t$$
$$= 20.9 \times 20.6$$
$$= 430 \text{ rad.}$$

Total work done $= T\theta = 30.5 \times 430 = 13\,100$ J.

Since the output starts from rest, its average speed during engagement is half that of the input; hence the angle it turns through is half that of the input, i.e. 215 rad. The work done against friction is the product of the torque and the difference in the angles turned through by input and output, i.e.

Energy converted into heat $= 30\cdot5 \times 215 = 6250$ J.

FRICTION BETWEEN TWO CONICAL SURFACES

For a new bearing or friction surface it may be assumed that the pressure is uniform so that, initially, the rate of wear will be a maximum at the greatest radius.

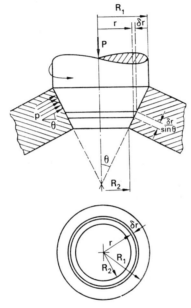

FIG. 225

Referring to Figure 225,

Total bearing surface $= \dfrac{\pi(R_1^2 - R_2^2)}{\sin \theta}$

Load = surface area × vertical component of pressure

$= \dfrac{\pi(R_1^2 - R_2^2)}{\sin \theta} \times p \sin \theta$

i.e. $P = \pi p(R_1^2 - R_2^2)$.

Hence $p = \dfrac{P}{\pi(R_1^2 - R_2^2)}$

and is evidently independent of θ.

Load on element $= p \times 2\pi r \left(\dfrac{dr}{\sin \theta}\right)$

Frictional force $= p \times 2\pi r \left(\dfrac{dr}{\sin \theta}\right) \times \mu$

Frictional torque on element $= p \times 2\pi r \left(\dfrac{dr}{\sin \theta}\right) \times \mu \times r$

$$= \dfrac{2\pi \mu p}{\sin \theta} r^2 \, dr.$$

The sum of all such elemental torques is the total torque, i.e.

$$T = \dfrac{2\pi \mu p}{\sin \theta} \int_{R_2}^{R_1} r^2 \, dr$$

$$= \dfrac{2\pi \mu p}{\sin \theta} \left(\dfrac{R_1^3 - R_2^3}{3}\right)$$

where

$$p = \dfrac{P}{\pi(R_1^2 - R_2^2)}$$

$\therefore$

$$T = \dfrac{2}{3} \dfrac{\mu P}{\sin \theta} \left[\dfrac{R_1^3 - R_2^3}{R_1^2 - R_2^2}\right].$$

Putting $\theta = 90°$ gives the result already obtained for annular surfaces.

For a worn bearing surface the pressure is no longer uniform and it is usual to assume a uniform rate of wear, i.e. that $pr = k$. It is left to the student to show that in this case the expression for the torque is that for an annular surface divided by $\sin \theta$, namely

$$T = \dfrac{1}{2} \dfrac{\mu P}{\sin \theta} (R_1 + R_2).$$

THE CENTRIFUGALLY ACTUATED CLUTCH

The essential components are shown in Figure 226. A spider S_1 is carried on the input shaft and has, usually, four equally spaced radial guides. Retained in each guide by a spring (under an initial compression) is a sliding shoe S_2 which extends beyond the guide and is lined with friction material on its outer face. The spider assembly is enclosed in a coaxial drum D mounted on the output shaft so that, if this assembly is rotated at increasing speed, the shoes will be on the point of moving outwards when the centripetal force required is on the point of exceeding the initial force in the spring. Increase in input speed beyond this point results in shoe movement, contact between shoe and drum (as shown) and transmission of torque. From this point onward there is no further increase in the spring force (see note on p. 253), while the centrifugal force increases as the square of the speed.

For one shoe:

Radial force between shoe and drum (i.e. engaging force)

$$= \text{centrifugal force} - \text{spring force}$$
$$= F - P.$$

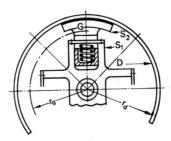

Fig. 226

If the mass of a shoe is M and the radius of rotation of its mass centre G (Fig. 227) is r_g, then the centrifugal force is given by

$$F = M\omega^2 r_g.$$

If the total compression of the spring in the engaged position is δ and the stiffness of the spring is λ then the spring force is given by

$$P = \lambda\delta.$$

Frictional force = $\mu \times$ engaging force
$$= \mu(F-P).$$

If the drum radius is r_d then

Frictional torque = $\mu(F-P) \times r_d$ per shoe.

If there are n shoes,

$$T = n\mu r_d(F-P).$$

As the lining material wears, the shoe must travel further to effect engagement, thereby increasing the spring force. It follows that, after wear has occurred, an increased centrifugal force (and hence a higher speed) must be attained before the drive is taken up. The radius of rotation of the mass centre of the shoe is also increased which helps to reduce this effect.

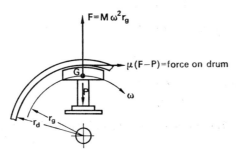

Fig. 227

Note: The value of P is reduced by centrifugal force acting on the mass of the spring itself. If this is not taken into account, the calculated speed at which engagement starts will be on the high side, i.e. the actual speed will be lower.

It is usual to provide some method of spring adjustment so that the engaging speed

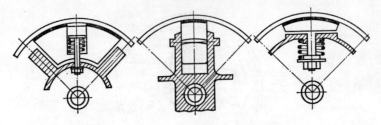

Fig. 228

may be varied. In this way starting torque difficulties associated with some types of electric motor may be avoided and the effects of lining wear minimised.

Alternative shoe arrangements are shown in Figure 228.

EXAMPLE. Each of the four shoes of a centrifugal clutch has a mass of 1·6 kg and, at the moment of engagement, the radius of the mass centres is 112 mm and the spring force on each shoe is 700 N. If the drum diameter is 280 mm and the friction coefficient is 0·3 determine the speed at which engagement begins. Find also the power which could be transmitted at 1480 rev/min. Neglect the effects of centrifugal force on the mass of the spring.

Solution. Engagement begins when centrifugal and spring forces are equal, i.e. when

$$1·6 \times \omega^2 \times 0·112 = 700$$

whence $\omega = 62·5$ rad/s

and $N = 597$ rev/min.

At 1480 rev/min, $\omega = \dfrac{2\pi \times 1480}{60} = 155$ rad/s

so that $F = 1·6 \times 155^2 \times 0·112 = 4300$ N.

Possible friction force per shoe $= \mu(F-P)$
$= 0·3(4300-700)$
$= 1080$ N.

Possible friction torque per shoe $= 1080\left(\dfrac{0·280}{2}\right)$
$= 151$ Nm.

∴ Torque for four shoes $= 4 \times 151$

i.e. $T = 604$ Nm.

Hence possible power $= T\omega = 604 \times 155 = 93\,600$ W $= 93·6$ kW.

BELT DRIVES

Belts are used to transmit power between two parallel shafts the axes of which are too far apart to be connected by gears. However, if the velocity ratio in such a situation is required to be exact, a chain drive is necessary.

Choice of material for the belt is influenced by environmental or climatic conditions and by service requirements. Reinforced fabrics (natural and synthetic, variously impregnated and of single or multi-ply—depending on the thickness required), rubber and steel, are in common use. Belt drives are cheap and reliable, have an efficiency of about 0·97 and have the added virtue of being able to slip in the event of sudden overload.

Friction between a Flat Belt and a Pulley

Consider two pulleys A and B, Figure 229(a), connected by a flat open belt (as distinct from a crossed belt) and let the initial tension be T_0. If B is prevented (by a resisting torque) from rotating and a torque is applied at A, Figure 229(b), then the tension at point N will rise to some value T_t while the tension at point M will fall to some value T_s. The torque

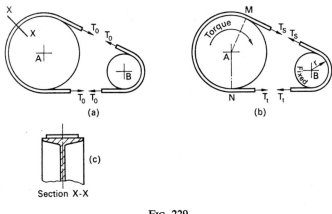

Fig. 229

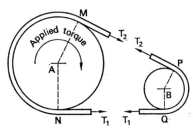

Fig. 230

on B due to these unequal tensions will then be $T_t r - T_s r$, or $(T_t - T_s)r$, and this will be a maximum when the belt is about to slip. (Although r is the radius of the neutral axis of the belt, the effect of belt thickness is usually neglected.)

Let the torque on A be increased until the belt is about to slip on B and suppose T_1 and T_2 to be the corresponding (i.e. maximum) values of T_t and T_s. Then, as shown in Figure 230,

Torque on B at instant of slipping $= (T_1 - T_2)r$.

Assuming the increase in length of NQ to be equal to the reduction in length of MP, i.e.

assuming the changes in the tensions to be the same, then, denoting such change by dT:

$$T_1 = T_0 + dT$$
and
$$T_2 = T_0 - dT$$
i.e.
$$T_1 + T_2 = 2T_0.$$

Thus the sum of the limiting tensions is fixed by the initial tension.

Now refer to Figure 231 and consider the element of belt on the stationary pulley B subtending an angle $d\theta$, subject to tensile forces T and $T+dT$ and about to slip in the direction of the latter.

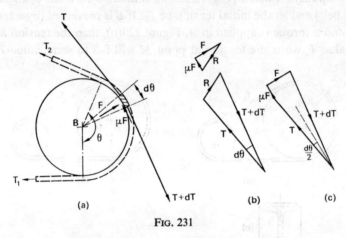

FIG. 231

The equilibriant of T and $(T+dT)$ is R, Figure 231(b), which itself is the resultant of the radial pulley reaction F and the force of friction μF between pulley and element, so that the complete force polygon is as shown in Figure 231(c).

Resolving parallel to F:

$$F = (T+dT)\sin\frac{d\theta}{2} + T\sin\frac{d\theta}{2}$$

$$= 2T\sin\frac{d\theta}{2} + dT\sin\frac{d\theta}{2}$$

$$= 2T\frac{d\theta}{2} + dT\frac{d\theta}{2}$$

since $d\theta$ is small.

The second term is the product of infinitely small quantities and may be neglected, so that

$$F = T.d\theta. \tag{1}$$

Resolving normal to F:

$$\mu F + T\cos\frac{d\theta}{2} = (T+dT)\cos\frac{d\theta}{2}$$

$$\mu F = dT\cos\frac{d\theta}{2}.$$

Since the angle $d\theta$ is infinitely small its cosine may be taken as unity, so that

$$\mu F = dT. \tag{2}$$

Hence $\quad \mu(T.d\theta) = dT \quad$ (substituting for F)

or $\quad \mu.d\theta = \dfrac{1}{T}.dT.$

The limits of $d\theta$ are 0 and θ (the angle of lap) while T varies round the pulley from T_1 to T_2. Thus, assuming μ constant and integrating both sides:

$$\mu \int_0^\theta d\theta = \int_{T_2}^{T_1} \dfrac{1}{T} dT$$

$$\mu \left|\theta\right|_0^\theta = \left|\log_e T\right|_{T_2}^{T_1}$$

$$\mu(\theta - 0) = \log_e T_1 - \log_e T_2$$

i.e. $\quad \mu\theta = \log_e \dfrac{T_1}{T_2}$

or $\quad \underline{\dfrac{T_1}{T_2} = e^{\mu\theta}.\ddagger}$

Thus, when the belt is on the point of slipping on the stationary pulley, the ratio of the tensions is fixed for a given angle of lap and increases with this angle. It is often referred to as the *limiting tension ratio*. Since slip will occur on the smaller pulley, the maximum value of θ in a simple drive is π. Note that the actual ratio of the tensions may have a lower value than the limiting one.

The relation between T_1 and T_2 also holds good when pulley B is rotating, i.e. when the belt is delivering power to it and is on the point of slipping. (For centrifugal effects see p. 260). It follows that the maximum power which can be absorbed by pulley B when driven at N rev/min is given by

$$\dfrac{2\pi N}{60} \times \text{possible torque}$$

or $\quad \dfrac{2\pi N(T_1 - T_2)r}{60}$

where $\quad T_1 = e^{\mu\theta} T_2.$

Alternatively, if the speed of the belt is u,

Possible input power = Work done/sec by belt
 = net tension × speed
 = $\underline{(T_1 - T_2)u.}$

If the angle of lap on the driven pulley exceeds 180° (i.e. π) then the above expressions give the possible output from the driving pulley for which, evidently, $\theta < 180°$.

‡ See Appendix 8 on p. 340.

The expression for the limiting tension ratio is also valid for a metal band when used for applying a braking torque to a rotating drum and for a wire rope when running on the bottom of a groove.

Note that, if the belt is crossed, the angle of lap is not only increased (enabling a greater torque to be transmitted) but is the same for each pulley so that, theoretically, slip occurs simultaneously on both. Crossing the belt reverses the direction of rotation of the driven pulley.

EXAMPLE. A mechanic wishes to lower a component having a mass of 165 kg from a trolley to the ground by means of a rope which he passes over a fixed horizontal pipe overhead. If he is to be able to control the motion with a force of 90 N or less on the free end of the rope, find the minimum number of times he must pass the rope round the pipe. Assume $\mu = 0.2$.

Solution.
$$\mu\theta = \log_e \frac{T_1}{T_2}$$

or
$$0.434\mu\theta = \log_{10} \frac{T_1}{T_2}$$

$\therefore$
$$\theta(0.434 \times 0.2) = \log_{10} \frac{165 \times 9.81}{90}$$

so that
$$\theta = \frac{\log_{10} 18}{0.868} = \frac{1.2553}{0.868} = 14.45 \text{ radians.}$$

Division by 2π gives a value of 2·3 turns. If the man were standing on the ground, the angle would probably be nearer to 2·5 turns.

EXAMPLE. Two parallel shafts 2·5 m apart are to be connected by a crossed belt for which the permissible load per mm width is 15 N. If the smaller pulley is to be driven at 500 rev/min and the nominal reduction ratio is to be 2·5 : 1, calculate, neglecting centrifugal effects, the width of belt required for the transmission of 24 kW and the necessary initial tension. Assume $\mu = 0.2$ and take the diameter of the smaller pulley as 0·3 m.

Solution. Referring to Figure 232,
$$R_2 = 2 \cdot 5 R_1 = 2 \cdot 5 \times 0 \cdot 15 = 0 \cdot 375 \text{ m}$$
$$\sin \alpha = \frac{0 \cdot 375 + 0 \cdot 15}{2 \cdot 5} = 0 \cdot 21$$
$$\alpha = 12 \cdot 6 \text{ deg} = 0 \cdot 22 \text{ rad.}$$

For each pulley,
$$\theta = \pi + 2\alpha$$
$$= 3 \cdot 14 + 0 \cdot 44$$
$$= 3 \cdot 58 \text{ rad.}$$

Speed of belt,
$$u = \omega_1 R_1 = \left(\frac{2\pi \times 500}{60}\right) 0 \cdot 15 = 7 \cdot 85 \text{ m/s.}$$

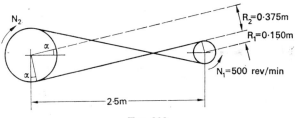

Fig. 232

Power transmitted $= (T_1 - T_2)u = 24\,000$

$\therefore \quad T_1 - T_2 = \dfrac{24\,000}{7\cdot 85} = 3060 \text{ N}.$

$\log_{10} \dfrac{T_1}{T_2} = 0\cdot 434\mu\theta = 0\cdot 434 \times 0\cdot 2 \times 3\cdot 58$

$= 0\cdot 311$

$\therefore \quad \dfrac{T_1}{T_2} = 2\cdot 05, \quad \text{i.e.} \quad T_1 = 2\cdot 05 T_2.$

Hence $\quad 2\cdot 05 T_2 - T_2 = 3060 \quad$ whence $\quad T_2 = 2910 \text{ N}$

and $\quad T_1 = 2910 \times 2\cdot 05$

$= 5970 \text{ N}.$

Required belt width $= \dfrac{5970}{15} = 398 \quad$ say 400 mm.

Initial tension, $\quad T_0 = \dfrac{T_1 + T_2}{2}$

$= \dfrac{5970 + 2910}{2}$

$= 4440 \text{ N}.$

Grooved Pulleys

In cases where a flat belt would be unable to cope with the power transmission requirements (say several hundred kW), a pulley having one or more grooves may be used in conjunction with either ropes or trapezoidal section belts, Figure 233.

Considering an element of belt, the reaction F of the pulley is equal to the sum of the radial components of the forces R exerted on the rope element by the sides of the groove, i.e. $\quad F = 2(R \sin \alpha)$

so that Normal force per side, $R = \dfrac{F}{2 \sin \alpha}$

Frictional force per side $= \mu R = \dfrac{\mu F}{2 \sin \alpha}$

and Total frictional force/element $= 2\left(\dfrac{\mu F}{2 \sin \alpha}\right) = \dfrac{\mu F}{\sin \alpha}.$

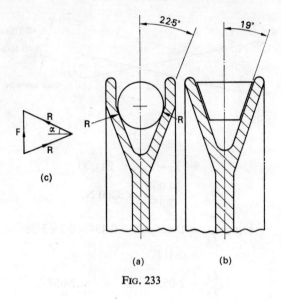

Fig. 233

Since the frictional force on an element of flat belt is μF it is evident that a rope resting on the sides of a groove having a semi-angle α is equivalent to a flat belt having a virtual coefficient of $\mu/\sin \alpha$ (which is greater than μ) so that in this case

$$\text{Limiting tension ratio,} \quad \frac{T_1}{T_2} = e^{\mu\theta/\sin \alpha}.$$

Thus the wedging effect of the groove increases the tension ratio (other things being equal) and makes possible the use of a relatively low initial tension.

Neglecting centrifugal effects, the power transmitted by a system having n ropes (or v-belts) is given by

$$P = n(T_1 - T_2)u.$$

The angle between the non-parallel sides of an unstrained standard belt section is 40° (see BS 1440 : 1962—Industrial Drives) and bending reduces this angle to somewhere near the usual pulley value of 38°.

Effects of Centrifugal Force

The centripetal force which is required to keep an element of belt in contact with the surface of a rotating pulley is provided by the radially inward component of what is known as the "centrifugal tension" T_c which is superimposed upon the "driving" tensions, the limiting values of which are denoted by T_1 and T_2. Thus, assuming slip to be imminent,

Tension on tight side, $\qquad T_t = T_1 + T_c$

and, Tension on slack side, $\qquad T_s = T_2 + T_c$.

While the belt is in contact with both pulleys it is justifiable to assume that, whatever the speed, the belt length and hence the mean tension $\frac{1}{2}(T_t + T_s)$ is constant and therefore equal to T_0.

Thus
$$T_0 = \frac{1}{2}(T_1+T_c+T_2+T_c)$$
$$= \frac{1}{2}\left(T_1+\frac{T_1}{e^{\mu\theta}}\right)+T_c$$

i.e.
$$T_0 = T_1\left(\frac{e^{\mu\theta}+1}{2e^{\mu\theta}}\right)+T_c.$$

Since T_0 is constant it is evident that, as T_c increases with speed, the available value of T_1 falls and with it the power transmission capacity.

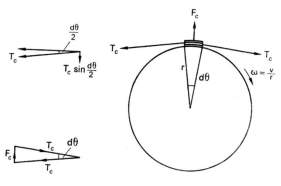

Fig. 234

Referring to Figure 234 and considering the forces (other than those producing rotation) on an element of belt subtending an angle $d\theta$ and having a mass per unit length of m,

Mass of element $= (r.d\theta)m$

Centrifugal force $= (r.d\theta)\,m\left(\dfrac{u^2}{r}\right)$

Centripetal force $=$ 2(radial component of T_c)
$$= 2\left(T_c \sin \frac{d\theta}{2}\right)$$
$$= 2T_c\left(\frac{d\theta}{2}\right) \text{ since } d\theta \text{ is small}$$
$$= T_c\,d\theta.$$

Equating centripetal to centrifugal force gives
$$T_c\,d\theta = (r.d\theta)\,m\left(\frac{u^2}{r}\right)$$

whence
$$T_c = mu^2.$$

This increases as the square of the speed and is independent of the diameter of the pulley.

Theoretical Speed for Maximum Safe Power

The safe value of T_t must be based on a knowledge of the properties of the belt material and the sum of T_1 and T_c must not exceed the value chosen. Since, however, the available value of T_1 is reduced as T_c increases, the tension difference (T_1-T_2) also falls. For this reason the power transmitted rises with speed only up to a certain point, after which it falls off.

Now
$$P = (T_1-T_2)u$$
where
$$T_2 = \frac{T_1}{e^{\mu\theta}}$$
i.e.
$$P = \left(T_1 - \frac{T_1}{e^{\mu\theta}}\right)u$$
$$= T_1\left(1 - \frac{1}{e^{\mu\theta}}\right)u.$$

The quantity in the brackets is constant. Denoting this by k gives

$$P = kuT_1$$
$$= ku(T_t - T_c)$$
where
$$T_c = mu^2$$
i.e.
$$P = kuT_t - kmu^3.$$

Since T_t is the arbitrary value decided upon for reasons of safety, durability and so on, the only variable in the above equation is the belt speed, u, so that if the expression is differentiated with respect to u and the result equated to zero, the value of u may be found at which the belt should be run to make the power a maximum.

Differentiating gives
$$\frac{dP}{du} = kT_t - km(3u^2).$$

At the given value of T_t the condition for maximum power is, therefore, that
$$kT_t = km(3u^2)$$
i.e. that
$$T_t = 3mu^2 = 3T_c.$$

Thus, in theory, maximum power is transmitted when the centrifugal tension has risen to one third of the figure laid down as the maximum permissible on the tight side. The belt speed which corresponds with this value of T_c is given by

$$u = \sqrt{\frac{T_c}{m}}$$

where
$$T_c = T_t/3.$$

It follows that, at this speed, $T_1 = \frac{2}{3}T_t$.

These deductions, though useful as a guide, should be treated with considerable reserve since they are based on the assumption that the initial tension has been adjusted to the value

given by
$$T_0 = \tfrac{1}{2}(T_t + T_s).$$

In practice the actual initial tension is easy neither to ascertain nor to maintain constant.

EXAMPLE. The V-section of a car fan belt is $130 \times 10^{-6} \text{m}^2$ and can withstand continuous operation at a stress of 3 MPa. The specific mass of the material is 1100 kg/m³, the groove angle is 38 deg, the effective pulley diameter is 100 mm and the coefficient of friction may be taken as 0·2. Find the theoretical maximum continuous power which may be delivered to the fan and the corresponding fan speed.

Solution.

Volume of belt per metre $\quad = 130 \times 10^{-6} \times 1 \cdot 0 = 0 \cdot 00013 \text{ m}^3$

Mass of belt per metre $\quad = 0 \cdot 00013 \times 1100 = 0 \cdot 143$ kg

Permissible maximum tension $\quad = 130 \times 10^{-6}(3 \times 10^6) = 390$ N

$\therefore$ Permissible centrifugal tension, $T_c = \tfrac{390}{3} = 130$ N

Corresponding belt speed, $u = \sqrt{\dfrac{T_c}{m}} = \sqrt{\dfrac{130}{0 \cdot 143}} = 30 \cdot 2$ m/s

Corresponding fan speed, $\omega = \dfrac{u}{r} = \dfrac{30 \cdot 2}{0 \cdot 05} = 604$ rad/s

$\therefore \qquad N = 604 \left(\dfrac{60}{2\pi} \right) = 5770$ rev/min.

Now, $\theta = \pi$ and $\alpha = 19$ deg, [so that $\dfrac{\mu \theta}{\sin \alpha} = \dfrac{0 \cdot 2\pi}{0 \cdot 326} = 1 \cdot 93$

$\therefore$ Limiting tension ratio, $\dfrac{T_1}{T_2} = e^{1 \cdot 93} = 6 \cdot 9$.

At the permissible maximum power, $T_1 = \tfrac{2}{3} \times 390 = 260$ N

so that $\qquad\qquad\qquad T_2 = \dfrac{260}{6 \cdot 9} = 37 \cdot 7$ N.

$\therefore$ Theoretical maximum power $= (260 - 37 \cdot 7)30 \cdot 2$

$\qquad\qquad\qquad\qquad\qquad = 6720$ W

$\qquad\qquad\qquad\qquad\qquad = 6 \cdot 72$ kW.

LUBRICATION

This is the general name given to the introduction of a film of fluid (lubricant) between two surfaces having relative motion, thereby reducing both the resistance to such motion and the heat generated. The continuous replacement of the film transfers about 85 per cent of the heat while the separation of the surfaces virtually eliminates wear.

In ball and roller bearings heat is generated by continuous external friction between the rolling elements and the cage and by intermittent internal friction due to repeated elastic straining of races and rolling elements. Although lubrication increases (marginally) the frictional torque, it prevents excessive rise in temperature and the attendant damage to highly polished surfaces. The presence of lubricant also goes some way towards protecting such surfaces against corrosion.

Effect of Temperature on Viscosity

Poiseuille showed in 1840 that the resistance offered by a capillary tube to the flow of water varies inversely as the temperature and that, at any temperature $T\,°C$, the dynamic viscosity is given by

$$\mu = \mu_0 \left(\frac{0\cdot0179}{1+0\cdot03368T+0\cdot000221T^2} \right) \text{ poises}$$

where μ_0 = viscosity at 0°C.

The viscosity of all lubricants varies in a similar manner, the lower the temperature the greater being the proportional variation.

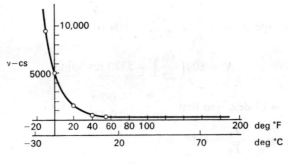

FIG. 235

Figure 235 shows how the kinematic viscosity varies with temperature for an SAE 20 lubricant as used in internal combustion motors, while Figure 236 is a commercial chart for a range of typical lubricants of the same family. Note that no numerical value of viscosity has any significance unless the temperature is specified.

The centistoke values ($cm^2/s \times 10^{-2}$) may be converted to SI (m^2/s) by dividing by 10^6.

With continued cooling, a point is reached at which the oil just ceases to flow. A temperature 5°F above this is defined as the *pour point* and is determined by cooling under prescribed conditions in a test apparatus.

The *cloud point* is the temperature at which visible insoluble wax is formed and, in the case of the paraffinic oils, is about 10°F above the pour point. Below this temperature the behaviour is no longer Newtonian.

At the other extreme, the *flash point* is the temperature at which the vapour pressure reaches the lower explosive limit and is determined by heating a test sample under prescribed conditions. Since each hydrocarbon constituent of an oil has its own vapour pressure, the volatility of the lightest element controls the flash point. It should be noted that the heaviest (least volatile) elements are relied upon to maintain the oil film on a hot surface such as a cylinder wall.

Finally, the viscosity of a mineral oil increases with pressure, the value at a given temperature being doubled (approximately) for every 35 MPa (5000 lbf/in²) increase. Such increase becomes of importance where surfaces are very heavily loaded, examples of this being the faces of gear teeth or cams. Compressibility of a fluid also becomes a factor to be considered where power is being transmitted via a fluid coupling and recent work has shown the existence of a close relation between compressibility and viscosity.

Viscosity Index (V.I.) Classification of Oils

This is an empirical number indicating the effect on kinematic viscosity of rise in temperature, a low index signifying a relatively large reduction and vice versa. By means of the V.I. function the slope of the v/T curve of the sample (v in centistokes and to a logarithmic scale) is interpolated between that of a highly paraffinic Pennsylvanian distillate (designated 100 V.I.) and that of a naphthenic Texas Coastal distillate (designated 0 V.I.) both of which so-called *reference oils* have the same viscosity at 210°F as the sample.

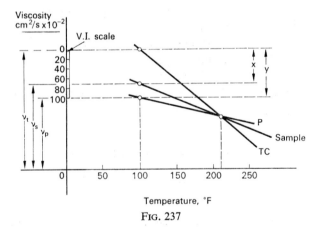

FIG. 237

Referring to Figure 237, if the viscosities at 100°F of Pennsylvanian (P), Texas Coastal (TC) and sample oils are denoted respectively by v_p, v_t and v_s then, by definition,

$$\text{Viscosity Index of sample} = \frac{x}{y} \times 100 = \left[\frac{v_t - v_s}{v_t - v_p}\right] 100.$$

The higher the index the less the reduction in viscosity with rise in temperature.

LUBRICANTS

Mineral oil was first struck in 1859 in Pennsylvania by a Colonel Drake and, in the refined (distilled) state of kerosine (known as paraffin in the UK), replaced colza or rapeseed oil as the principal lamp fuel. Further simple processing resulted in the production of more volatile fractions (e.g. petrol which, until the advent of the I.C. engine *circa* 1900, was wasted) and of a lubricating oil superior to and cheaper than the vegetable and animal oils it replaced. Although the vegetable oils—linseed, castor, rape, etc.—and animal oils (lard, sperm, neatsfoot, etc.) share the common disadvantage of instability in the presence of oxygen, especially at high temperature, their satisfactory performance in the early days was due in large measure to a molecular structure favouring the formation of a tenacious adsorbed layer on the strength of which the state of boundary lubrication depends. This is the only respect in which mineral oils are inferior to "fixed" oils, so called because they cannot be distilled.

Since a mineral oil is a blend of many hydrocarbons its performance under given operating conditions will be influenced by the behaviour of the constituents. As such conditions may demand more than just the ability to separate the surfaces and transfer the heat (the lower the efficiency of a machine, the more heat to be carried away) the actual performance may diverge from that predicted as a result of specified laboratory tests, i.e. the tests may not give an accurate picture of other qualities. A specification based on a limited number of physical and chemical characteristics, although laying down minimum standards, may therefore prove an unreliable guide to actual performance, may limit the possible sources of supply and may lead to the exclusion of a superior product. It is now sound practice to consult one of the major oil companies, each of which develops its products in collaboration with machinery manufacturers, particularly as, although a given oil may be satisfactory from the point of view of price, bearing temperature and annual consumption, its use may result in the formation of excessive fouling and consequent increased frequency of overhaul. In the last resort a lubricant must be judged according to the sum of running *and* maintenance costs over a considerable period.

Oil Contamination and Bearing Failure

When oil is in a state of fine dispersion in a heated atmosphere there is a pronounced oxidising effect. This may lead to

(a) the formation of acids, some of which are volatile and may corrode the bearing surfaces, the corrosion products then contaminating the oil; provision of adequate sump ventilation will minimise this;

(b) the precipitation of black deposits which may (by lodging in oilways, coolers and filters) interfere with the circulation so reducing heat transfer and contributing to eventual bearing failure.

Contamination of the lubricant by water (from leaks or condensation) leads to what is known as *emulsification*. This is the formation of an *emulsion* or intimate mixture of oil and water, the latter existing as particles so fine as to be in permanent suspension. Such emulsification is aggravated by the presence of dust.

[*Note:* An emulsion of milky appearance is also formed when minute particles of oil are suspended in water, e.g. the cooling fluid supplied to metal cutting tools. Emulsions are non-Newtonian in behaviour since their viscosity falls with increase in the rate of shear.]

Contamination of the lubricant by the products of incomplete combustion increases the viscosity and the quantity of sludge deposited, while the entry into the sump of the condensate of unburnt fuel reduces the viscosity and produces *thinning*.

Bearing failure in internal combustion engines may also be initiated by a fatigue crack resulting (usually) from the presence of foreign matter (dirt, swarf, etc.) between the steel back of the bearing shell and its housing. (The resulting lack of intimate contact between the two over the whole area affects the rigidity of the shell and the transfer of heat from it.) The crack develops into a small area of loose metal which, when eventually flushed away, leaves a shallow *pit*. The multiplication of such pits as the surface disintegrates increases the pressure on the remaining surface, so hastening the process until, eventually, hydrodynamic lubrication (see p. 269) can no longer be sustained. The onset of such failure is hastened by any of the types of contamination mentioned.

The arrangement of oil pipes is also of importance; anchorage to a hot casing (causing unnecessary heating) and discharge of the return above sump level (causing aeration and frothing) are both to be avoided.

Oiliness

Different liquids having the same dynamic viscosity offer different resisting torques at the same bearing under identical conditions. The lubricant offering the lowest resistance is said to be the most *oily*. The property of *oiliness* is responsible for the adherence to a metal surface (adsorption) of the *boundary layer* of molecules which is able to provide lubrication of a sort when there is no full separating film, e.g. at the start and finish of relative motion. The molecules of very stable fluids (e.g. medicinal paraffin) have little chemical affinity with a metallic surface and so provide negligible boundary lubrication. A tenacious boundary layer is also resistant to corrosion when machinery is at rest.

Additives

It was early discovered that, by adding as little as 5 per cent of fixed oil, the adsorption capacity (or oiliness) of a mineral oil could be made to approach that of the "additive". (Such blends—known as *compounded oils*—are still made for use in situations where boundary conditions are likely to occur.) A further consequence of such an addition to a mineral oil is a considerable reduction in its natural tendency to remain distinct from water so that, particularly where there is agitation, contamination by water will lead to the formation of a stubborn emulsion. Such an emulsion possesses reasonable lubricating properties and is often useful in unavoidably humid (e.g. marine) situations.

The early and successful attempts at modification of the natural properties of a lubricant have been followed by the development of a range of agents each designed, when incorporated in the oil, either to impart or improve some desirable property, or to inhibit or reduce some undesirable tendency. The main categories are as follows.

1. *Dispersant*. Hard carbon in the piston-ring region of I.C. engines is formed partly by the combustion of the cylinder wall oil film (which is composed of the least volatile elements and is renewed every revolution) and partly by the trapping of combustion soot behind the rings. Although the formation of such carbon is unavoidable it is possible to

modify its physical character by means of a dispersant (detergent) additive. This is adsorbed on to the insoluble carbon particles and, by keeping them in suspension, prevents their agglomeration (coagulation) or deposition as sludge. The oil becomes black but when drained off takes the carbon with it.

2. *Anti-oxidant*. The oxidation encouraged by high-temperature agitation under aerated conditions is aggravated by the presence of certain metals (such as copper) which act as catalysts, i.e. which assist oxidation. Such oxidation produces acidic petroleum oxides (which are corrosive) together with surface lacquer and sludge. The latter reduces heat transfer not only by choking filters and oilways but also by increasing viscosity.

With new lubricant there is an inception (or delay) period before oxidation begins seriously (the oxidation rate then increasing roughly as the square of the time) and one type of additive—called an oxidation inhibitor—lengthens this delay period by a factor of about four. Anti-oxidants proper may be designed either to interfere directly with the oxidation reaction or to counter the catalytic action already mentioned.

3. *Extreme pressure*. It may happen that, at points of extremely high pressure (e.g. on the nose of a cam), the oil film may be disrupted causing local high temperature. At such points the additive decomposes and releases the E.P. element which then reacts with the oxide layer on the bearing surface to form a film of inorganic compound having anti-welding properties.

4. *Corrosion inhibiting*. The water formed by the combustion of the hydrogen in a hydrocarbon (having a condensed volume more than equalling that of the fuel) contains acidic compounds which, as condensate, attack ferrous parts (e.g. cylinder walls) when cold. The additive surrounds the water globules and prevents contact until the temperature has risen to the point where water and acid are driven off as vapours.

Anti-corrosion agents may take the form either of organic acids preventing contact with water by preferential adsorption on sliding surfaces, or of alkaline substances able to neutralise, for example, the acids formed by the combustion of high sulphur-content diesel fuels.

Lubrication of a Plain Bearing

The behaviour of the lubricant in a simple bearing under load was investigated successfully (in 1882–5) by Beauchamp Tower who was commissioned by a "Committee on Friction" set up in 1878 by the Institution of Mechanical Engineers. He discovered that, when a shaft dipping into an oil bath (Fig. 238) was rotated, oil was carried up into the convergent space between shaft and bearing and separated them completely, as a result of which the frictional torque was nearly independent of the load. (In 1886 Reynolds presented to the Royal Society a mathematical analysis of Tower's work and showed that a necessary condition for the generation of pressure in fluid between sliding surfaces is that they must converge in the direction of motion.)

By measuring the pressure at each of a series of drilled holes, Tower also showed that the pressure distribution in the oil film varied as shown in Figure 239, the maximum pressure being more than six times the mean bearing pressure (i.e. than load/projected bearing area) and offset in the direction of rotation. This is known as *hydrodynamic lubrication* and, for it to be sustained oil must be supplied at or near to the position of minimum load, i.e. of maximum clearance.

If the load is applied, not to the bearing (as in Fig. 238) but to the shaft, then the latter adopts the attitude shown in Figure 240

When stationary the shaft (journal) is in contact with the bearing (bush) at point P and separated from it only by the boundary layer. If the clearance is fed with oil, then, as the

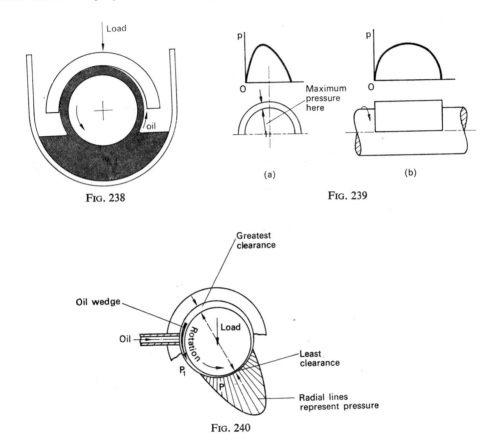

FIG. 238

FIG. 239

FIG. 240

shaft begins to rotate, it rolls up the bearing (against the direction of rotation) to some point P_1 at which slipping begins. Between the now moving boundary layer on the shaft and the similar (stationary) layer on the bearing, other fluid layers move in the direction of rotation (at intermediate velocities) into the convergent space, so lifting the shaft on a wedge-shaped film of oil. As soon as this film is complete the shaft adopts the attitude shown, the state now being one of hydrodynamic lubrication.

Due to the absence of metallic contact, hydrodynamic bearings under steady load have an almost unlimited life (i.e. there is negligible wear) and are practically noiseless. They can withstand shock loads, and have a load capacity which increases with speed.

During its passage through the bearing the oil rises in temperature (the viscosity falling) and transfers upwards of 85 per cent of the heat generated, the operating temperature being controlled by the rate of flow.

Resisting Torque at a Plain Bearing

If the shaft is unloaded it may be assumed concentric with the bearing, i.e. the thickness of the film of lubricant may be assumed uniform as shown in Figure 241.

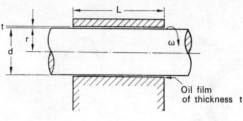

Fig. 241

Since the clearance, i.e. the film thickness, is very small, the flow will be laminar so that, as already shown,

$$\text{Shear stress, } q = \mu \frac{du}{dy}$$

where $du = \omega r$ and $dy = t$,

i.e.
$$q = \mu \frac{\omega r}{t}.$$

The area wetted by the film of fluid is πdL so that, neglecting end effects,

$$\text{Shear force at radius } r = \pi dL \left(\frac{\mu \omega r}{t} \right)$$

i.e.
$$\text{Resisting torque} = \pi dL \left(\frac{\mu \omega r}{t} \right) \times r$$

$$= \frac{\pi \mu \omega \, dr^2 L}{t} \quad \text{and} \quad r^2 = \frac{d^2}{4}$$

so that
$$T = \frac{\pi \mu \omega \, d^3 L}{4t}.$$

For a given set of conditions the amount of "lift" (i.e. the film thickness) and the load capacity increase with the viscosity. As T also increases with μ, an oil should be chosen which has just sufficient viscosity at its outlet temperature to float the shaft under load with a reasonable margin of safety. When the shaft is under load it is no longer concentric with the housing so that the expression for the resisting torque is then only approximately true.

EXAMPLE. A plain bearing 100 mm long supports a shaft 75 mm diameter, the total clearance being 0·076 mm. Estimate the resisting torque due to viscous friction at 3000 rev/min on light load given that, at the delivery temperature, the lubricant has a dynamic viscosity of 0·0372 kg/ms. Hence find the approximate power loss at the bearing.

Solution.

Angular velocity of shaft, $\omega = \dfrac{2\pi \times 3000}{60} = 314 \cdot 2$ rad/s

Film thickness, $t = \dfrac{1}{2}\left(\dfrac{0 \cdot 076}{1000}\right) = 38 \times 10^{-6}$ m

Viscosity of lubricant, $\mu = 0 \cdot 0372$ kg $\left(\dfrac{1}{\text{ms}}\right)$ or $\dfrac{\text{Ns}^2}{\text{m}}\left(\dfrac{1}{\text{ms}}\right)$ or $\dfrac{\text{Ns}}{\text{m}^2}$

and
$$T = \dfrac{\pi \mu \omega\, d^3 L}{4t}$$

$$= \dfrac{\pi(0 \cdot 0372) 314 \cdot 2 (0 \cdot 075^3) 0 \cdot 1}{4(38 \times 10^{-6})} \quad \left(\dfrac{\text{Ns}}{\text{m}^2}\right)\dfrac{\text{m}^4}{\text{ms}}$$

$\simeq 10$ Nm, assuming the viscosity not to change.

(In fact the viscosity falls during the passage of the oil through the bearing.)

Hence, Power absorbed, $P = T\omega$
$$= 10 \times 314 \cdot 2 \text{ W}$$
$$= 3 \text{ kW approx.}$$

About 15 per cent of this (in the form of heat) would travel along the shaft and through the bush, the remainder being absorbed by the lubricant.

Resisting Torque at a Collar Bearing

Let an axial load be applied to a shaft as shown in Figure 242 and let the thrust be taken by the annular surface $\pi(R_1^2 - R_2^2)$, assumed flat and parallel to the bearing surface.

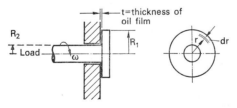

Fig. 242

Let the shear (viscous) stress at radius r be q, and assume the thickness of the oil film to be maintained constant. Since, within the radial distance, t, the fluid velocity rises from zero to ωr,

$$\dfrac{du}{dy} = \dfrac{\omega r}{t}$$

i.e.
$$q = \mu \frac{du}{dy} = \mu \frac{\omega r}{t}.$$

Tangential force on element $= q \times 2\pi r\, dr$
$$= \frac{2\pi\mu\omega}{t} r^2\, dr$$

Resisting torque on element $= \frac{2\pi\mu\omega}{t} r^2\, dr \times r$
$$= \frac{2\pi\mu\omega}{t} r^3\, dr$$

so that
$$T = \frac{2\pi\mu\omega}{t} \int_{R_2}^{R_1} r^3\, dr$$

i.e.
$$T = \frac{\pi\mu\omega}{2t}(R_1^4 - R_2^4).$$

$$\text{Power loss} = T\omega = \frac{\pi\mu\omega^2}{2t}(R_1^4 - R_2^4)\ \text{W}.$$

EXAMPLE. Estimate the heat transferred per minute by the lubricant supplied to a flat pivot 150 mm diameter when rotating at 600 rev/min given that the viscosity is 0·3 poise and that the clearance is maintained at 0·125 mm.

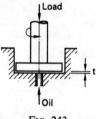

FIG. 243

Solution. Referring to Figure 243 and to the table on p. 176,

$\mu = 0.3 \times 0.1$
$\quad = 0.03$ kg/ms or Ns/m²,

$\omega = \dfrac{2\pi \times 600}{60}$

$\quad = 62.84$ rad/s

$R_2 = 0, \quad R_1 = 0.075$ m $= R$, say, and $t = 0,000125$ m.

Hence
$$T = \frac{\pi\mu\omega}{2t} R^4$$
$$= \frac{\pi(0.03)62.84}{2(0.000125)}(0.075^4) \quad \frac{\text{Ns}}{\text{m}^2} \frac{\text{m}^4}{\text{sm}}$$
$$= 0.75\ \text{Nm}.$$

Heat transferred per minute = 60 $(T\omega)$
= 60(0·75×62·84)
= 2830 J.

EXAMPLES 9

1. A rough casting having a mass of 2000 kg rests on parallel rails inclined at 20° to the horizontal. Calculate the force required to drag it up the incline by means of a rope acting away from the plane and inclined at 20° to it. Assume $\mu = 0\cdot25$. Check the result by drawing the force triangle (13 050 N).
2. At the outer end of a horizontal piece of string a mass of 2·27 kg is made to describe a circle of 610 mm radius, the speed of rotation being 18 rev/min. Calculate the radius of the circle described by the inner end of the string assuming $\mu = 0\cdot25$ (68·5 mm).
3. The slide of a bolt has a mass of 0·91 kg and is constrained by vertical guides. It rests on an actuating wedge having an included angle of $\tan^{-1} 0\cdot125$, the lower surface being horizontal. Assume $\mu = 0\cdot2$ for all surfaces and find the horizontal force required on the wedge (a) to raise, (b) to lower the slide (5·07 N, 2·4 N).
4. The saddle of a lathe has a mass of 254 kg and is traversed by a single start square thread lead screw of 12·7 mm pitch and 38 mm mean thread diameter. Assume that $\mu = 0\cdot1$ for both screw and lathe bed and estimate the torque required to rotate the screw. Neglect friction at the screw thrust collar (0·99 Nm).
5. Gas at a pressure of 1·38 MPa is controlled by means of a valve 50 mm diameter, the spindle of which is 38 mm external diameter and carries a square thread of 3·18 mm pitch. Estimate the torque required to close the valve taking $\mu = 0\cdot25$ and neglecting other sources of friction. Calculate also the efficiency of the screw (14·2 Nm, 10 per cent approx.).
6. A load of 227 kg is raised by rotating a nut on a square-threaded screw 19 mm pitch, 75 mm mean thread diameter, the screw axis being vertical. The bearing face of the nut is annular, the radii being 50 mm and 100 mm. Take $\mu = 0\cdot15$ for both bearing face and thread and find the torque required on the nut (46 Nm).
7. A vertical shaft assembly has a mass of 45 kg and ends in a flat pivot 50 mm diameter. Estimate the heat carried away per minute by the lubricant at 330 rev/min assuming the pressure to be uniform and taking $\mu = 0\cdot04$ (623 J).
8. 75 kW is to be transmitted at 4000 rev/min by a single-plate clutch using both sides of the friction ring. Assume $\mu = 0\cdot25$ and determine suitable ring diameters given that the pressure when fully engaged is not to exceed 104 kPa. Assume a diameter ratio of 1·5 (224 and 336 mm).
9. Part of a recording instrument has a mass of 0·017 kg and is supported by a vertical spindle 0·457 mm radius. The lower end of the spindle is conical, the cone angle being 120°. Assume $\mu = 0\cdot02$ and calculate:
(a) the normal pressure on the cone face (253 kPa);
(b) the frictional torque (1·15×10^{-6} Nm);
(c) the power loss at 12 rev/min (14·7×10^{-6} W).
10. Part of a machine tool having a moment of inertia of 0·678 kg m² and initially at rest is connected via a cone clutch to a motor rotating uniformly at 1480 rev/min. Assume uniform wear and estimate the duration of the engagement period and the heat produced during it, given that $\mu = 0\cdot35$. The cone angle is 30°, the engaging force is 225 N and the greatest and least diameters of the faces are 150 and 100 mm respectively (5·5 s, 8180 J).
11. The lining diameters of a multi-plate clutch are 140 and 90 mm and there are six active surfaces. Assume that wear is uniform and that $\mu = 0\cdot3$ and determine:
(a) the engaging force necessary for the transmission of 30 kW at 3000 rev/min (9350 N);
(b) the pressures at the inner and outer edges of the linings (65·5 and 41·4 kPa).
12. When the four shoes of a centrifugal clutch are at rest, the radial clearance between them and the drum is 6·35 mm, the radius of their mass centres is 200 mm and the spring force on each is 535 N. Assume a coefficient of friction of 0·3 and estimate the power which could be transmitted at 660 rev/min given that the mass per shoe is 7·25 kg, the spring stiffness is 42 kN/m and the drum diameter is 455 mm. Find, for the same speed, the effect on this power of 1·25 mm radial wear on the linings (120 kW, no change).
13. A centrifugal clutch is to engage at 500 rev/min and is to transmit 21 kW at 750 rev/min. If each of the four shoes is to have its mass centre at 25 mm from the contact surface (which is of 165 mm radius) calculate:
(a) the normal force between shoe and drum on full load (1610 N);
(b) the required mass per shoe (3·38 kg);
(c) the force exerted by the spring on the adjusting screw (1285 N).

If the spring stiffness is to be 168 kN/m find, assuming no adjustment, the reduction in the power which would result from 1·6 mm radial lining wear (3 kW).

14. A flat leather belt is to transmit 30 kW from a pulley 760 mm diameter when running at 500 rev/min, the angle of lap being 160°. Calculate the belt section necessary to limit the stress to 2·15 MPa taking $\mu = 0.3$ and $\varrho = 970$ kg/m³ (1.29×10^{-3} m²).

15. A pulley is to transmit 75 kW at 300 rev/min to an equal pulley by means of ropes 25 mm diameter. The grooves are to have an effective diameter of 915 mm and included angle of 45°. Estimate neglecting centrifugal effects, the minimum number of ropes required (9).

16. Show that, if μ is taken as 0·23, the approximate number of times that a rope must be passed round a cylindrical surface to give a limiting tension ratio of n is given by $N = (5/\pi) \log_{10} n$. A winch is to haul a train of 30 000 kg up an incline of 1 in 30. Find, using the above formula, the least value of N at the start of winding which will limit the force on the cable anchorage to 178 N. Assume a rolling resistance of 45 N for each 1000 kg of load (3).

17. A crossed belt is to connect two parallel shafts and run at 27 m/s, the common angle of lap being 200°. The belt is 200 mm wide and 7·5 mm thick and the material has a specific mass of 1000 kg/m³. Calculate the value of the centrifugal tension (1160 N). Estimate the greatest power which safely may be transmitted at this speed assuming a permissible stress of 1·72 MPa and taking $\mu = 0.2$ (20 kW).

18. At what speed will a flat belt 100 mm wide and having a mass per metre of 5·85 kg transmit maximum power if the permitted tension per mm width is 12 N? If the limiting tension ratio is 2·1, estimate the value of this maximum power (26·5 m/s, 11·5 kW).

19. A rope drive transmits 82 kW between equal pulleys having a groove angle of 40°. There are ten ropes and the mass per metre is 0·415 kg. If $\mu = 0.2$ and the rope speed at the given power is 15 m/s find the tension in each rope, assuming that the power is equally divided and that slip is imminent (740 and 205 N).

20. A flat pulley on which the angle of lap is 150° can absorb 10 kW when the belt speed is 20 m/s and the belt tension in the tight side is 1350 N. Find the tension in the slack side given that the belt has a mass per metre of 0·75 kg. If the angle of lap is increased by 60° (by means of a jockey pulley) find, for the same speed and tension in the tight side, the new value of the tension in the slack side. Hence find the increase in power which the pulley can absorb (840 N, 720 N, 2·5 kW). (N.B. It is not necessary to know the value of μ.)

21. A rope is given three complete turns round the drum of a powered capstan at an effective radius of 230 mm, and a force of 90 N is applied manually to the "slack" side. Assume a coefficient of friction of 0·3 between the drum and the rope when slipping and estimate the force which would be exerted on the "tight" side of the rope and the speed which would be imparted to a ship having a mass of 5×10^6 kg, starting from rest, if this force were applied to it for one minute. Find also the power required to drive the capstan at 50 rev/min (25·5 kN, 0·3 m/s, 30 kW).

22. The grooves on the pulleys of a multiple rope drive are to have an angle of 45° and are to accommodate ropes 38 mm diameter having a mass per metre of 1·04 kg for which a safe operating stress of 2·4 MPa has been laid down. The smaller pulley is to have a lap angle of 165° and is to rotate at 300 rev/min and the velocity ratio is to be 1·5. If the power to be transmitted is to be a maximum, find suitable diameters for the pulleys. If this maximum is to be 220 kW, find the least number of ropes needed. Take $\mu = 0.28$ (1·9 m, 2·84 m, 5).

23. A v-belt has a section of 645×10^{-6} m² and connects two equal pulleys 300 mm effective diameter. Find the power which can be transmitted between the pulleys at 1000 rev/min if the permissible stress is 4·13 MPa and $\mu = 0.15$. The groove angle is 45° and the mass per metre of belt may be taken as 0·75 kg (27 kW).

24. A rope drive is to transmit 600 kW from a pulley 3·8 m effective diameter when rotating at 90 rev/min. If the groove angle is 45°, $\mu = 0.28$, the allowable tension per rope is 2200 N and the belt has a mass per metre of 1·35 kg, find the least necessary number of ropes (21).

25. Explain the meaning of the following terms:
    Boundary layer, velocity gradient, emulsification, centipoise, cloud point, viscosity index, oxidation.

26. At atmospheric pressure and 0°C the dynamic viscosity of water is 1·793 cP. Express this value in
    (a) lbf s/ft² (37.5×10^{-6});
    (b) gf s/cm² (18.3×10^6).

27. Describe how and explain why the position of a loaded shaft changes relative to the bearing as it increases speed from rest, given that the bearing is fed correctly with lubricant.

28. A shaft having a nominal diameter of 75 mm is supported in a plain bearing 75 mm long, the clearance being 0·127 mm. Estimate the power absorbed at 700 rev/min if the lubricant has a dynamic viscosity of 0·056 kg/ms (420 W).

TRIBOENGINEERING

29. Comment on the following statements:
   (a) The upper temperature limit is about 60°C if a lubricant is to withstand continuous use for several years.
   (b) A specification based upon a limited number of physical characteristics is an unreliable guide to the suitability of an oil for a particular duty.
   (c) Since a mineral oil is a blend of many hydrocarbons its performance under given operating conditions will be influenced by the behaviour of the constituents.
   (d) No numerical value of viscosity has any significance unless the corresponding temperature is specified.
30. A single plain collar bearing has diameters of 120 mm and 160 mm and the oil supply is such as to maintain under load an oil film 0·3 mm thick. Estimate the power loss at 1250 rev/min if the lubricant has a mean dynamic viscosity of 120 cP (112 W).

10

VIBRATION

Principles of Kinematic Constraint

The position in space of a point on a body is defined, usually, by reference to three mutually perpendicular axes (called a *frame of reference*) say XX, YY and ZZ, Figure 244. Their common intersection is called the *origin* and the distances to the point are measured from this origin along each axis. If the *orientation* of the body is also stated (relative to the same axes) then its position in space is also defined.

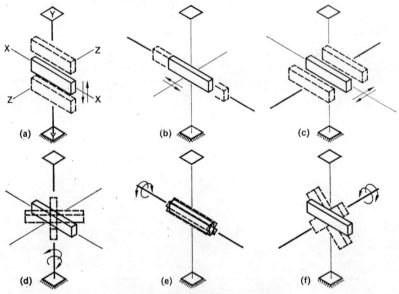

Full line represents equilibrium position
Dotted lines represent extremes of vibration

FIG. 244

The motion of the body in space may consist of components of linear velocity (translation) along each axis together with components of angular velocity (rotation) about each axis, each of these possibilities corresponding with what is known as a *degree of freedom*. Any circumstance which nullifies a degree of freedom is called a *constraint* so that, for a body to be fixed (rigid) relative to the frame of reference (i.e. all motion prevented) a total of six constraints must be provided.

276

VIBRATION

If the frame of reference is anchored to the earth (an effectively "infinite" mass) and the body is mounted upon it in such a way that the restoring force (or torque) is proportional to the linear (or angular) displacement from the equilibrium position (linear or angular) for each axis—i.e. there is an elastic constraint for each degree of freedom—then six "modes" of simple harmonic motion (vibration) are possible, one about each equilibrium position. These six modes are illustrated in Figure 244.

The body may also vibrate in any combination of these modes, e.g. the motions of Figures 244(a) and 244(d) may occur simultaneously. In this case, if the frequencies are equal, the path in space of any point is evidently helical.

Considering the axis YY, the constraint acts in a torsional manner in Figure 244(d), in a longitudinal (or axial) manner in Figure 244(a) and in a transverse (or flexural) manner in Figures 244(b) and 244(c). In this chapter will be considered torsional, longitudinal and transverse vibrations in systems having one degree of freedom only.

Free and Forced Vibration

When part of an engineering component is forcibly deflected, i.e. displaced from its equilibrium position relative to its frame of reference (the elastic limit being not exceeded), the work done by the force in producing the deflection is stored as elastic strain energy. If the deflecting force is removed suddenly, the component begins to resume its original shape under the action of the restoring force and at the instant it achieves this the whole of the strain energy has been converted into kinetic energy. The motion continues therefore beyond the equilibrium position until this kinetic energy has been reconverted into strain energy. Thus, although the external force has ceased to act, an unhindered or "free" vibration has been set up. Further, since the restoring force is proportional to the displacement (as is always the case with elastic bodies) the resulting acceleration is also, showing that the motion is simple harmonic. This motion occurs at what is called the *natural frequency* or at various multiples of it. (In practice the motion is hindered by the surrounding medium and by molecular friction so that the amplitude of the vibration is progressively reduced—or "damped"—until the component comes to rest in its equilibrium position.)

If the component is subjected to a harmonic deflecting force, the resulting vibration is said to be *forced* and the frequency is, necessarily, that of the disturbance. As this *exciting frequency* approaches any of the natural values, the amplitude increases, the component being said to *resonate* when this frequency coincides with any natural value. If the damping forces are small enough, the amplitude during such a state of resonance may become large enough to bring about rapid failure; quite small amplitudes may be, if prolonged, sufficient to cause failure in fatigue. (Besides which, any resonance is usually accompanied by unacceptable noise.) Evidently a system should be arranged to have, if possible, a natural frequency appreciably different from that of any harmonic exciting force to which it might be subjected.

Mathematical Equation for Linear s.h.m.

Let a radius OP, Figure 245, rotate at ω rad/s about point O so that point Q (i.e. the projection of P) executes s.h.m. along the horizontal diameter. OP is called a *generating vector*.

Suppose that at any instant t sec after passing point P_0 the generating vector makes an angle θ with OP_0. Then $\theta = \omega t$. Then, from the centre of oscillation (or equilibrium position),

$$\text{Displacement,} \quad x = r\cos\theta = r\cos\omega t.$$

(This is negative when ωt lies between 90° and 270°.)

$$\text{Velocity,} \quad v = \frac{dx}{dt} = -\omega^2 r \sin\omega t$$

$$\text{Acceleration,} \quad f = \frac{d^2x}{dt^2} = -\omega^2 r\cos\omega t = -\omega^2 x.$$

(This is positive when x is negative, i.e. when ωt lies between 90° and 270°. Thus f is proportional to x and is of opposite sense (i.e. directed towards O) so that the motion is, by definition, simple harmonic.)

Transposing the last equation gives what is known as the differential equation of s.h.m., viz.

$$\frac{d^2x}{dt^2} + \omega^2 x = 0.$$

If the point Q is replaced by a body of mass M and this is mounted elastically, then the acceleration corresponding with a displacement x (i.e. $-\omega^2 x$) is imparted by the restoring force consequent upon such displacement. If the stiffness (restoring force per unit displacement) is denoted by λ, then this force is $-\lambda x$. (It must have the same sign as the acceleration it produces, i.e. it is always towards the equilibrium position.)

Since Restoring force = Mass × linear acceleration,

$$\therefore \quad -\lambda x = M(-\omega^2 x)$$

$$\text{i.e.} \quad \omega^2 = \frac{\lambda}{M}.$$

Thus if λ is known, the value of ω can be found for a given vibrating mass. The periodic time is then obtainable from

$$T = \frac{2\pi}{\omega} \text{ sec/cycle.}$$

The frequency (in cycles/sec) is the reciprocal of this, i.e.

$$n = \frac{1}{T} = \frac{1}{2\pi}\sqrt{\frac{\lambda}{M}} \text{ c/s.}$$

[*Note:* The unit of frequency (one cycle per second) will be denoted in future by the abbreviation Hz, after Heinrich Hertz, 1857–94.]

Mathematical Equation for Angular s.h.m.

This is identical in form to that for linear s.h.m. If the angular displacement from an equilibrium position is θ at t sec after passing that position, then

$$\frac{d^2\theta}{dt^2} + \omega^2\theta = 0.$$

(In this case it is impossible to represent the generating vector.)

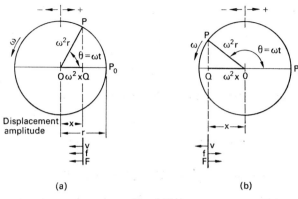

Fig. 245

If I is the moment of inertia of the body about the axis on which it is mounted elastically, then the angular acceleration corresponding with a displacement θ (i.e. $-\omega^2\theta$) is imparted by the restoring torque consequent upon such displacement. If the torsional stiffness (restoring torque per unit angular displacement) is denoted by q, then this torque is $-q\theta$. (It must have the same sign as the acceleration it produces, i.e. it is always towards the equilibrium position.)

Since Restoring torque = Moment of inertia × angular acceleration,

$$-q\theta = I(-\omega^{12}\theta)$$

i.e.

$$\omega^2 = \frac{q}{I}.$$

Thus if q is known, the value of ω can be found for a given vibrating mass. The periodic time and frequency are given by

$$T = \frac{2\pi}{\omega} \text{ sec}$$

and

$$n = \frac{\omega}{2\pi} = \frac{1}{2\pi}\sqrt{\frac{q}{I}} \text{ Hz.}$$

Note that in the case of linear s.h.m.

$$\text{Stiffness} \times x = \text{Mass} \times f$$

i.e.

$$\frac{f}{x} = \frac{\text{stiffness}}{\text{mass}} = \frac{\lambda}{M} = \omega^2,$$

so that

$$n = \frac{1}{2\pi}\sqrt{\frac{f}{x}}.$$

Similarly, for angular s.h.m.

$$\text{Stiffness} \times \theta = \text{Moment of inertia} \times \alpha$$

i.e.

$$\frac{\alpha}{\theta} = \frac{\text{stiffness}}{\text{moment of inertia}} = \frac{q}{I} = \omega^2,$$

so that
$$n = \frac{1}{2\pi}\sqrt{\frac{\alpha}{\theta}}.$$

In general therefore,
$$n = \frac{1}{2\pi}\sqrt{\frac{\text{acceleration}}{\text{displacement}}}.$$

Various applications of the foregoing will now be considered.

TORSIONAL VIBRATION

Two-mass System, One Mass being Finite

Imagine a disk of finite mass to be attached by an elastic uniform shaft to a fixed anchorage (e.g. to the earth, a relatively "infinite" mass), as shown in Figure 246. The application to the disk of a torque, T, will (from the torsion equation) produce an angular displacement given by

$$\theta = \frac{TL}{GJ}.$$

Hence, Shaft stiffness,
$$q = \frac{T}{\theta} = \frac{GJ}{L}$$

where
$$J = \frac{\pi d^4}{32}.$$

If the disk is released it will have an angular acceleration given by

$$\alpha = \frac{T}{I}$$

where I = polar moment of inertia of disk,
T = restoring torque = $GJ\theta/L$.

Hence:
$$\frac{\text{Acceleration}}{\text{Displacement}} = \frac{\alpha}{\theta} = \frac{T}{I}\frac{GJ}{TL} = \frac{GJ}{IL} = \text{a constant, say } k.$$

Since $\alpha = k\theta$, the angular motion after release is simple harmonic, the frequency being given by

$$n = \frac{1}{2\pi}\sqrt{\frac{GJ}{IL}} \text{ Hz.}$$

$$\left(\text{Alternatively,} \quad n = \frac{1}{2\pi}\sqrt{\frac{q}{I}} \quad \text{where} \quad q = \frac{GJ}{L}.\right)$$

Two-mass System, Both Masses Finite

If the "fixed" anchorage of Figure 246 is replaced by a second disk, the system becomes that shown in Figure 247. Imagine the disks to have polar moments of inertia I_1 and I_2 and to be given simultaneous angular displacements in opposite directions (so twisting the shaft) and then to be released simultaneously. At the instant of release the angular velocity

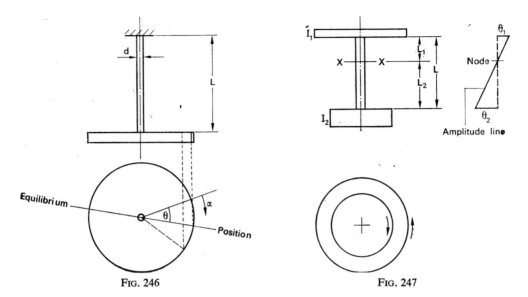

Fig. 246   Fig. 247

of all parts is zero so that the system then possesses zero angular momentum. In the absence of an externally applied torque, this momentum must remain zero so that the angular momenta of the disks (each in association with part of the shaft) must be equal and opposite at all times, i.e. the disks must move at all times in opposite directions. Now angular momentum is the product of moment of inertia and angular velocity so that if the displacements from the equilibrium position at time t are, respectively, θ_1 and θ_2, then

$$I_1 \frac{d\theta_1}{dt} = -I_2 \frac{d\theta_2}{dt}$$

i.e. $\quad I_1\theta_1 = -I_2\theta_2$ (integrating with respect to time)

so that
$$\frac{\theta_1}{\theta_2} = -\frac{I_2}{I_1} = \text{a constant.}$$

For the ratio of the angular displacements to be constant, it is necessary for some section of the shaft to have zero amplitude, i.e. to remain at rest. In Figure 247 this section—called a *node*—is denoted by XX.

It follows that, relative to the node, each part of the shaft behaves as though it were anchored to an infinite mass. Hence the frequency of I_1 is given by

$$n_1 = \frac{1}{2\pi}\sqrt{\frac{GJ}{I_1 L_1}}$$

and the frequency of I_2 is given by

$$n_2 = \frac{1}{2\pi}\sqrt{\frac{GJ}{I_2 L_2}}.$$

Since $n_1 = n_2$ ($= n$ say), it follows that

$$\frac{1}{I_1 L_1} = \frac{1}{I_2 L_2}$$

MECHANICAL TECHNOLOGY FOR HIGHER ENGINEERING TECHNICIANS

i.e. that
$$L_1 = \frac{I_2}{I_1} L_2 = \frac{I_2}{I_1}(L-L_1) = \frac{I_2}{I_1} L - \frac{I_2}{I_1} L_1$$

$$\therefore \quad L_1 + \frac{I_2}{I_1} L_1 = \frac{I_2}{I_1} L$$

whence
$$L_1 = \left(\frac{I_2}{I_1+I_2}\right) L.$$

This gives the distance of the node from I_1. Substituting for L_1 in the expression for the frequency:

$$n = \frac{1}{2\pi} \sqrt{\frac{GJ}{I_1 L_1}}$$

$$= \frac{1}{2\pi} \sqrt{\frac{GJ}{I_1}\left(\frac{I_1+I_2}{LI_2}\right)}$$

$$= \frac{1}{2\pi} \sqrt{\frac{GJ}{L}\left(\frac{I_1+I_2}{I_1 I_2}\right)} \quad \text{and} \quad \frac{GJ}{L} = q$$

so that
$$n = \frac{1}{2\pi} \sqrt{q\left(\frac{I_1+I_2}{I_1 I_2}\right)}.$$

The amplitude of the vibration is evidently proportional to the distance from the node and may be represented, as shown in Figure 247, by the distance between the axis and a straight line passing through the node.

Torsionally Equivalent Length

If a torque, T, is applied to a uniform solid shaft as shown in Figure 248(a), the resulting angle of twist will be given by

$$\theta = \frac{TL}{GJ}$$

where
$$J = \frac{\pi d^4}{32}$$

i.e.
$$\theta = \frac{32T}{\pi G}\left(\frac{L}{d^4}\right).$$

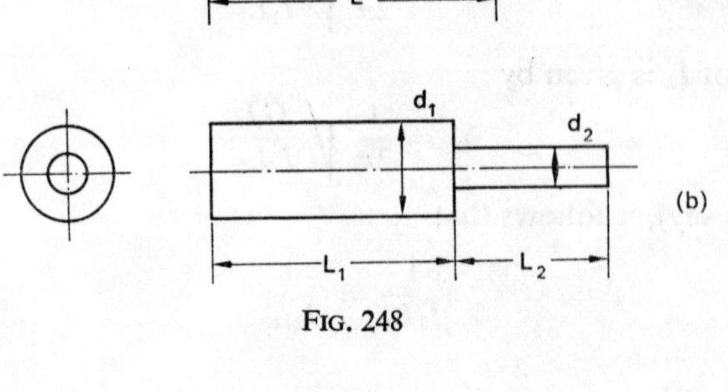

Fig. 248

282

VIBRATION

If the shaft is stepped as shown in Figure 248(b) and the same torque applied, the total twist will be given by

$$\theta = \theta_1 + \theta_2 = \frac{32T}{\pi G}\left(\frac{L_1}{d_1^4} + \frac{L_2}{d_2^4}\right).$$

If the overall twist is to be the same in the two cases, i.e. the shafts are to be *torsionally equivalent*, then

$$\frac{L}{d^4} = \frac{L_1}{d_1^4} + \frac{L_2}{d_2^4}$$

i.e.
$$L = L_1\left(\frac{d}{d_1^4}\right) + L_2\left(\frac{d}{d_2^4}\right).$$

This gives the value of L (i.e. the length of uniform shaft) which is torsionally equivalent to the stepped shaft. If, for convenience, an equivalent shaft of diameter d_1 is required, then writing d_1 for d we obtain

$$L = L_1 + L_2\left(\frac{d_1}{d_2}\right)^4.$$

This expression may be extended to cover a third length of shaft of diameter d_3, viz.

$$L = L_1 + L_2\left(\frac{d_1}{d_2}\right)^4 + L_3\left(\frac{d_1}{d_3}\right)^4 \quad \text{and so on.}$$

EXAMPLE. Figure 249 represents a motor-generator set. If $I_g = 6\cdot 55$ kg m² and $I_m = 21\cdot 0$ kg m², estimate the natural frequency of torsional vibration assuming $G = 80\,000$ MPa. Neglect the inertia of the shafts and coupling.

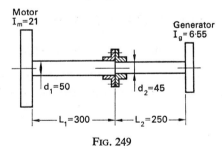

FIG. 249

Solution.

Equivalent length of shaft of diameter d_1

$$L = L_1 + L_2\left(\frac{d_1}{d_2}\right)^4 = 300 + 250\left(\frac{50}{45}\right)^4 = 300 + 381 = 681 \text{ mm.}$$

The assembly reduces therefore to that shown in Figure 250.

283

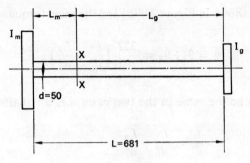

Fig. 250

Distance of node, XX, from motor,

$$L_m = \left(\frac{I_g}{I_m + I_g}\right)L$$

$$= \left(\frac{6 \cdot 55}{21 + 6 \cdot 55}\right)681$$

$$= 162 \text{ mm.}$$

Polar second moment, $\quad J = \dfrac{\pi}{32}(d_1)^4 = \dfrac{\pi}{32}(0 \cdot 05)^4 = 0 \cdot 613 \times 10^{-6} \text{ m}^4$

Natural frequency, $\quad n = \dfrac{1}{2\pi}\sqrt{\dfrac{GJ}{I_m L_m}}$

$$= \frac{1}{2\pi}\sqrt{\frac{(80\,000 \times 10^6)}{21 \times 0 \cdot 162}\left(\frac{0 \cdot 613}{10^6}\right)}$$

$$= 19 \cdot 1 \text{ Hz.}$$

FLEXURAL VIBRATION

Transverse Vibration of a Uniform Shaft

Let an element of shaft, dx (or of uniformly loaded beam), Figure 251, vibrate with amplitude a about its statically deflected position with frequency n. Then if m = mass/unit length,

Maximum acceleration of element $= \omega^2 a$

where $\quad\quad\quad\quad\quad\quad\quad \omega = 2\pi n,$

and Maximum inertia force $= (m\,dx)\omega^2 a$

$$= (\text{a constant}) \times ma.$$

Thus the dynamic loading is proportional to the product ma and not, as with static loading, proportional to m. This means that the profile during vibration is different from the static deflection curve. Now the intensity of dynamic loading (i.e. the inertia force per unit length)

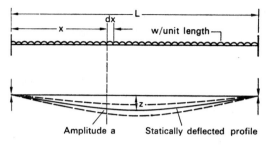

Fig. 251

is $(m)\omega^2 a$ so that from the theory of simple bending

$$(m)\omega^2 a = EI \frac{d^4 a}{dx^4},$$

or, rearranging,

$$\frac{d^4 a}{dx^4} - (m)\left(\frac{\omega^2}{EI}\right) a = 0$$

whence, putting $\quad (m)\dfrac{\omega^2}{EI} = k^4: \quad \dfrac{d^4 a}{dx^4} - k^4 a = 0.$

The solution to this equation is

$$a = A \cos kx + B \sin kx + C \cosh kx + D \sinh kx. \tag{1}$$

Successive differentiation gives

$$\frac{da}{dx} = -kA \sin kx + kB \cos kx + kC \sinh kx + kD \cosh kx \tag{2}$$

$$\frac{d^2 a}{dx^2} = -k^2 a \cos kx - k^2 B \sin kx + k^2 C \cosh kx + k^2 D \sinh kx \tag{3}$$

Since $a = 0$ when $x = 0$, substitution of these values in eq. (1) shows that $C = -A$. But, since $d^2 a/dx^2 = 0$ when $x = 0$, a similar substitution in eq. (3) shows that $C = +A$. It follows that both C and A must be zero. Again, since $a = 0$ when $x = L$, substitution of these values in eq. (1) shows that $D \sinh kL = -B \sin kL$. But, since $d^2 a/dx^2 = 0$ when $x = L$, a similar substitution in eq. (3) shows that $D \sinh kL = +B \sin kL$ also. Hence

$$2D \sinh kL = 0.$$

Since $\sinh kL$ cannot be zero, it follows that $D = 0$.

Hence, from above, $\quad B \sin kL = 0$

i.e. $\quad\quad\quad\quad \sin kL = 0 \quad (\text{since } B \neq 0)$

so that $\quad\quad\quad\quad kL = \pi, \ 2\pi, \ 3\pi \ \text{ and so on.}$

Now $\quad\quad\quad\quad (m)\dfrac{\omega^2}{EI} = k^4$

so that $\quad\quad\quad\quad \omega^2 = k^4\left(\dfrac{EI}{m}\right).$

285

The minimum possible (or *fundamental*) frequency corresponds with the lowest possible value of ω and this, in turn, corresponds with the lowest of the many values of k, namely π/L.

In this case, therefore, $\quad \omega^2 = \left(\dfrac{\pi}{L}\right)^4 \left(\dfrac{EI}{m}\right) = \pi^4 \left(\dfrac{EI}{mL^4}\right) = \pi^4 g \left(\dfrac{EI}{wL^4}\right)$

where $\quad\quad\quad\quad w = mg =$ weight per unit length.

Now, for a simply supported beam weighing w per unit length, the static deflection at mid-span is given by

$$\boxed{z = \dfrac{5}{384}\left(\dfrac{wL^4}{EI}\right)}$$

so that $\quad\quad\quad\quad \dfrac{EI}{wL^4} = \dfrac{5}{384}\left(\dfrac{1}{z}\right).$

Substitution of this value in the previous equation gives

$$\omega^2 = \left(\dfrac{5\pi^4}{384}\right)\dfrac{g}{z} = 1\cdot 267\left(\dfrac{g}{z}\right).$$

Hence $\quad\quad\quad\quad \omega = \sqrt{\left(1\cdot 267\dfrac{g}{z}\right)}$

and $\quad\quad\quad\quad n = \dfrac{1}{2\pi}\sqrt{\left(1\cdot 267\dfrac{g}{z}\right)}.$

Normal Modes of Vibration

Vibration as shown in Figure 251 is described as the *1st Normal Mode* and this corresponds with $k = \pi/L$. With suitable initiation it is possible for the shaft to vibrate in a number of other ways (two of which are shown in Fig. 252), all parts having the same frequency and being in the same or opposite phase. Figure 252(a) represents what is known as the *2nd*

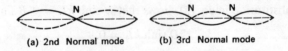

(a) 2nd Normal mode   (b) 3rd Normal mode

FIG. 252

Normal Mode and this corresponds with $kL = 2\pi$, i.e. with $k = 2(\pi/L)$. Similarly, when $k = 3(\pi/L)$ the vibration is represented by Figure 252(b) and called the *3rd Normal Mode*.

Since, in general, $\quad\quad\quad\quad \omega^2 = k^4\left(\dfrac{gEI}{w}\right),$

$\therefore \quad\quad\quad\quad 2\pi n = k^2 \sqrt{\left(\dfrac{gEI}{w}\right)}.$

The general expression for the natural frequency is, therefore,

$$n = k^2 \frac{1}{2\pi} \sqrt{\left(\frac{gEI}{w}\right)}$$

where $\quad k^2 = \left(\frac{\pi}{L}\right)^2, \quad 4\left(\frac{\pi}{L}\right)^2, \quad 9\left(\frac{\pi}{L}\right)^2,$ etc.

Substitution of successive values of k^2 gives

for 1st Normal Mode, $\quad n_1 = \left(\frac{\pi}{L}\right)^2 \frac{1}{2\pi} \sqrt{\left(\frac{gEI}{w}\right)},$

for 2nd Normal Mode, $\quad n_2 = 4\left(\frac{\pi}{L}\right)^2 \frac{1}{2\pi} \sqrt{\left(\frac{gEI}{w}\right)} = 4n_1,$

for 3rd Normal Mode, $n_3 = 9n_1$ and so on.

Thus the frequency increases rapidly with the number representing the mode while the amplitude falls in like manner. Note that, excluding the ends, the number of nodes is one less than the number representing the mode.

Shaft with Concentrated Loads—Approximate (Energy) Method

Let a uniform shaft carry a concentrated load W_1 producing a static deflection y_1 (Fig. 253) and let the amplitude at this point be a_1. Neglecting the inertia of the shaft and

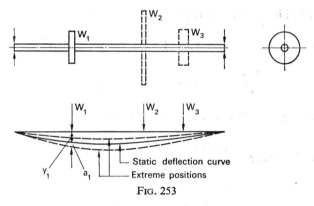

Fig. 253

the rotation of the disk about a horizontal diameter during the vibration, and assuming that the shape of the shaft during vibration is identical to the statically deflected shape (i.e. that $a = Cy$, where C is some constant), then

Additional load to produce $a_1 =$ (Load/unit deflection)a_1

$$= \left(\frac{W_1}{y_1}\right)a_1 \quad \text{where} \quad a_1 = Cy_1$$

Additional strain energy $= \frac{1}{2}\left(\frac{W_1}{y_1}\right)a_1 \times a_1$

$$= \frac{C^2}{2}(W_1 y_1) \quad \text{since} \quad a_1^2 = C^2 y_1^2.$$

The maximum velocity (and hence the maximum kinetic energy) occurs at the equilibrium position and is given by

$$u_{max} = \omega a_1$$
$$= 2\pi n a_1$$

where n = frequency.

$$\text{Maximum kinetic energy} = \frac{1}{2}\left(\frac{W_1}{g}\right)(2\pi n a_1)^2$$
$$= \frac{C^2}{2}\left(\frac{4\pi^2 n^2}{g}\right)(W_1 y_1^2)$$

since $a_1^2 = C^2 y_1^2$.

Equating strain and kinetic energies:

$$\frac{C^2}{2}\left(\frac{4\pi^2 n^2}{g}\right)(W_1 y_1^2) = \frac{C^2}{2}(W_1 y_1)$$

whence
$$4\pi^2 n^2 = \frac{g}{y_1}$$

i.e.
$$n = \frac{1}{2\pi}\sqrt{\frac{g}{y_1}}.$$

If there are additional loads W_2 and W_3 (shown dotted in Figure 253) then

$$\text{Additional strain energy} = \frac{C^2}{2}(W_1 y_1) + \frac{C^2}{2}(W_2 y_2) + \frac{C^2}{2}(W_3 y_3)$$
$$= \frac{C^2}{2}\Sigma(Wy).$$

Note that y_1, y_2 and y_3 are the deflections under each load due to all the loads acting together. Also

$$\text{Maximum kinetic energy} = \frac{C^2}{2}\left(\frac{4\pi^2 n^2}{g}\right)(W_1 y_1^2 + W_2 y_2^2 + W_3 y_3^2)$$
$$= \frac{C^2}{2}\left(\frac{4\pi^2 n^2}{g}\right)\Sigma(Wy^2).$$

Equating strain and kinetic energies:

$$\frac{C^2}{2}\left(\frac{4\pi^2 n^2}{g}\right)\Sigma(Wy^2) = \frac{C^2}{2}\Sigma(Wy)$$

i.e.
$$\frac{4\pi^2 n^2}{g} = \frac{\Sigma(Wy)}{\Sigma(Wy^2)}$$

whence
$$n = \frac{1}{2\pi}\sqrt{\left(g \cdot \frac{\Sigma(Wy)}{\Sigma(Wy^2)}\right)}.$$

Note that the effect of the inertia of the shaft itself can be allowed for, approximately, by including a proportion of its weight in each load.

VIBRATION

Dunkerley's Empirical Equation

This states that, if concentrated loads $W_1 W_2 W_3$ and so on produce static deflections (when acting *individually*) of $y_1 y_2 y_3$ and so on, corresponding with natural frequencies of $n_1 n_2 n_3$ and so on, when mounted on a uniform shaft of weight W_s, the maximum deflection and frequency of which are respectively y_s and n_s, then the natural frequency, n, of the system is obtainable from the equation

$$\frac{1}{n^2} = \frac{1}{n_1^2} + \frac{1}{n_2^2} + \frac{1}{n_3^2} + \ldots + \frac{1}{n_s^2}.$$

Now
$$n_s = \frac{1}{2\pi}\sqrt{\left(1{\cdot}267\,\frac{g}{y_s}\right)}$$

so that
$$\frac{1}{n_s^2} = \left(\frac{4\pi^2}{1{\cdot}267 g}\right) y_s$$

and
$$n_1 = \frac{1}{2\pi}\sqrt{\frac{g}{y_1}}$$

so that
$$\frac{1}{n_1^2} = \left(\frac{4\pi^2}{g}\right) y_1.$$

Similarly
$$\frac{1}{n_2^2} = \left(\frac{4\pi^2}{g}\right) y_2 \quad \text{etc.}$$

Hence
$$\frac{1}{n^2} = \left(\frac{4\pi^2}{g}\right)\left(y_1 + y_2 + y_3 + \ldots + \frac{y_s}{1{\cdot}267}\right)$$

i.e.
$$n^2 = \left(\frac{g}{4\pi^2}\right)\frac{1}{(y_1 + y_2 + y_3 + \ldots + y_s/1{\cdot}267)}.$$

If the deflections are quoted in millimetres, and g is taken as $9{\cdot}81 \times 10^3$ mm/s² then $g/4\pi^2 = (9{\cdot}81 \times 10^3)/4\pi^2 = 248{\cdot}5$. The expression for the natural frequency then reduces to the more useful form

$$n = \frac{15{\cdot}75}{\sqrt{(y_1 + y_2 + \ldots + y_s/1{\cdot}267)}} \text{ Hz.}$$

EXAMPLE. Two self-aligning bearings 1·52 m apart support a 50 mm diameter shaft on which are mounted masses as shown in Figure 254.

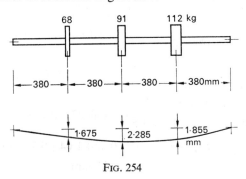

Fig. 254

MECHANICAL TECHNOLOGY FOR HIGHER ENGINEERING TECHNICIANS

Neglect the effects of the inertia of the shaft and use the energy method to estimate the natural frequency of the system given that the static deflections (with all loads acting) are as shown. Next, assume for the shaft a specific mass of 7750 kg/m³ and an elastic modulus of 206 000 MPa, and make a second estimate of the frequency based on Dunkerley's Equation.

Solution. The loads in newtons are 667, 893 and 1100, and for these alone

$$n = \frac{1}{2\pi} \sqrt{\left(g \frac{\Sigma Wy}{\Sigma Wy^2}\right)}$$

$$= \frac{1}{2\pi} \sqrt{\left[(9 \cdot 81) \frac{(667 \times 0 \cdot 00167) + (893 \times 0 \cdot 00229) + (1100 \times 0 \cdot 00185)}{(667 \times 0 \cdot 00167^2) + (893 \times 0 \cdot 00229^2) + (1100 \times 0 \cdot 00185^2)}\right]}$$

$$= \underline{11 \cdot 2 \text{ Hz.}}$$

For the shaft section, $I = \dfrac{\pi}{64}(0 \cdot 05)^4 = 0 \cdot 3065 \times 10^{-6} \text{ m}^4$

The weight of the shaft, $W_s = \dfrac{\pi}{4}(0 \cdot 05)^2 1 \cdot 52(7750 \times 9 \cdot 81)$

$$= 226 \cdot 5 \text{ N.}$$

The deflection of the shaft at mid-span *due to its own weight* is given by

$$y_s = \frac{5}{384} \frac{W_s L^3}{EI} \quad \text{(referring to Fig. 255(d))}$$

$$= \frac{5 \times 226 \cdot 5 \times 1 \cdot 52^3 \times 10^6}{(206\,000 \times 10^6) 0 \cdot 3065} = 0 \cdot 000164 \text{ m} = 0 \cdot 164 \text{ mm}$$

so that $\dfrac{y_s}{1 \cdot 267} = 0 \cdot 13.$

For a non-central concentrated load, the deflection at the load point is given by

$$y = \frac{W a^2 b^2}{3EIL}$$

where a and b are the distances from the load,

$$a + b = L = \text{span.}$$

Referring to Figure 255, the individual static deflections are

$$y_1 = \frac{667 \times 0 \cdot 38^2 \times 1 \cdot 14^2 \times 10^6}{3(206\,000 \times 10^6) 0 \cdot 3065 \times 1 \cdot 52} = 0 \cdot 00042 \text{ m} = 0 \cdot 42 \text{ mm,}$$

$$y_2 = \frac{1}{48} \frac{WL^3}{EI} = \frac{893 \times 1 \cdot 52^3}{48(206\,000 \times 10^6) 0 \cdot 3065} = 0 \cdot 00104 \text{ m} = 1 \cdot 04 \text{ mm,}$$

$$y_3 = \frac{1100 \times 1 \cdot 14^2 \times 0 \cdot 38^2 \times 10^6}{3(206\,000 \times 10^6) 0 \cdot 3065 \times 1 \cdot 52} = 0 \cdot 00072 \text{ m} = 0 \cdot 72 \text{ mm.}$$

$$\therefore \quad y_1 + y_2 + y_3 + (y_s/1 \cdot 267) = 0 \cdot 42 + 1 \cdot 04 + 0 \cdot 72 + 0 \cdot 13$$

$$= 2 \cdot 31 \text{ mm.}$$

VIBRATION

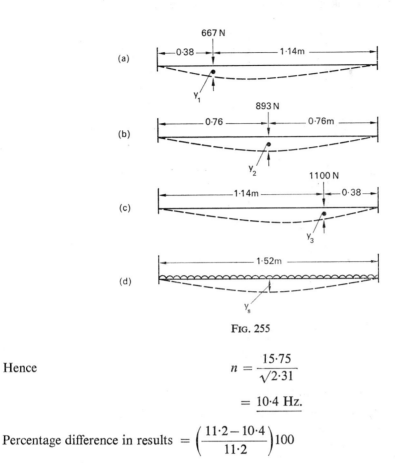

Fig. 255

Hence
$$n = \frac{15 \cdot 75}{\sqrt{2 \cdot 31}}$$
$$= 10 \cdot 4 \text{ Hz.}$$

Percentage difference in results $= \left(\dfrac{11 \cdot 2 - 10 \cdot 4}{11 \cdot 2}\right)100$

$= 7$ approx.

CRITICAL (WHIRLING) SPEED OF A SHAFT

The mass centre of a shaft is *always* displaced from the axis of rotation due to one or more of the factors listed below:

1. Initial curvature resulting from imperfect manufacture.
2. Variation in density of material.
3. Rough handling.
4. Bending under gravitational force or unbalanced magnetic forces.

The mass centre is subjected therefore to centripetal acceleration so that the resulting centrifugal reaction bends the shaft still further and increases the radius of rotation of the mass centre. The effect is thus cumulative.

Suppose a disk of mass M and weight W to be mounted on an elastic shaft as shown in Figure 256 and let its mass centre, G, have an initial eccentricity, h, when stationary. If y is the dynamic displacement of G at some speed ω, and if λ is the lateral stiffness of the

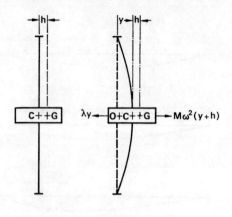

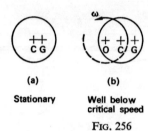

(a) Stationary (b) Well below critical speed

Fig. 256

shaft, then

$$\text{Centrifugal reaction} = \text{centripetal force}$$

i.e. $M\omega^2(y+h) = \lambda y$ (neglecting the inertia of the shaft itself)

or $M\omega^2 y + M\omega^2 h - \lambda y = 0$

so that $y(M\omega^2 - \lambda) = -M\omega^2 h$,

or $y(\lambda - M\omega^2) = M\omega^2 h$ (changing signs)

whence $$\frac{y}{h} = \frac{M\omega^2}{\lambda - M\omega^2}$$

or $$\frac{y}{h} = \frac{1}{(\lambda/M\omega^2) - 1}. \tag{1}$$

Now ω can be increased from zero until the term $\lambda/M\omega^2$ approaches unity and makes the ratio y/h very large indeed. Evidently the ratio y/h is infinite, theoretically, when this term is unity. The angular velocity is then said to be *critical* and the shaft is said to *whirl*. Denoting this velocity by ω_c and putting $M\omega_c^2 = \lambda$ we obtain

Critical angular velocity, $\omega_c = \sqrt{\dfrac{\lambda}{M}}$ rad/s

and Corresponding speed, $n_c = \dfrac{1}{2\pi}\sqrt{\dfrac{\lambda}{M}}$ rev/s.

Since $\sqrt{(\lambda/M)}$ is the angular velocity of the generating vector corresponding to the harmonic motion of a mass M when mounted on an elastic support of stiffness λ (see p. 278), it follows that the speed in rev/s at which the system in Figure 256 becomes unstable (i.e. at which y becomes infinite theoretically) is identical numerically to the frequency at which the system vibrates transversely. A similar conclusion would result from the analysis of any other system so that, in general, any expression giving the natural frequencies of transverse vibration gives also the speeds at which the system will "whirl" when rotated.

Note that the static deflection produced by the load *at the load point* when the system is horizontal is given by

$$z = \frac{W}{\lambda} \quad (\text{where } W = Mg)$$

i.e. that

$$\frac{\lambda}{W} = \frac{1}{z}$$

or

$$\frac{\lambda}{M} = \frac{g}{z}.$$

Hence, Critical speed, $n_c = \dfrac{1}{2\pi}\sqrt{\dfrac{g}{z}}$ rev/s.

Putting $g = 9\cdot81 \times 10^3$ mm/s² gives, for a static deflection quoted in mm,

$$n = \frac{15\cdot75}{\sqrt{z}} \text{ rev/s.}$$

Writing ω_c^2 for λ/M in eq. (1) gives

$$\frac{y}{h} = \frac{1}{(\omega_c/\omega)^2 - 1} = \frac{1}{(n_c/n)^2 - 1} \quad (\text{putting } \omega = 2\pi n \text{ and } \omega_c = 2\pi n_c). \qquad (2)$$

As the speed, n, rises, y not only increases but lags behind h by an increasing angle, φ, as shown in Figure 257.

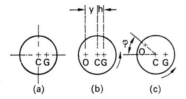

(a) Stationary (b) Well below n_c (c) Approaching n_c

FIG. 257

At the critical speed $\varphi = 90°$ and y is very large relative to h. (The theoretically infinite value is not attained in practice owing to restraints and damping.) As soon as the speed exceeds the critical value ($n > n_c$) the quantity n_c/n becomes less than unity so that, as is evident from eq. (2), the ratio y/h becomes negative. If follows that at this point the dynamic deflection, y, must change sign, the physical meaning of this being that the radii CO and CG are approaching.

Since the quantity $(n_c/n)^2$ tends to zero as the speed rises above the critical value, the ratio y/h approaches -1, i.e. the dynamic deflection tends towards $-h$. This indicates that the axis of rotation tends to pass through G as shown in Figure 258(a).

MECHANICAL TECHNOLOGY FOR HIGHER ENGINEERING TECHNICIANS

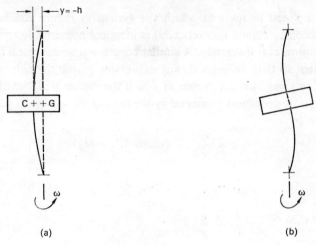

Fig. 258

If the speed increase is continued and account is taken of the inertia of the shaft, the system will enter a further range of instability corresponding to the 2nd Normal Mode of transverse vibration of the shaft itself, the configuration then being as shown in Figure 258(b).

EXAMPLE. A shaft is to be supported in self-aligning bearings 50 mm and 100 mm respectively from the plane of a disk mounted on it and forming part of a brake. If the disk assembly has a mass of 0·68 kg and the 1st Critical Speed is to be at least 3000 rev/min, estimate the minimum necessary diameter of shaft required, neglecting the inertia of the shaft itself. Hence find the lowest critical speed of the shaft itself given that the material has a specific mass of 7900 kg/m³ and an elastic modulus of 206 000 MPa.

Solution.

2nd Moment of shaft section, $I = \dfrac{\pi d^4}{64} = 0{\cdot}049 d^4$ m⁴.

Deflection of shaft at disk, $y_d = \dfrac{Wa^2b^2}{3EIl}$ (see p. 35),

$$= \frac{(0{\cdot}68 \times 9{\cdot}81) 0{\cdot}05^2 \times 0{\cdot}01^2}{3(206\,000 \times 10^6)(0{\cdot}049 d^4) 0{\cdot}15}$$

$$= \frac{3{\cdot}68}{10^{14} d^4} \text{ m.}$$

Neglecting the inertia of the shaft, the 1st Critical Speed is given by

$$n_{cd} = \frac{1}{2\pi}\sqrt{\frac{g}{y_d}} = \frac{1}{2\pi}\sqrt{\left(\frac{9{\cdot}81 \times 10^{14} d^4}{3{\cdot}68}\right)} = \frac{1}{2\pi}\sqrt{d^4(2{\cdot}66 \times 10^{14})}$$

∴ $n_{cd}^2 = \left(\dfrac{2{\cdot}66 \times 10^{14}}{4\pi^2}\right) d^4 = (6{\cdot}75 \times 10^{12}) d^4.$

VIBRATION

Putting $n_{cd} = \dfrac{3000}{60} = 50$ rev/s gives $d^4 = \dfrac{50^2}{6.75 \times 10^{12}} = \dfrac{371}{10^{12}}$

so that
$$d^2 = \dfrac{19.3}{10^6}$$

i.e.
$$d = \dfrac{4.4}{10^3} = 0.0044 \text{ m} = 4.4 \text{ mm.}$$

Weight of shaft $= [(\pi/4)d^2]0\cdot15(7900\times9\cdot81) = 9120d^2$ N.
Deflection of shaft at centre under its own weight

$$y_s = \dfrac{5}{384} \dfrac{WL^3}{EI} \quad \text{(see p. 38)},$$

$$= \dfrac{5(9120d^2)0\cdot15^3}{384(206\,000\times10^6)0\cdot049d^4} = \dfrac{0\cdot397}{10^{10}d^2} \text{ m.}$$

The 1st Critical Speed of the shaft itself is given by

$$n_{cs} = \dfrac{1}{2\pi}\sqrt{\left(1\cdot267\dfrac{g}{y_s}\right)} = \dfrac{1}{2\pi}\sqrt{\left(\dfrac{1\cdot267\times9\cdot81\times10^{10}d^2}{0\cdot397}\right)}$$

$$= \dfrac{1}{2\pi}\sqrt{d^2(31\cdot4\times10^{10})}$$

∴ $\quad n_{cs}^2 = \left(\dfrac{31\cdot4\times10^{10}}{4\pi^2}\right)d^2 = (0\cdot794\times10^{10})d^2$

i.e. $\quad n_{cs} = (0\cdot89\times10^5)d$

$= 0\cdot89\times0\cdot0044\times10^5$

$= 393$ rev/s

$= 23\,500$ rev/min.

Referring to Figure 258(b), the second critical speed of the combination would be, therefore, $23\,500\times 4$ or $94\,000$ rev/min.

VISCOUS DAMPING

If an oscillatory motion is opposed by a force proportional to velocity, i.e. to dx/dt, the motion is said to be *viscous damped*. Since the restoring force has to overcome such opposition before producing any acceleration, the acceleration is less than with unopposed motion, the equation being

$$\text{Mass}\times\text{acceleration} = \text{Restoring force} - \text{damping force}$$

i.e.
$$M\dfrac{d^2x}{dt^2} = -\lambda x - b\dfrac{dx}{dt}$$

where b is a constant,

so that
$$\frac{d^2x}{dt^2} + \left(\frac{b}{M}\right)\frac{dx}{dt} + \left(\frac{\lambda}{M}\right)x = 0.$$

Writing ω^2 for λ/M and $2k$ for b/M we obtain

$$\frac{d^2x}{dt^2} + 2k\frac{dx}{dt} + \omega^2 x = 0.$$

The auxiliary equation is therefore

$$m^2 + 2km + \omega^2 = 0$$

whence $\qquad m = -k \pm \sqrt{(k^2 - \omega^2)}$

and writing p for $\sqrt{(k^2 - \omega^2)}$ gives $\quad m = -k \pm p$.

There are three possibilities:

1. $k > \omega$ (p is positive).

   The solution is
   $$x = Ae^{m_1 t} + Be^{m_2 t}$$

where $\qquad m_1 = -k+p$ and is negative since $p < k$,

and $\qquad m_2 = -k-p$ and is evidently positive.

Since both these terms have negative indices, the displacement x is reduced exponentially to zero so that there is no oscillation, i.e. the motion is *aperiodic*. The damping is said to be *heavy* and the system is said to be *overdamped*.

2. $k = \omega$ (p is zero).

   As before, $x = Ae^{m_1 t} + Be^{m_2 t}$, but now $m_1 = -k = m_2$
   so that $x = e^{-kt}(At + B)$.

In this case the damping coefficient, b, is just sufficient to bring the system to rest after a disturbance without oscillation. The damping is said to be *critical*.

3. $k < \omega$.

   As before, $m = -k \pm \sqrt{(k^2 - \omega^2)}$
   $\qquad\qquad = -k \pm j\sqrt{(\omega^2 - k^2)}$
   $\qquad\qquad = -k \pm jp$

where $\qquad p = \sqrt{(\omega^2 - k^2)}$.

Substituting for m gives
$$x = Ae^{(-k+jp)t} + Be^{(-k-jp)t}$$
$$= e^{-kt}(Ae^{jpt} + Be^{-jpt})$$

where $\qquad e^{jpt} = \cos pt + j \sin pt$

and $\qquad e^{-jpt} = \cos pt - j \sin pt$

so that $\qquad x = e^{-kt}[A(\cos pt + j \sin pt) + B(\cos pt - j \sin pt)]$
$\qquad\qquad = e^{-kt}[(A+B)\cos pt + j(A-B)\sin pt]$.

VIBRATION

Since x is a real quantity it cannot be equal to a complex expression, so that A and B must be conjugate complex numbers, i.e. $(A+B)$ and $j(A-B)$ must be real numbers. Putting $A+B = L$ and $j(A-B) = M$ gives

$$x = e^{-kt}(L \cos pt + M \sin pt).$$

The motion represented by this equation is oscillatory, the term e^{-kt} (called the *damping factor*) indicating that successive amplitudes decrease exponentially. In other words, when the system is released suddenly from a displaced position, the damping is insufficient to prevent overshooting of the equilibrium position. Such damping is said to be *sub-critical* and the system is said to be *underdamped*, the graph of displacement against time having the form shown in Figure 259, x being measured at time t after release from an initial displacement r_0.

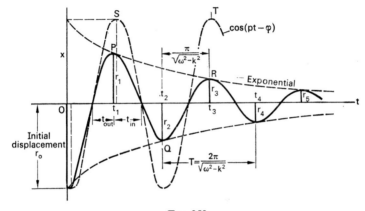

FIG. 259

It can be shown that* the equation for the displacement can be written in the alternative form

$$x = \frac{\omega r_0}{\sqrt{(\omega^2 - k^2)}} e^{-kt} \cos [\sqrt{(\omega^2 - k^2)}t - \varphi]$$

where $\varphi = \tan^{-1} \dfrac{k}{\sqrt{(\omega^2 - k^2)}}.$

From this it is evident that the periodic time $(2\pi/p)$ is given by

$$T = \frac{2\pi}{\sqrt{\left[\dfrac{\lambda}{M} - \left(\dfrac{b}{2M}\right)^2\right]}} \quad \text{since} \quad \frac{\lambda}{M} = \omega^2 \quad \text{and} \quad \frac{b}{2M} = k.$$

When there is no damping, $b = 0$, i.e. $k = 0$ and $p = \omega$, so that then

$$T = \frac{2\pi}{\omega} \quad \text{instead of} \quad \frac{2\pi}{\sqrt{(\omega^2 - k^2)}}.$$

* See pp. 124 et seq., *Mechanics of Machines*, by the same author.

MECHANICAL TECHNOLOGY FOR HIGHER ENGINEERING TECHNICIANS

Since $\omega > \sqrt{(\omega^2 - k^2)}$ it is evident that damping increases the periodic time. However, with light damping, such change is usually insignificant.

The frequency in Hz (i.e. $1/T$) is given by

$$n = \frac{1}{2\pi}\sqrt{\left(\frac{\lambda}{M} - \left[\frac{b}{2M}\right]^2\right)}.$$

It can also be shown that the ratio of any amplitude to the one preceding it is constant. Referring to Figure 259,

$$\frac{r_5}{r_4} = \frac{r_4}{r_3} = \frac{r_3}{r_2} = \frac{r_2}{r_1} = -e^{-(k/2)T}$$

where
$$k = \frac{b}{2M}.$$

This ratio is negative because r_2 is negative while r_1 is positive, i.e. the amplitudes are successively of opposite sign.

Similarly, the ratio of any amplitude to the one preceding it on the same side of the equilibrium position is also constant.

Thus
$$\frac{r_5}{r_3} = \frac{r_3}{r_1} = e^{-kT}.$$

This is positive because the amplitudes are of the same sign.

Finally, the ratio of the nth to the initial amplitude is given by

$$\frac{r_n}{r_0} = e^{-k(n-1)T}.$$

But the time for $(n-1)$ oscillations is $(n-1)T$ so that this can also be written

$$\frac{r_n}{r_0} = e^{-kt}$$

where $t = (n-1)T$ = interval between 1st and nth amplitudes.
Since kt is the logarithm (to the base e) of this ratio, it is called the *logarithmic decrement*.

EXAMPLE. Figure 260 represents a mass of 4·08 kg suspended from a close-coiled spring and coupled to a coaxial dashpot. If the measured time for a total of forty damped oscillations was 16 sec and the measured ratio of the initial displacement to the amplitude of the sixth displacement on the same side of the equilibrium position was 2·25, calculate:
  (a) the damping force per unit velocity (value of b);
  (b) the stiffness of the spring.

Solution.

$$\text{Amplitude ratio,} \quad \frac{r_n}{r_0} = e^{-k(n-1)T}.$$

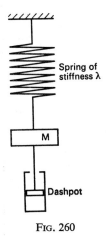

Fig. 260

The given ratio is the reciprocal of this, i.e. r_0/r_n, so that,

when $T = \frac{16}{40}$,
$$2 \cdot 25 = e^{+k(6-1)0 \cdot 4}$$
$$= e^{2k}.$$

Hence
$$2k = \log_e 2 \cdot 25$$
from which
$$k = 0 \cdot 405.$$

Damping coefficient,
$$b = (2M)k = 2 \times 4 \cdot 08 \times 0 \cdot 405$$
$$= 3 \cdot 3 \text{ N per m/s}.$$

But,
$$T = \frac{2\pi}{\sqrt{(\omega^2 - k^2)}}$$

so that
$$\sqrt{(\omega^2 - k^2)} = \frac{2\pi}{0 \cdot 4} = 15 \cdot 7.$$

Hence
$$\omega^2 - k^2 = 15 \cdot 7^2 = 246 \cdot 8$$
i.e.
$$\omega^2 = 246 \cdot 8 + 0 \cdot 405^2$$

so that
$$\frac{\lambda}{M} = 247$$

whence
$$\lambda = 247 \times 4 \cdot 08$$
$$= 1010 \text{ N/m}.$$

FORCED VIBRATION WITH VISCOUS DAMPING

In the previous paragraph, consideration was given to a spring-mass system such as is shown in Figure 260, the mass being released suddenly from a position below the equilibrium position and the subsequent motion analysed.

Now suppose the spring support to be given s.h.m. of amplitude A about the equilibrium position OO, Figure 261. The problem is to determine the amplitude a of the load about the equilibrium position oo and its phase relative to the applied amplitude.

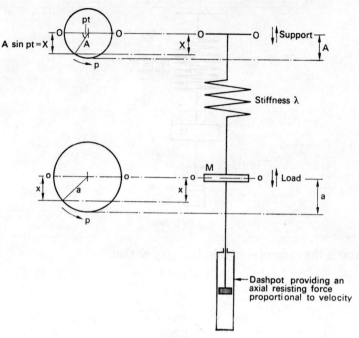

Fig. 261

If the corresponding instantaneous displacements of lower and upper ends at time t are, respectively, x and X where $X = A \sin pt$ and $x > X$, then, taking downward displacements as positive,

Instantaneous upward spring force on load $= -\lambda(x-X)$
$\qquad\qquad\qquad\qquad\qquad\qquad\qquad\quad = -\lambda x + \lambda X.$

As before,

Mass × acceleration = Restoring force − Damping force so that

$$M\frac{d^2x}{dt^2} = -\lambda x + \lambda X - b\frac{dx}{dt}$$

where M = inertia constant,
$\qquad\lambda$ = restoring constant,
$\qquad b$ = damping constant.

Rearrangement gives

$$\frac{d^2x}{dt^2} + \left(\frac{b}{M}\right)\frac{dx}{dt} + \left(\frac{\lambda}{M}\right)x = \left(\frac{\lambda}{M}\right)X.$$

Writing ω^2 for λ/M, $2k$ for b/M and $A \sin pt$ for X:

$$\frac{d^2x}{dt^2} + 2k\frac{dx}{dt} + \omega^2 x = \omega^2 A \sin pt.$$

The solution to this equation will be in two parts, i.e. the expression for x will be the sum of a Particular Integral (any particular solution which satisfies the whole equation) and the Complementary Function. The latter—obtained by solving the equation $d^2x/dt^2 + 2k(dx/dt) + \omega^2 x = 0$—will give the expression for x when the external disturbing force (producing the motion of the support) is zero. As already shown, this is

$$x = e^{-kt}[L \cos \sqrt{(\omega^2 - k^2)}t + M \sin \sqrt{(\omega^2 - k^2)}t]$$

or,
$$x = Ce^{-kt} \cos [\sqrt{(\omega^2 - k^2)}t - \varphi]$$

where
$$C = \sqrt{(L^2 + M^2)}$$

and
$$\varphi = \tan^{-1} M/L.$$

If $k < \omega$ this equation represents a damped free vibration, the damping factor, e^{-kt}, indicating an exponential dying away of amplitude with time, i.e. that the vibration is *transient*. The constants C and φ depend on the position of the load at the instant corresponding to zero time and can be determined from the initial conditions.

Assuming $x = P \cos pt + Q \sin pt$ as a particular integral, differentiating twice, substituting the values of dx/dt and d^2x/dt^2 so obtained in the equation of motion, simplifying and equating coefficients, we obtain

$$P = -\frac{2kp\omega^2 A}{(\omega^2 - p^2)^2 + 4k^2 p^2} \quad \text{and} \quad Q = +\frac{\omega^2 A(\omega^2 - p^2)}{(\omega^2 - p^2)^2 + 4k^2 p^2}.$$

Now the assumed equation may also be written

$$x = R\left(\frac{P}{R} \cos pt + \frac{Q}{R} \sin pt\right)$$

or
$$x = R \cos (pt - \alpha)$$

where
$$\alpha = \tan^{-1} \frac{Q}{P} \quad \text{and} \quad R = \sqrt{(P^2 + Q^2)}.$$

Substitution of the values obtained for P and Q gives

$$R = \frac{\omega^2 A}{\sqrt{[(\omega^2 - p^2)^2 + 4k^2 p^2]}}$$

so that the particular solution can be written

$$x = \frac{\omega^2 A}{\sqrt{[(\omega^2 - p^2)^2 + 4k^2 p^2]}} \cos [\sqrt{(\omega^2 - k^2)}t - \alpha].$$

Now
$$\tan \alpha = \frac{Q}{P} = -\frac{\omega^2 - p^2}{2kp}$$

where α lies between 90° and 180°,

$$\therefore \quad \tan (\alpha - 90) = +\frac{2kp}{\omega^2 - p^2}.$$

Writing β for $(\alpha - 90)$ and referring to Figure 262 we have $\alpha = 90 + \beta$ so that
$$\cos (pt - \alpha) = \sin (pt - \beta).$$

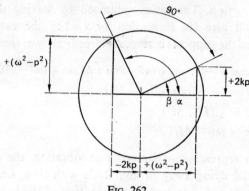

Fig. 262

Thus the particular solution may be written in the same form as the equation for the applied motion, namely

$$x = \frac{\omega^2 A}{\sqrt{[(\omega^2-p^2)^2 + 4k^2p^2]}} \sin[\sqrt{(\omega^2-k^2)}\,t - \beta].$$

This represents a steady vibratory motion of frequency

$$n = \frac{p}{2\pi} = \frac{\sqrt{(\omega^2-k^2)}}{2\pi}$$

and amplitude

$$a = \frac{\omega^2 A}{\sqrt{[(\omega^2-p^2)^2 + 4k^2p^2]}}$$

lagging the applied motion by the angle (lying between 0° and 90°)

$$\beta = \tan^{-1}\frac{2kp}{\omega^2-p^2}.$$

The complete solution is the sum of the particular solution just found and the complementary function, viz.

$$x = Ce^{-kt}\cos[\sqrt{(\omega^2-k^2)}\,t - \varphi] + \frac{\omega^2 A}{\sqrt{[(\omega^2-p^2)^2 + 4k^2p^2]}} \sin[\sqrt{(\omega^2-k^2)}\,t - \beta].$$

|_____Transient_____| |_____Steady State_____|

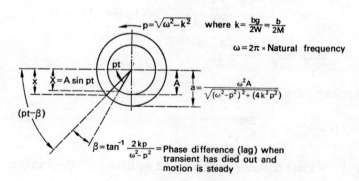

Fig. 263

The phase difference, β, between the applied and resulting motions is illustrated in Figure 263 which should now be compared with Figure 261.

Since, after the application of the disturbing force, the transient affects only the first few amplitudes, it is usual to neglect it, i.e. to omit it from the equation for the displacement.

Dynamic Magnification Factor

This is the ratio of the amplitude of the load, Figure 261, to the amplitude of the mounting and is denoted by Q.

Thus
$$Q = \frac{a}{A}$$

$$= \frac{1}{A} \times \frac{\omega^2 A}{\sqrt{[(\omega^2 - p^2)^2 + 4k^2 p^2]}}$$

which reduces to

$$Q = \frac{1}{\sqrt{\left[\left(1 - \frac{p^2}{\omega^2}\right)^2 + \left(\frac{2k}{\omega}\right)^2 \left(\frac{p}{\omega}\right)^2\right]}}.$$

For an undamped system b (and therefore k) is zero, in which case

$$Q = \frac{1}{1 - \left(\frac{p}{\omega}\right)^2}.$$

Evidently when $p = \omega$ (i.e. when the applied frequency is equal to the natural frequency) the value of Q—and therefore the load amplitude—is theoretically infinite. The system is then said to be in a state of *resonance*. Although k may be very small it is in fact never zero so that the load amplitude merely becomes very large.

Thus, when $p = \omega$ and k is not zero,

$$Q = \frac{1}{2k/\omega} = \frac{\omega}{2k}$$

so that
$$a_{max} = \frac{\omega A}{2k}.$$

Now
$$a = \omega^2 A (\omega^4 - 2\omega^2 p^2 + p^4 + 4k^2 p^2)^{-1/2}$$
$$= \text{amplitude of steady damped motion.}$$

Differentiating with respect to p and equating to zero shows that this is a maximum when

$$p = \sqrt{(\omega^2 - 2k^2)}.$$

This is less than ω, but not much less when k is small.

The phase lag, β (which is given by $\tan^{-1} 2kp/(\omega^2 - p^2)$), increases rapidly as p approaches ω until, when $p = \omega$:

$$\beta = \tan^{-1} \infty = \frac{\pi}{2} = 90°.$$

This is true whether or not $k = 0$.

If the applied frequency is increased beyond this point (making $p > \omega$), then $2kp/(\omega^2 - p^2)$ becomes negative in which case $\beta > \pi/2$. The value of Q falls rapidly and the phase lag increases considerably beyond $\pi/2$ until, when $p \simeq 1\cdot4\omega$, the two amplitudes (a and A) are approximately equal (though about 170° out of phase) for values of $2k/\omega$ between 0·1 and 0·2.

When $2k/\omega = 0\cdot2$, the approximate relation between Q, β and the ratio p/ω is shown in Figure 264.

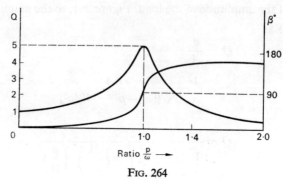

FIG. 264

Note that, the smaller the ratio $2k/\omega$, the greater the value of Q and the nearer its maximum value is to resonance.

Consideration has been given previously to the response of a damped spring-mass system—as represented in Figure 261— to a periodic displacement of the end of the spring remote from the mass. Now suppose this end to be fixed and a periodic disturbing *force* to be applied to the mass itself, a very common occurrence in engineering and one which may be exemplified by an elastically mounted electric motor, Figure 265, having an out-of-balance rotor.

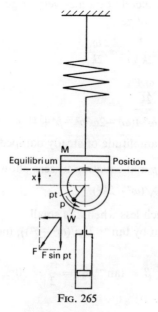

FIG. 265

If m = mass of rotor,
   r = radius of rotation of mass centre,
   p = angular velocity of rotation,

then Unbalanced reaction to centripetal force, $F = mp^2r$
and Axial (i.e. vertical) disturbing force $= F \sin pt$.

Taking the downward direction as positive, the equation of vertical motion is therefore

$$M \frac{d^2x}{dt^2} = -\lambda x + F \sin pt - b \frac{dx}{dt}.$$

Rearrangement gives

$$\frac{d^2x}{dt^2} + \left(\frac{b}{M}\right)\frac{dx}{dt} + \left(\frac{\lambda}{M}\right)x = \left(\frac{F}{M}\right) \sin pt.$$

Writing ω^2 for λ/M and $2k$ for b/M gives

$$\frac{d^2x}{dt^2} + 2k\frac{dx}{dt} + \omega^2 x = \left(\frac{F}{M}\right) \sin pt.$$

Now, for a given angular velocity, p, the maximum value of $F \sin pt$ is F. Suppose F to produce an axial spring deflection A when applied statically, i.e. $F = \lambda A$. Then the right-hand side of the equation becomes $(\lambda/M)A \sin pt$ and, since $\lambda/M = \omega^2$, the equation can be written

$$\frac{d^2x}{dt^2} + 2k\frac{dx}{dt} + \omega^2 x = \omega^2 A \sin pt.$$

Since this is identical to the displacement equation corresponding to harmonic disturbance of the mounting (fixed end of the spring), the other equations already derived apply equally to harmonic disturbance of the mass itself, provided only that A is taken as the deflection resulting from a static application of F.

In this case the dynamic magnification factor may be defined as the ratio

Amplitude of load/Static deflection correspondig to maximum disturbing force.

EXAMPLE. A motor-driven reciprocating pump is to be mounted on a rigid frame, the estimated total mass being 295 kg. The frame is to rest on an elastic packing which is required to limit the fluctuation in ground load to ± 45 N at 750 rev/min. Assume negligible damping and find the required stiffness of the packing given that at this speed the primary unbalanced vertical harmonic force has a maximum value of 355 N. Hence find the amplitude of the vibration. Find also the amplitude at half speed.

Solution. The system may be represented as in Figure 266:

Angular velocity of rotor, $\quad p = \dfrac{2\pi N}{60} = \dfrac{2\pi \times 750}{60} = 78.5$ rad/s,

so that $\quad p^2 = 6160.$

Also $\quad \omega^2 = \dfrac{\lambda}{M} = \dfrac{\lambda}{295} = 0.00339\lambda.$

305

MECHANICAL TECHNOLOGY FOR HIGHER ENGINEERING TECHNICIANS

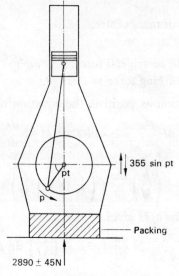

FIG. 266

The equation for the viscous-damped motion of an elastically mounted body when subjected to a direct harmonic force is

$$\frac{d^2x}{dt^2} + 2k\frac{dx}{dt} + \omega^2 x = \omega^2 A \sin pt$$

and the steady state displacement is given by

$$= \frac{\omega^2 A}{\sqrt{[(\omega^2 - p^2)^2 + 4k^2 p^2]}} \sin[\sqrt{(\omega^2 - k^2)}\, t - \beta]$$

which, when the motion is undamped ($k = 0$), reduces to

$$x = \frac{\omega^2 A}{\omega^2 - p^2} \sin(\omega t - \beta).$$

Hence, Steady state undamped amplitude $= \dfrac{\omega^2 A}{\omega^2 - p^2}$.

But A = static deflection corresponding to maximum disturbing force

$$= \frac{F}{\lambda} = \frac{F}{\omega^2 M} = \frac{F/M}{\omega^2}$$

where $F = 355$ N.

$\therefore$ Steady undamped amplitude $= \dfrac{\omega^2}{\omega^2 - p^2}\left(\dfrac{F/M}{\omega^2}\right) = \dfrac{F}{M}\left(\dfrac{1}{\omega^2 - p^2}\right)$

$$= \frac{355}{295}\left(\frac{1}{0\cdot00339\lambda - 6160}\right).$$

Now the maximum variation in the force on the ground is to be ± 45 N so that the change in the deflection of the packing (above or below the static position) must be limited to $\pm 45/\lambda$. Since this is the permitted amplitude, we have

$$\pm \frac{45}{\lambda} = \frac{355}{295}\left(\frac{1}{0.00339\lambda - 6160}\right)$$

whence $\qquad \lambda = 204\,500 \text{ N/m} = \text{required stiffness.}$

Hence, at 750 rev/min,

$$\text{Amplitude} = \pm \frac{45}{\lambda}$$

$$= \pm \frac{45}{204\,500} = 0.00022 \text{ m} = 0.22 \text{ mm.}$$

Again, at half speed (375 rev/min),

$$p^2 = \frac{6160}{2^2} = 1540$$

and $\qquad \omega^2 = 0.00339\lambda = 0.00339 \times 204\,500 = 693.$

Since the harmonic force is proportional to the square of the speed,

$$F = \frac{355}{2^2} = 88.7 \text{ N.}$$

Hence $\qquad$ New amplitude $= \pm \dfrac{F}{M}\left(\dfrac{1}{\omega^2 - p^2}\right)$

$$= \pm \frac{88.7}{295}\left(\frac{1}{693 - 1540}\right)$$

$$= \pm 0.000355 \text{ m}$$

$$= 0.355 \text{ mm.}$$

Viscous Damping of Torsional Vibration

Figure 267 represents an elastic shaft of torsional stiffness, q, supporting a disk of polar moment of inertia I and fitted with a coaxial damper which provides an opposing torque proportional to angular velocity.

If the disk is displaced from the equilibrium position by some angle θ_0 and released suddenly, the equation of motion will be

Accelerating (inertia) torque = Restoring torque − damping torque

or $\qquad I\dfrac{d^2\theta}{dt^2} = -q\theta - b\dfrac{d\theta}{dt}$

where b = damping constant.

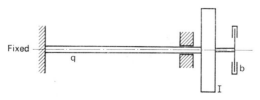

Fig. 267

Rearranging gives

$$\frac{d^2\theta}{dt^2} + \left(\frac{b}{I}\right)\frac{d\theta}{dt} + \left(\frac{q}{I}\right)\theta = 0.$$

Writing ω^2 for q/I and $2k$ for b/I we obtain

$$\frac{d^2\theta}{dt^2} + 2k\frac{d\theta}{dt} + \omega^2\theta = 0.$$

This equation is identical in form to that for viscous-damped linear motion (p. 296) so that, when $k < \omega$, the displacement equation is

$$\theta = \frac{\omega\theta_0}{\sqrt{(\omega^2-k^2)}} e^{-kt} \cos\left[\sqrt{(\omega^2-k^2)}t - \tan^{-1}\frac{k}{\sqrt{(\omega^2-k^2)}}\right],$$

while the frequency of vibration is given by

$$n = \frac{1}{2\pi}\sqrt{\left[\frac{q}{I} - \left(\frac{b}{2I}\right)^2\right]} \text{ Hz}.$$

The graph of angular displacement against time will be identical in form to that shown in Figure 259.

Damping Ratio

Critical damping has been defined as that which will just suffice to bring a system to rest without oscillation after sudden release from a displaced position. In other words, when $k = \omega$ there is just no overshoot of the equilibrium position, the displacement–time curve having the form shown in Figure 268.

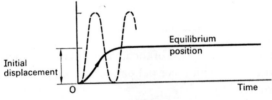

Fig. 268

Now, Damping constant, $b = 2k$ (Inertia constant), so that if b_c is the value of b when the damping is critical—i.e. when $k = \omega$—then

$$b_c = 2\omega \text{ (Inertia constant)}.$$

Evidently

$$\frac{b}{b_c} = \frac{k}{\omega}.$$

The ratio of the damping constants for non-critical and critical conditions (b/b_c) is called the *damping ratio* and is denoted by ζ (zeta). It is dimensionless.

Thus

$$k = \zeta\omega$$

where $\zeta = 1.0$ for critical damping.

Now the periodic time is given by $T = \dfrac{2\pi}{\sqrt{(\omega^2 - k^2)}}$

so that, writing $\zeta\omega$ for k we obtain

$$T = \dfrac{2\pi}{\sqrt{(\omega^2 - \zeta^2\omega^2)}} = \dfrac{2\pi}{\sqrt{\omega^2(1-\zeta^2)}} = \dfrac{2\pi}{\omega\sqrt{(1-\zeta^2)}}$$

and

$$n = \dfrac{1}{T} = \dfrac{\omega\sqrt{(1-\zeta^2)}}{2\pi}.$$

Inspection of this reveals that the frequency falls as the damping becomes heavier.

If the first overshoot is expressed as a percentage of the undamped value and is plotted against damping ratio, the curve obtained will have the form shown in Figure 269.

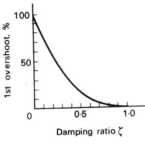

Fig. 269

EXAMPLE. Find the value of damping ratio which will limit the first overshoot of an oscillation to 5 per cent of the initial displacement.

Solution.

First overshoot, $\theta_1 = \left(\dfrac{5}{100}\right)\theta_0$

$\therefore \quad \dfrac{\theta_0}{\theta_1} = 20.$

Ratio of successive amplitudes, $\dfrac{\theta_1}{\theta_0} = e^{-(k/2)T}$

$\therefore \quad \dfrac{\theta_0}{\theta_1} = e^{+(k/2)T}$

$\therefore \quad \left(\dfrac{k}{2}\right)T = \log_e 20 = 2\cdot 99.$

Writing $\dfrac{2\pi}{\omega\sqrt{(1-\zeta^2)}}$ for T and $\zeta\omega$ for k:

$$\left(\dfrac{\zeta\omega}{2}\right)\dfrac{2\pi}{\omega\sqrt{(1-\zeta^2)}} = 2\cdot 99$$

i.e. $\pi^2\zeta^2 = 2\cdot 99^2(1-\zeta^2)$

309

so that
$$\zeta^2 = \frac{2\cdot 99^2}{\pi^2 + 2\cdot 99^2} = 0\cdot 475$$

i.e. Required damping ratio, $\zeta = 0\cdot 69$.

EXAMPLES 10

1. When suspended from a helical spring, a load of 9·1 kg is found to vibrate vertically with a periodic time of 0·75 s. Determine, for an amplitude of 51 mm:
   (a) the angular velocity of the generating vector (8·37 rad/s);
   (b) the maximum acceleration (3·57 m/s²);
   (c) the acceleration when the displacement is 127 mm (2·67 m/s²);
   (d) the stiffness of the spring (633 N/m);
   (e) the static deflection in the spring (140 mm);
   (f) the maximum force in the spring (122 N).

2. Determine the permissible speed of the flywheel shown in Figure 270 if the acceleration of the oscillating rod is not to exceed 6·1 m/s² (74 rev/min).

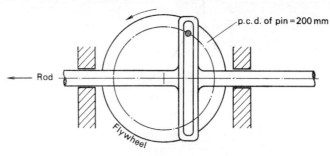

FIG. 270

3. A cube is placed on a table which is then made to oscillate horizontally with s.h.m. of frequency 0·4 Hz. If the coefficient of friction is 0·26, determine the permissible amplitude of the motion if no sliding is to occur (405 mm). (It is unnecessary to know the mass of the cube.)

4. A sieve is to have a vertical travel of 76 mm and to be vibrated with s.h.m. If the material resting on it is just to lose contact, estimate the least necessary frequency of vertical vibration (2·55 Hz).

5. A component having a mass of 22·7 kg oscillates harmonically between extremes 1·22 m apart, the periodic time being 2·0 s. Find:
   (a) the angular velocity of the generating vector (π rad/s);
   (b) the maximum acceleration (6 m/s²);
   (c) the velocity at a point 152 mm from the centre of oscillation (1·86 m/s);
   (d) the time taken to travel a distance of 76 mm from an extremity (0·16 s).

6. The periodic time of vertical vibration of a load of 4·5 kg when suspended from a helical spring was found to be 0·25 s. Find:
   (a) the stiffness of the spring (2850 N/m);
   (b) the maximum tension at an amplitude of 51 mm (185 N).

7. Mercury is contained in a uniform-section U-tube, the arms of which are vertical. If the liquid level in one arm is depressed below the position of equilibrium and released suddenly, show that the mercury will oscillate about that position with s.h.m. If such an oscillation has a periodic time of 1·012 s, determine the length of the column of liquid (0·5 m).

8. Determine the accelerating force exerted by a tappet moving with s.h.m. on a valve of mass 0·45 kg and having a lift of 5·2 mm when the camshaft speed is 1500 rev/min, the cam having no dwell period (29 N).

9. Estimate the natural frequency of vertical oscillation of a four-wheeled vehicle of mass 1500 kg assuming uniform mass distribution and a static deflection of 50 mm in each spring (2·175 Hz).

10. A steel wire 3·18 mm diameter and 1·83 mm long is fixed at its upper end and loaded coaxially by a cylinder 255 mm diameter, 50 mm long. If the specific mass of the cylinder material is 7200 kg/m³ and the

rigidity modulus of the wire material is 80 000 MPa, estimate the frequency of free torsional vibration (0·277 Hz).

11. Two gears having moments of inertia of 17·5 and 9·9 kg m² respectively are keyed 76 mm apart on a shaft 50 mm diameter. Assume $G = 80\,000$ MPa and determine:
    (a) the position of the node (28 mm from the larger);
    (b) the natural frequency of torsional vibration (53 Hz).

12. A hollow shaft 76 mm o/d, 57 mm i/d, carries at one end a flywheel of mass 272 kg and radius of gyration 610 mm. It is coupled to a solid shaft 50 mm diameter, 915 mm long at the other end of which is a gear of mass 218 kg and radius of gyration 457 mm. Find the length of hollow shaft required in order that the node of free torsional vibration shall be at the coupling. Assume $G = 80\,000$ MPa and find the frequency under these conditions (1·42 m, 5·56 Hz).

13. A shaft connecting two identical coaxial disks each having a mass of 455 kg and a radius of gyration of 380 mm is stepped as follows: 254 mm of 76 mm diameter, 100 mm of 127 mm diameter, 254 mm of 90 mm diameter. Assume $G = 80\,000$ MPa and find:
    (a) the torsionally equivalent length of 76 mm diameter shaft (405 mm);
    (b) the position of the node (202 mm from the disk on the 76 mm part);
    (c) the frequency of free torsional vibration (22·8 Hz).

14. A joist having a flexural rigidity (EI) of $7·15 \times 10^6$ Nm² and having a mass per metre of 31·3 kg is simply supported over a span of 3·66 m. Calculate the lowest frequency of free transverse vibration (56 Hz).

15. An I-section joist for which $I = 15 \times 10^{-6}$ m⁴ and $E = 206\,000$ MPa has a mass per metre of 22·4 kg. Find the fundamental frequency of a simple span when loaded transversely with 1000 kg at points 915 mm from each end (95 Hz).

16. A crane of mass 3000 kg travels along two identical parallel joists bridging a span of 9·15 m. Each joist has a mass per metre of 80·5 kg and a second moment of area of 154×10^{-6} m⁴. Assume $E = 206\,000$ MPa and estimate the value of the lowest natural frequency of transverse vibration of the system. Assume simple supports (5·18 Hz).

17. The first critical speed of a horizontal shaft carrying at its midpoint a disk of mass M is 700 rev/min. If the dynamic (centrifugal) deflection at the load point is not to exceed 0·25 mm at twice this speed, find the permissible initial eccentricity of the mass centre of the disk (0·19 mm). If the first critical speed is to be raised to 850 rev/min, find the required percentage increase in the second moment of area of the shaft section (47 per cent). What percentage increase in diameter does this represent? (10 per cent).

18. The self-aligning bearings supporting a steel shaft are respectively 0·305 and 2·15 m on each side of a gear of mass 18 kg. If the shaft material has a specific mass of 7750 kg/m³ and an elastic modulus of 206 000 MPa, estimate the shaft diameter necessary to give a critical speed in excess of 2400 rev/min (118 mm).

19. A shaft 12·7 mm diameter is made of steel having a specific mass of 7750 kg/m³ and an elastic modulus of 206 000 MPa. It is supported in self-aligning bearings 1·2 m apart and carries identical disks of mass 0·9 kg at 410 and 890 mm from the left-hand bearing. Find the lowest critical speed allowing for the inertia of the shaft (595 rev/min).
    Estimate the mass of a single disk which, when substituted for the other two and mounted at the centre, would give the same critical speed (1·22 kg).
    Find the vaue of the second critical speed for the modified system (4175 rev/min⁶). (Remember that the disk is now at the node and has no effect on the frequency.)

20. The amplitude of an elastically mounted component falls to half its initial displacement after 10 complete cycles. If the stiffness of the mounting is 16 750 N/m and the component has a mass of 54·5 kg, find the value of the damping coefficient (1900 N per m/s).

21. Two parallel pipes each 100 mm diameter and having a second moment of area about a diameter of 208×10^{-6} m⁴ are built into a wall to form a cantilever which supports at the free end an electric hoist of mass 22·7 kg. The armature of the hoist motor has a mass of 9·06 kg and has its axis horizontal, parallel to the wall and 1·52 m from it. If the mass centre of the rotor is 0·25 mm from the axis of rotation, estimate the speed at which a maximum bending stress of 166 MPa will be induced in the pipe material. Assume $E = 206\,000$ MPa (1710 rev/min).

11

CONTROL

GENERAL

Mechanisms, singly or in series, must be adjusted or *controlled* in order that they do what is required of them. In the language of control, the result of the act of adjustment (i.e. of pressing a button, pulling a lever, closing a valve, etc.) is called a *command* while the subsequent behaviour of the system is called a *response*.

The automatic or self-actuated adjustment of mechanisms is no new thing, long-standing examples being the centrifugally actuated ball governor linked to a valve (throttle) and the rotation of a windmill head by a wind-actuated vane (fantail) so as to maintain the plane of the sails normal to the wind direction.

The so-called process industries (foodstuffs, paper, chemicals and so on), where production is characterised by continuous flow through consecutive treatments, are particularly amenable to control by automatic methods since these make, in turn, an important contribution to the uniformity and quality of the product. Further, by reducing the time spent in "machine watching", automatic control either releases a proportion of the labour force for more useful work or makes possible a reduction in the wage bill. Alternatively, shift working may be introduced with the same amount of labour.

It is convenient here to define *process control* as "the automatic control of variables (temperature, thickness, composition and so on) encountered in industrial flow production".

CLOSED LOOP CONTROL

This is the name given to a system in which the actual value of the controlled quantity is measured and compared continuously with the desired value by a differential device called an *error detector*. In its simplest (!) form the detector is a man ("pink box"), but more conveniently a "black box" of some kind is used.

In the first case a calibrated regulator such as a knob, lever or handwheel is set manually—as a result of experience usually—to give, say, a desired machine response. The machine must then be watched (monitored) continuously and with concentration and the regulator adjusted to cope with changes in the load or in any other variable. By combining monitoring with adjustment the operator is said to close the *information loop*. An example of this simple method is the adjustment of room temperature with the aid of a thermometer and an electric heater fitted with an on/off switch. Temperature control within fine limits is impossible because, due to the time lag between cause and effect (i.e. between switch operation and heat transfer), the operator tends to over-correct, that is, he operates the switch too late. Although his judgement may improve with practice he will certainly become bored. His concentration will then suffer and the control will not only become erratic but will cease altogether when, eventually, he falls asleep.

In the second case the human operator is replaced by a detector which uses the error (deviation from the desired result) to energise an external source of power—called an *actuator* or *servomotor*—connected to the input of the controlled machine in such a way as to reduce the error. Thus correction is achieved by using the error to reduce itself.

[*Note:* An actuator is a device which produces a limited motion—angular or linear—and may be mechanical, pneumatic, electric or hydraulic. A servomotor is usually electric or hydraulic and has a continuous output.]

FEEDBACK

The use of a measured output for purposes of comparison—i.e. its feeding back to the input—is called, naturally enough, *feedback* and, when such use results in a reduction in the error, is called *negative feedback*. (Evidently positive feedback has the opposite effect.) In the same way the use of the error in energizing a servomotor constitutes *feedforward*.

OPEN LOOP CONTROL

If the behaviour of the output has no influence on the input (i.e. there is no feedback) the so-called information loop does not exist. Such a system is said to have *open loop control* and a typical example is the switching on of street lighting at a predetermined time by means of a time switch. The fact that the sun sets at a different time each day has no effect on the setting of the timing mechanism. (Eventually the information that the illumination is out of phase with requirements will be fed back to the time switch by the human being altering the setting!)

Referring to the simple method of adjusting room temperature already considered (in which the loop is closed by visual observation) an improved result may be obtained by substituting for the man a temperature-sensitive switch, i.e. one which "knows" when to operate and can itself close the loop. This may be done by mounting one of the contacts on the unrestrained end of a bi-metallic cantilever so arranged that the change in its curvature resulting from a fall in temperature below the desired value brings the fixed and moving contacts together, so switching on the power—and vice versa. Such a switch is known as a *thermostat* (patented by Andrew Ure in 1830) and the closed loop control it provides, although discontinuous (on/off or stop/go), is automatic.

AMPLIFICATION

If the output from the error detector (called the *error signal*) is small, as is usual, the detector itself has insufficient power to actuate the correcting mechanism direct so that some form of power amplifier is necessary. The need of considerable correcting power is exemplified by, say, a rocket launching platform—weighing several tons and having a large moment of inertia—which is required to follow a *remote position control* or RPC. If the power input to the platform were made proportional to the error, i.e. to the difference in the directions of control and platform (or to some function of it), it is evident that the momentum of the platform would not be zero at the instant of zero error signal, so that the platform would overshoot the required position and thereafter oscillate about it. Such oscillation is unwanted and must be damped out as rapidly as possible.

As will be seen later, the rapidity with which the error is reduced and the new position adopted depends partly upon the inertia of the moving parts and partly on the way in which the resistance to motion varies while the parts accelerate.

Note that the torque or thrust of the actuating mechanism itself may be "amplified" by a gearbox, lever system or hydraulic/pneumatic relay as the case may be.

THE TRANSDUCER

The measurement of the quantity to be controlled often involves its conversion into different units, e.g. angular velocity into potential difference. The converting device is called a *transducer*—in the case just mentioned, a tachogenerator—and upon its accuracy depends that attainable by the system of control. Of the many other devices for transduction may be mentioned the Bourdon Tube (pressure change into angular rotation), strain gauge (change in length into resistance change), and thermocouple (temperature change into e.m.f.).

It may be remarked that, due to the relative ease with which minute electric signals can, by solid state electronic means, be detected, transmitted, amplified and manipulated generally, it is usual to convert the error signal into an electrical quantity. However, where reliability is essential under adverse operating conditions—prevalent high temperature, corrosive or dust-laden atmosphere, exposure to radiation and so on—pneumatic equipment may be preferable.

THE BLOCK DIAGRAM

As already explained the output from the load transducer is fed back, subtracted from the required quantity—in the same units—by the error detector, the difference transformed into a signal, amplified as necessary and fed forward to the actuating mechanism. Such a sequence may be illustrated by what is known as a *block diagram*. In such a diagram the various components or operations are represented by rectangles while the connecting arrows show either the flow of working medium (fluid, electrons, air) or of information.

CONTROL

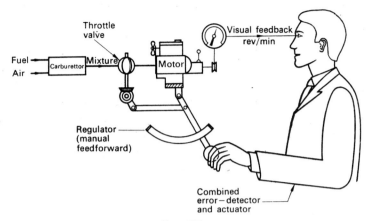

Fig. 271

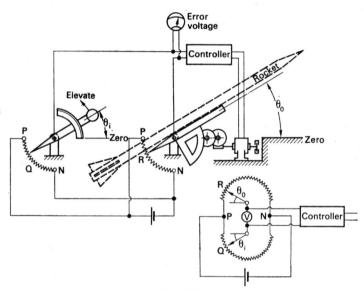

Fig. 272

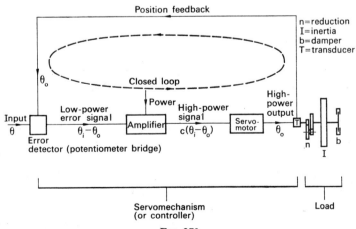

Fig. 273

Figure 271 illustrates in block form the speed control of a petrol motor using visual error detection and manual control. Figure 272 shows how the components of the launching system mentioned earlier might be arranged so as to give remote control of position, say elevation. Figure 273 is the corresponding block diagram.

The combination of error-detector, amplifier and actuator is known as a *servomechanism* (Latin: *servus*, a slave). Alternatively a servomechanism—or "servo" for short—may be defined as an "error-actuated device which causes a system output to follow automatically an input of the same sort prescribed by a human operator".

LAG IN RESPONSE

In the example of temperature control quoted earlier it was explained that the tendency to over-correct on the part of the operator was due to the time lag between cause and effect. In *any* system there are various inherent causes of delay in response so that in practice it is difficult to measure the input and output simultaneously. For example, due to torsional strain there is an angular lag of one end of a shaft behind the other, this lag being greatest during periods of acceleration. Again, due to inductive effect there is a delay in the build up of current—and hence of torque—after switching on a motor, while, due to inertia effect, a torque must act for a definite time before a required velocity is attained. An undesirable (but unavoidable) effect of such lags is to raise the order of the differential equation of motion and so increase the possibility of unstable operation.

TYPES OF INPUT

The input to a system may vary in any way, e.g. it may be subject to uniform change, have a high or a low rate of change or be subject to sinusoidal change of long (days) or short (milliseconds) periodic time. The first step therefore in the evaluation of a servo is to examine its reaction—i.e. its *response*—to each of the input forms mentioned.

STEP FUNCTION POSITION INPUT

Suppose the operator to rotate rapidly the handlever of Figure 272 through some angle θ_i from the zero position and hold it in the new position. The graph of θ_i against time will have the form shown in Figure 274.

Due to its inertia the platform response is not instantaneous so that an error is detected and amplified. Torque (or thrust) is applied to the platform and causes it to accelerate. As it approaches the required position the error and actuating force diminish and both are zero in that position. However, since the platform has been accelerated from rest, its kinetic energy causes it to overshoot so that an error and actuating force of opposite sign then bring it to rest before accelerating it in the reverse direction. The platform is evidently in a state of oscillation about the required position. If the motion is undamped the oscillation

CONTROL

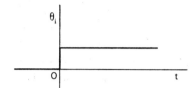

FIG. 274

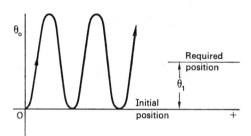

FIG. 275

is steady and the amplitude constant, the graph of output displacement against time being as shown in Figure 275.

Such a response—known as "hunting"—cannot be tolerated and requires the introduction of some form of damping.

DAMPING

By this is meant the introduction of a torque or force in opposition to the motion. In the case of viscous damping (see p. 295) the opposition is (theoretically) proportional to relative velocity and is therefore zero in the steady state. Approximately viscous damping may be provided by some kind of *dashpot* or *eddy current brake* and whichever is chosen should be made capable of accurate adjustment.

Since viscous damping is present at lubricated plain bearings it assists in reducing (attenuating) the amplitude of any torsional vibration of a motor armature and its load (two-mass system, see p. 281) initiated by the sudden increase in strain energy in the connecting shaft at the instant of switching on.

PROPORTIONAL CONTROL

Referring again to Figures 272 and 274,

let θ_i = step input (i.e. angle turned through by lever)
and θ_o = angle turned through by output (platform response).

Then, by definition,
　　Instantaneous error = $\theta_i - \theta_o$.

If the actuating torque on the platform is required to be proportional to this error and viscous damping is assumed—i.e. the damping torque is proportional to $d\theta_o/dt$—then, since the torque required to accelerate the platform is proportional to $d^2\theta_o/dt^2$ we have:

$$\text{Inertia torque} + \text{damping torque} = \text{actuating torque}$$

so that if I = inertia constant (moment of inertia of platform)
b = damping constant
c = actuating constant,

we have
$$I\frac{d^2\theta_o}{dt^2} + b\frac{d\theta_o}{dt} = c(\theta_i - \theta_o).$$

Rearrangement gives
$$\frac{d^2\theta_o}{dt^2} + \left(\frac{b}{I}\right)\frac{d\theta_o}{dt} + \left(\frac{c}{I}\right)\theta_o = \left(\frac{c}{I}\right)\theta_i.$$

Writing ω^2 for c/I and $2k$ for b/I we obtain
$$\frac{d^2\theta_o}{dt^2} + 2k\frac{d\theta_o}{dt} + \omega^2\theta_o = \omega^2\theta_i.$$

The left-hand side of this equation is identical in form to the left-hand side of the equation for forced damped vibration (see p. 300) so that putting it equal to zero will enable the type of transient motion to be identified. Thus

1. If $k < \omega$ the damping is sub-critical, i.e. the system is *underdamped* and there is

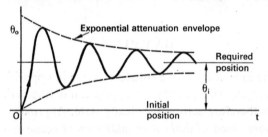

Fig. 276

transient oscillation of the output about the desired position. The graph of θ_o against time has the form shown in Figure 276, the solution being

$$\theta_o = \frac{\omega\theta_i}{\sqrt{(\omega^2 - k^2)}} e^{-kt} \cos\left[\sqrt{(\omega^2 - k^2)}\,t - \tan^{-1}\frac{k}{\sqrt{(\omega^2 - k^2)}}\right].$$

Dividing the right-hand side of the differential equation by the coefficient of θ_o to obtain the particular integral (i.e. the value of θ_o when the transient has died away) we obtain, for the steady state in which both $d\theta_o/dt$ and $d^2\theta_o/dt^2$ are zero:

$$\theta_o = \frac{\omega^2\theta_i}{\omega^2} = \theta_i.$$

2. If $k = \omega$ the damping is *critical*, i.e. there is just no overshoot. The graph of θ_o

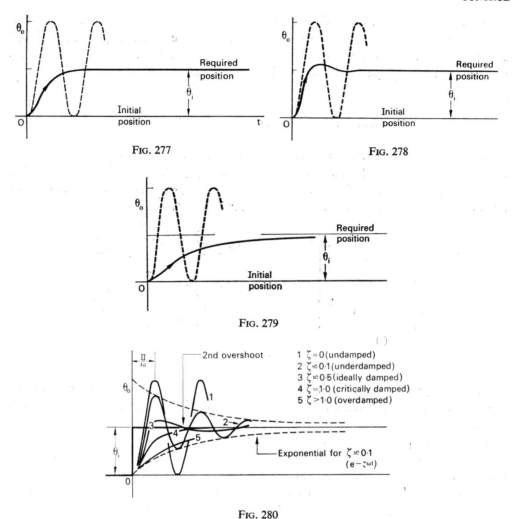

Fig. 277

Fig. 278

Fig. 279

Fig. 280

against time has the form shown in Figure 277, the solution being

$$\theta_o = e^{-kt}(At+B).$$

(The student should compare this with Figure 268.)

If slight overshoot is permissible (i.e. the graph of θ_o against time may have the form shown in Figure 278) a value of ζ of about 0·5 (or $k \simeq 0\cdot5\omega$) will give the most satisfactory response, such response being even more rapid than that corresponding to critical damping conditions.

3. If $k > \omega$, i.e. $\zeta > 1\cdot0$, the system is *overdamped* and the response is sluggish. The graph of θ_o against time has the form shown in Figure 279, the solution being

$$\theta_o = Ae^{[-k+\sqrt{(k^2-\omega^2)}]t} + Be^{[-k-\sqrt{(k^2-\omega^2)}]t}.$$

The effect on step response (i.e. on output response to a step function input) of increasing the value of the damping ratio from zero (undamped state) is best shown by superimposing the curves of Figures 275–9 to give Figure 280. Referring to the equation of motion it is evident that the rapidity of response depends upon the value of ω, i.e. upon the ratio c/I, so that the smaller the value of I and the larger that of c the higher the natural frequency of the system and the quicker a new position is adopted.

RAMP FUNCTION POSITION INPUT

If the lever of Figure 272 is replaced by a handwheel and this is suddenly given a uniform angular velocity instead of an angular displacement—i.e. if $d\theta_i/dt$ is made a step function instead of θ_i—the graph of θ_i against time will have the form shown in Figure 281(a), the relation being called a *ramp function*.

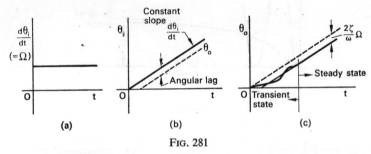

Fig. 281

Inertia torque causes a delay in response and hence an error-actuated correcting torque as before but tending, in this case, to bring the *speed* of the platform up to that of the handwheel.

If viscous friction is applied to the output, the transient variation in speed is similar to the variation in displacement just discussed, the speeds being identical when the transient has died away. However, the residual friction torque on the output at uniform speed requires an error $(\theta_i - \theta_o)$ proportional to it in order that power shall be supplied to overcome it. Hence, in the steady state (speeds equal) there is a fixed difference in the angular positions of input and output. This lag is given the name *steady state error* and is represented in Figure 281(b) by the vertical distance between the full and dotted lines. Figure 281(c) shows the actual graph of output displacement against time for the case where $\zeta \simeq 0.5$. In addition to introducing this error, the friction torque required for damping out the transient also results in unwanted heat so that, evidently, not all the actuating torque is usefully employed.

An expression for this error will now be found. Referring to Figure 281(b),

$$\text{Slope} = \frac{\theta_i}{t} = \frac{d\theta_i}{dt} = \Omega, \quad \text{say (assumed constant),}$$

or

$$\theta_i = \Omega t.$$

Substituting for θ_i in the equation of motion (p. 318) and putting $k = \zeta\omega$ gives

$$\frac{d^2\theta_o}{dt^2} + 2\zeta\omega\frac{d\theta_o}{dt} + \omega^2\theta_o = \omega^2\Omega t.$$

Since $d^2\theta_o/dt^2$ is zero in the steady state, the particular integral is obtainable from

$$2\zeta\omega \frac{d\theta_o}{dt}+\omega^2\theta_o = \omega^2\Omega t$$

so that, writing D for d/dt, we obtain

$$(2\zeta\omega D+\omega^2)\theta_o = \omega^2\Omega t,$$

i.e.
$$\left(\frac{2\zeta}{\omega}D+1\right)\theta_o = \Omega t \quad \text{(dividing by } \omega^2\text{)}.$$

The platform position in the steady state therefore is given by

$$\theta_o = \Omega t\left(1+\frac{2\zeta}{\omega}D\right)^{-1}$$

$$= \Omega t\left[1+(-1)\left(\frac{2\zeta}{\omega}D\right)+\frac{(-1)(-2)}{2!}\left(\frac{2\zeta}{\omega}\right)^2 D^2,\ \text{etc.}\right]$$

$$= \Omega t\left[1-\frac{2\zeta}{\omega}D+\frac{4\zeta^2}{\omega^2}D^2,\ \text{and so on}\right]$$

$$= \Omega t - \Omega\frac{2\zeta}{\omega}\frac{d(t)}{dt},$$

i.e.
$$\theta_o = \theta_i - \frac{2\zeta}{\omega}\Omega.$$

It is also clear that, $\quad\dfrac{d\theta_o}{dt} = \dfrac{d\theta_i}{dt}.$

Thus the speeds are equal—i.e. there is no error in velocity—but the output lags the input by the fixed angle $(2\zeta/\omega)\Omega$. Evidently this positional lag (known, unfortunately, as *velocity lag*) increases with the degree of damping as well as with the velocity.

EXAMPLE. A rotor having a moment of inertia of 0·054 kg m² is driven via a flexible coupling the torsional stiffness of which is 540 Nm per radian. Neglect the inertia of the connecting shaft and find the natural frequency of the combination, assuming the input side of the coupling to be held stationary. Determine the effect on this frequency of adding a viscous damper giving a damping ratio of 0·6. If the assembly is driven at 145 rev/min estimate the value of the steady-state positional lag.

Solution

Natural undamped frequency, $\quad n = \dfrac{\omega}{2\pi} = \dfrac{1}{2\pi}\sqrt{\dfrac{q}{I}}$

$$= \frac{1}{2\pi}\sqrt{\frac{540}{0{\cdot}054}} = 15{\cdot}9\ \text{Hz}.$$

When $\quad \zeta = 0{\cdot}6$:

$$n = \frac{1}{2\pi}\omega\sqrt{(1-\zeta^2)} = \frac{1}{2\pi}100\sqrt{(1-0{\cdot}6^2)} = 12{\cdot}7\ \text{Hz}.$$

MECHANICAL TECHNOLOGY FOR HIGHER ENGINEERING TECHNICIANS

At 145 rev/min:
$$\frac{d\theta_i}{dt} = \Omega = \frac{2\pi \times 145}{60} = 15 \cdot 2 \text{ rad/s}$$

∴ Positional lag, $\frac{2\zeta}{\omega}\Omega = \left(\frac{2 \times 0 \cdot 6}{100}\right) 15 \cdot 2$

$$= 0 \cdot 182 \text{ radian}$$
$$\simeq \underline{10 \cdot 5 \text{ deg.}}$$

EFFECT ON POSITIONAL LAG OF UNIFORM LOAD TORQUE

If a uniform opposing torque, T, is applied to the output of the system represented by the previous equation of motion, the modified equation is

$$I\frac{d^2\theta_o}{dt^2} + \left(b\frac{d\theta_o}{dt} + T\right) = c(\theta_i - \theta_o)$$

which reduces to

$$\frac{d^2\theta_o}{dt^2} + 2\zeta\omega\frac{d\theta_o}{dt} + \omega^2\theta_o = \omega^2\Omega t - \frac{T}{I}.$$

Reasoning similar to that used previously shows that the output position in the steady state is obtainable from

$$\theta_o = \theta_i - \left(\frac{2\zeta}{\omega}\Omega + \frac{T}{I\omega^2}\right)$$

where
$$I = \frac{c}{\omega^2}.$$

Thus the positional lag now is $(2\zeta/\omega)\Omega + T/c$.

VELOCITY FEEDBACK

The equation of motion for proportional control may be transposed to read

$$I\frac{d^2\theta_o}{dt^2} = c(\theta_i - \theta_o) - b\frac{d\theta_o}{dt}.$$

Inspection of this equation reveals that an identical step velocity response will be obtainable *without* friction damping of the output if the term $b(d\theta_o/dt)$ represents a signal proportional to output velocity, i.e. if the actuating torque is made proportional to the error less such a signal. By simulating viscous damping in this way (known as *stabilisation by velocity feedback* or *velocity damping*) none of the actuating torque is, theoretically, wasted. This leads therefore to an improvement in efficiency.

Further inspection of the equation shows that, while $d\theta_o/dt$ is changing, i.e. while the system is in the transient state, the accelerating torque on the output is not only zero when

$b(d\theta_o/dt) = c(\theta_i - \theta_o)$, i.e. before overshoot, *but is negative* (reversed) *and increasing* from that point onward. Thus the response is now similar to that of an operator anticipating overshoot and reversing the correcting torque in good time. Moreover, the kinetic energy of the system at and during overshoot, instead of being converted into heat, may be conserved by returning it to the supply via the motor since this can be arranged to function as a generator.

To obtain a signal proportional to the derivative of the output (i.e. to $d\theta_o/dt$) a d.c. generator having a linear speed-voltage characteristic is driven from the output. Its voltage is then subtracted from the error signal, the difference amplified and the result fed forward. Figure 282 shows a suitable block diagram.

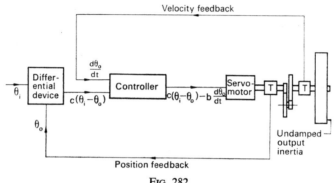

Fig. 282

In the steady state $d^2\theta_o/dt^2$ is zero so that then

$$c(\theta_i - \theta_o) = b\frac{d\theta_o}{dt} \quad \text{and} \quad \frac{d\theta_o}{dt} = \frac{d\theta_i}{dt} = \Omega,$$

$$\therefore \quad \theta_i - \theta_o = \left(\frac{b}{c}\right)\Omega$$

$$= \frac{2kI}{I\omega^2}\Omega \quad \text{where} \quad k = \zeta\omega$$

so that

$$\theta_i - \theta_o = \frac{2\zeta}{\omega}\Omega.$$

Thus, although velocity feedback eliminates the energy loss associated with viscous damping of the output, the position error remains unchanged. This is because, in the steady state (constant speeds), the then constant velocity signal must be opposed by an equal and opposite constant error signal and this requires a constant error. Since, inevitably, there is also some output resistance due to both viscous and Coulomb friction, the actual lag is somewhat greater than $(2\zeta/\omega)\Omega$.

Since a signal proportional to the derivative of the output is subtracted from a signal proportional to the error in position, the terms *derivative compensation* and *derivative of output feedback* are also in use.

The student should note that, since θ_o need not necessarily be an angle—i.e. $d\theta_o/dt$ need not necessarily be a velocity—this type of feedback, using a signal proportional to the rate of change in the output quantity, can be applied to the control of *any* variable and is in fact in wide use. He should also note that a fall in actuating torque when the speed is rising—and vice versa—may be achieved more easily by using a motor having a suitable speed-torque characteristic.

Velocity feedback may of course be used in conjunction with a viscous-damped output, the equation of motion then being

$$I\frac{d^2\theta_o}{dt^2} + b\frac{d\theta_o}{dt} = c(\theta_i - \theta_o) - b_1 \frac{d\theta_o}{dt}$$

where b and b_1 are, respectively, the viscous damping constant and the tachogenerator constant. Rearrangement gives

$$I\frac{d^2\theta_o}{dt^2} + (b+b_1)\frac{d\theta_o}{dt} + c\theta_o = c\theta_i.$$

Evidently in this case the positional lag is increased because the damping constant is increased from b to $(b+b_1)$. Note that, if the polarity of the tachogenerator is reversed—whether accidentally or otherwise—the constant b_1 becomes negative. There is then zero damping when $b_1 = b$ and negative damping when $b_1 > b$. In the latter case the system is unstable.

ERROR RATE DAMPING (DERIVATIVE CONTROL)— PHASE ADVANCE NETWORK

It has been shown that, although energy loss in a speed control system can be reduced by eliminating viscous damping of the output and simulating it by means of velocity feedback, a steady state error in position of $(2\zeta/\omega)\Omega$ results from the then constant signal from the output tachogenerator. Now, if this value of the feedback signal could be opposed by an equal and opposite signal—i.e. if $(2\zeta/\omega)\Omega$ or $(b/c)\Omega$ could be added in the steady state to the input angle—the only input to the controller when the transient has disappeared would be the position error signal and this could then be zero. This state of affairs is realisable by driving a second tachogenerator from the input, adjusting its steady state voltage to be equal and of opposite polarity to that of the first and feeding it *forward* as shown in Figure 283.

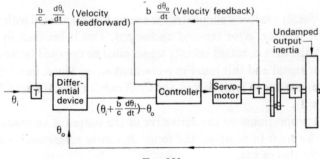

Fig. 283

This addition to the system requires the addition of $b(d\theta_i/dt)$ to the right-hand side of the equation of motion for velocity feedback so giving

$$I\frac{d^2\theta_o}{dt^2} = c(\theta_i - \theta_o) - b\frac{d\theta_o}{dt} + b\frac{d\theta_i}{dt}$$

$$= c\left(\theta_i - \theta_o + \frac{b}{c}\frac{d\theta_i}{dt}\right) - b\frac{d\theta_o}{dt}$$

$$= c\left[\left(\theta_i + \frac{b}{c}\frac{d\theta_i}{dt}\right) - \theta_o\right] - b\frac{d\theta_o}{dt}.$$

This represents the adding of $\dfrac{b}{c}\dfrac{d\theta_i}{dt}$ to θ_i before measuring the error so that the equation may be written

$$I\frac{d^2\theta_o}{dt^2} + b\frac{d\theta_o}{dt} + c\theta_o = c\left(\theta_i + \frac{b}{c}\frac{d\theta_i}{dt}\right).$$

In the steady state $d^2\theta_o/dt^2 =$ is zero while $d\theta_o/dt = \Omega = d\theta_i/dt$
so that
$$b\Omega + c\theta_o = c\theta_i + b\Omega,$$
i.e.
$$\theta_o = \theta_i.$$

Thus the lag in position has now disappeared.

The ratio b/c is known as the *velocity coefficient* of a system and its value is theoretically constant for all values of $d\theta_i/dt$ (i.e. of input velocity). Since, however, other parameters are not in practice constant, a best value has to be found by trial and error.

When the transient has died away the arrangement just described gives a servomotor output proportional only to the positional lag. Velocity feedback then exists only at output speeds differing from the steady state speed.

Now the two tachogenerators of Figure 283 together supply the controller with a signal proportional to $(d\theta_i/dt - d\theta_o/dt)$, that is, proportional to $d/dt(\theta_i - \theta_o)$. This is the rate of change in the error so that the required state of affairs (zero position error) is realisable by substituting for them a single device capable of emitting a signal proportional to the derivative of the error, or *error rate*, and feeding it forward to the amplifier as shown in Figure 284.

This method, called *error rate damping*, results during the transient period in a servomotor torque proportional both to the error itself and to the rate at which it is changing, i.e. to the rapidity with which the output speed is approaching/leaving the desired value.

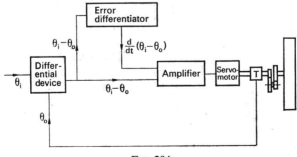

FIG. 284

In other words the amplifier input contains an anticipatory element which is greatest at the start of overshoot when the error is increasing most rapidly. In the steady state both the error and its derivative are zero.

Since the derivative of the error is difficult to obtain by mechanical means, it is usual to employ the network shown in Figure 285 (known variously as a *stabilising, transitional lead* or *phase advance* network) connected between the error detector and the amplifier.

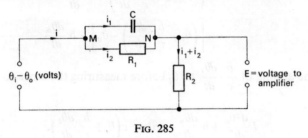

FIG. 285

Suppose a step function voltage to be applied to the control system. Then

$$\text{Voltage across MN} = \frac{1}{C}\int i_1 \, dt = i_2 R_1$$

or, using operator D, $\quad \dfrac{1}{C}\dfrac{1}{D}(i_1) = i_2 R_1,$

whence $\quad i_1 = R_1 CD(i_2) \quad$ and $\quad R_1 C = T,$

so that $\quad i_1 = TD(i_2).$

Hence $\quad \dfrac{i_2 R_1}{i_1 + i_2} = \dfrac{i_2 R_1}{TD(i_2) + i_2} = \dfrac{R_1}{1 + TD}.$

Voltage ratio, $\quad \dfrac{E}{\theta_i - \theta_o} = \dfrac{(i_1 + i_2)R_2}{i_2 R_1 + (i_1 + i_2)R_2}$

$$= \dfrac{R_2}{\dfrac{i_2 R_1}{i_1 + i_2} + R_2}$$

$$= \dfrac{R_2}{\dfrac{R_1}{1 + TD} + R_2} \quad \text{(from above)}$$

$$= \dfrac{R_2(1 + TD)}{R_1 + R_2(1 + TD)}$$

$$= \dfrac{\dfrac{R_2}{R_1 + R_2}(1 + TD)(R_1 + R_2)}{(R_1 + R_2) + R_2 TD}$$

$$= \dfrac{\dfrac{R_2}{R_1 + R_2}(1 + TD)}{1 + \left(\dfrac{R_2}{R_1 + R_2}\right)TD}.$$

Writing γ for the ratio $R_2/(R_1+R_2)$ we obtain

$$\frac{E}{\theta_i-\theta_o} = \frac{\gamma(1+TD)}{1+\gamma TD}.$$

If γ is small (in practice it approximates to 0·1) the denominator approaches unity in which case the voltage across the amplifier may be taken as given by

$$E = \gamma(1+TD)(\theta_i-\theta_o)$$

or

$$E = \gamma(\theta_i-\theta_o) + \gamma T \frac{d}{dt}(\theta_i-\theta_o).$$

The input to the amplifier is therefore the sum of a component proportional to the error and a component proportional to its derivative. That the second component is of opposite sign to the direction of change in θ_o (negative when θ_o is increasing—and vice

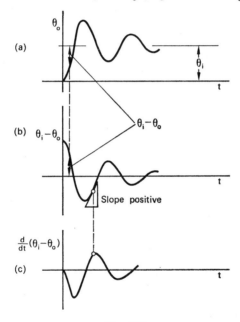

Fig. 286

versa) and has maximum values when the error is zero (i.e. at overshoot) is evident from Figure 286. This shows (a) the step response, (b) the corresponding variation in error with time, and (c) the corresponding variation in the slope of the error curve.

INTEGRAL CONTROL

A simpler method of compensation not dependent on the value of the ratio b/c enables positional (velocity) lag to be eliminated against a steady state output torque by using an additional component of actuating torque proportional to the time integral of the error.

Suppose the equation to be

$$I\frac{d^2\theta_o}{dt^2} + b\frac{d\theta_o}{dt} = c(\theta_i - \theta_o) + c_1 \int (\theta_i - \theta_o)\, dt$$

where c_1 is the constant of the integrating component. Then

$$I\frac{d^2\theta_o}{dt^2} + b\frac{d\theta_o}{dt} + c\theta_o = c\theta_i + c_1 \int (\theta_i - \theta_o)\, dt.$$

Differentiation of this equation gives the 3rd Order equation

$$I\frac{d^3\theta_o}{dt^3} + b\frac{d^2\theta_o}{dt^2} + c\frac{d\theta_o}{dt} = c\frac{d\theta_i}{dt} + c_1(\theta - \theta_o).$$

In the steady state

$$\frac{d\theta_o}{dt} = \frac{d\theta_i}{dt} = \Omega$$

so that $\dfrac{d^2\theta}{dt^2}$ and $\dfrac{d^3\theta}{dt^3}$ are zero.

Hence
$$c\Omega + c_1\theta_o = c\Omega + c_1\theta_i$$
so that
$$\theta_o = \theta_i = \Omega t.$$

The positional lag in a system having the equation assumed would therefore be zero.

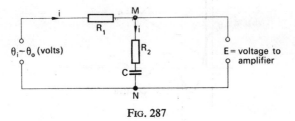

FIG. 287

The time integral of the error is obtained, usually, by means of the network shown in Figure 287—known as a *transitional lag* network. If a step function voltage is applied to the control system, reference to Figure 287 shows that the voltage across MN is

$$E = iR_2 + \frac{1}{C}\int i\, dt$$

$$= iR_2 + \frac{1}{C}\frac{1}{D}(i) \quad \text{using operator } D,$$

$$= i\left(R_2 + \frac{1}{CD}\right).$$

Integrator input voltage, $\theta_i - \theta_o = iR_1 + E$

$$= iR_1 + i\left(R_2 + \frac{1}{CD}\right)$$

$$= i\left(R_1 + R_2 + \frac{1}{CD}\right).$$

The ratio between the voltage across the amplifier to the input voltage is, therefore,

$$\frac{E}{\theta_i - \theta_o} = \frac{i[R_2 + 1/(CD)]}{i[R_1 + R_2 + 1/(CD)]}$$

$$= \frac{(R_2 CD + 1)/CD}{(R_1 CD + R_2 CD + 1)/CD}$$

$$= \frac{R_2 CD + 1}{(R_1 + R_2)CD + 1}$$

$$= \frac{R_2 CD + 1}{R_2 CD + R_2/(R_1 + R_2)} \times R_2/(R_1 + R_2).$$

So that, putting $R_2/(R_1 + R_2) = \gamma$ and $R_2 C = T$ we obtain

$$\frac{E}{\theta_i - \theta_o} = \frac{\gamma(1 + TD)}{TD + \gamma}.$$

If γ is small the denominator approaches TD in which case the voltage across the amplifier may be taken as

$$E = \gamma \left(1 + \frac{1}{TD}\right)(\theta_i - \theta_o)$$

or

$$E = \gamma(\theta_i - \theta_o) + \frac{\gamma}{T}\int (\theta_i - \theta_o)\,dt.$$

The input to the amplifier is therefore the sum of a component proportional to the error and a component proportional to its integral. Referring to the supposed equation it is seen that $\gamma = c$ and that $\gamma/T = c_1$.

This method is known as *integral-of-error feedforward* and the relevant block diagram is shown in Figure 288.

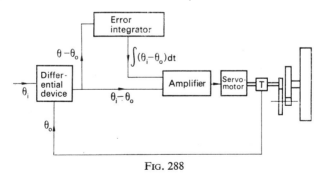

Fig. 288

SUMMARY

Three distinctive forms of control have been considered and the equations are repeated below for comparison.

1. Proportional Control

$$I\frac{d^2\theta_o}{dt^2} + b\frac{d\theta_o}{dt} + c\theta_o = c\theta_i.$$

If θ_i is an angular velocity, there is a steady state position lag.

2. Derivative Control

$$I\frac{d^2\theta_o}{dt^2} + b\frac{d\theta_o}{dt} + c\theta_o = c\left(\theta_i + \frac{b}{c}\frac{d\theta_i}{dt}\right).$$

The output is the sum of components proportional to the error and to its rate of change and, where θ_i is an angular velocity, the steady state lag in position is eliminated.

3. Integral Control

$$I\frac{d^2\theta_o}{dt^2} + b\frac{d\theta_o}{dt} + c\theta_o = c\theta_i + c_1\int(\theta_i - \theta_o)\,dt.$$

The output is the sum of components proportional to the error and to its duration and, where θ_i is an angular velocity, the steady state position lag is eliminated.

APPENDICES

1

NOTES ON METHODS OF ASSESSING POWER OUTPUT FOR I.C. MOTORS PRIOR TO THE INTRODUCTION OF SI

JAMES WATT first measured the sustained performance of mill horses and arrived at a figure of 32 400 ft lbf/min. By measuring the energy output of a stationary steam engine in the same units he was able to estimate the number of horses which it replaced. By 1783 he had modified this "standard" rate of doing work to 33 000 ft lbf/min (745·7 watts) and called it the "horsepower".

Where petrol motors are concerned the output rises with both speed and torque so that any method of power calculation which neglects to take either into account is of little value. However, such a formula was devised by the Royal Automobile Club (for fiscal purposes) early in this century and was, until recently, known as the *RAC Rating*. According to it

$$\text{(Fiscal) h.p.} = \frac{\text{(No. of cylinders) (cylinder diameter in inches)}^2}{2\cdot 5}.$$

The French equivalent is obtained from

(Fiscal) c.v. = (No. of cylinders) (cylinder diameter in cm)^2 (stroke in cm)30

and is given to the nearest whole number.

(Neither of the above is a true measure of power.)

Under the metric system the horsepower is defined as a rate of working equivalent to 75 kgm/s. Since this is equal to 735·5 watts it is evident that the metric h.p. (c.v. in French, p.s. in German) is about 1·5 per cent smaller than the British. It should be noted that, in the automobile industry, there are in common use[*] three official systems of stating the power developed by a petrol motor, none of which specifies a minimum time for which the rated output must be sustained.

1. SAE (American). This figure gives what is usually meant by the gross output (in h.p.), i.e. the output at the flywheel with the generator/alternator, fan, fuel pump and water pump separately driven, with the ignition timing and fuel mixture set to give optimum

---

[*] At the time of writing.

APPENDIX 1

performance and with the clutch, air filter and silencer disconnected. Under operating conditions the input to the transmission may be as little as 80 per cent of the quoted figure.

2. DIN (German). This figure gives the output (in p.s.) at the flywheel with all auxiliaries connected and is a reliable guide to what may be expected under ideal operating conditions.

3. CUNA (Italian). This figures gives the output (in p.s.) at the flywheel with all auxiliaries connected except air filter and silencer and is between 5 and 10 per cent higher than the DIN rating. Both DIN and CUNA values must be multiplied by 0·985 to convert to h.p.

2

DIMENSIONS

MEASUREMENT is the comparison of a quantity—height, inductance, viscosity—with a standard quantity of the same kind called a "unit", the size of which may be arbitrary and is usually chosen for convenience. There exist several systems of measurement for most quantities and, within any system, multiples of units having various names. Between the units of different quantities there may or may not be a relation.

Quantities of the same kind are said to have the same *dimension*, e.g. distances may be measured using different units (feet, kilometres), the shared dimension in this case being *length*. For algebraic purposes this is denoted by L and, evidently, the dimensions of area and volume are represented by L^2 and L^3 respectively.

If the dimension of time is represented by T, then the dimensions of velocity and acceleration are represented by L/T (or LT^{-1}) and L/T^2 (or LT^{-2}) respectively. In the same way if the dimension of mass is represented by M, then the dimensions of force (mass × acceleration) and moment of inertia (mass × k^2) are respectively MLT^{-2} and ML^2. Thus it is seen that a relation exists between the dimensions of different kinds of quantity.

Since the dimensions of any engineering quantity may be expressed in terms of M, L and T, these are called the *basic* or *fundamental* dimensions. Evidently the ratio of two quantities having the same dimension has itself *no* dimensions, i.e. it is a pure number (e.g. the radian). Such a ratio is said to be *dimensionless*. It follows that multiplication of any quantity by such a number does not alter the dimensions of that quantity. It also follows that the two sides of an equation *must* have the same dimensions, i.e. that the relation between quantities must be *dimensionally homogeneous*. The student should get into the habit of checking empirical or unfamiliar equations for dimensional correctitude. Such a check will reveal the existence of any quantity concealed, say, by the inclusion of a constant.

3

SOME BOOKS FOR FURTHER READING

Fisher, *Carburation*, Chapman and Hall.
Rogers and Mayhew, *Engineering Thermodynamics*, Longmans, 1967.
Thomas and Hunt, *Applied Heat*, Heinemann, 1970.
Campbell, *The Sports Car Engine*, Chapman and Hall, 1963.
Wrangham, *Elements of Heat Flow*, Chatto and Windus.
Smith, *Exhaust and Intake Systems*, Foulis, 1962.
Young and Pryor, *The Testing of I.C. Engines*, EUP.
Howarth, *Design of High Speed Diesel Engines*, Constable, 1966.
Black, *Strength of Materials*, Pergamon, 1966.
Duncan Thorn and Young, *Mechanics of Fluids*, Arnold.
Pearson, *Applications of Valves and Fittings*, Pitman, 1968.
Goodwin, *Power Hydraulics*, Cleaver-Hume, 1963.
Black, *Mechanics of Machines*, Pergamon, 1967.
Morris, *Control Engineering*, McGraw-Hill, 1968.
Pearson, *Technology of Instrumentation*, EUP, 1957.
Corrosion in Action, International Nickel Co., 1955.
Hydraulic Oils, Wakefield Castrol Group Ltd., 1960.
The Application of Lubricants, Shell I.P. Co. Ltd., 1965.
Designers' Handbooks, Nos. 1–9, The Glacier Metal Co.
Biggs, *Mechanical Behaviour of Engineering Materials*, Pergamon, 1965.
Lubrication of Industrial Gears, Shell I.P. Co. Ltd., 1964.
T. Potma, *Strain Gauges: Theory and Application*, Iliffe.
Bryan, *Control Systems for Technicians*, U.L. Press, 1967.
Redford, *Mechanical Engineering Design*.
Mayall, *Industrial Design for Engineers*, Iliffe, 1967.
Petersen, *Stress Concentration Design Factors*, Wiley, 1953.
Martin, *Handbook of Industrial Pipework*, Pitman, 1961.
Giles, *Vehicle Operation and Testing*, Iliffe, 1969.

4

SOME MORE CONVERSION FACTORS

	1 lbf	= 4·45 N
	1 lbf/in	= 175 N/m
	1 lbf/ft	= 14·57 N/m
(a)	1 lbf/in²	= 6900 Pa
	1 lbf/ft²	= 48 Pa
	1 lbf in	= 0·113 Nm
(b)	1 lbf ft	= 1·356 Nm
(c)	1 in lbf	= 0·113 J
	1 in	= 0·0254 m
	1 in²	= 645×10⁻⁶ m²
	1 in³	= 16·4×10⁻⁶ m³
(d)	1 in⁴	= 0·416×10⁻⁶ m⁴
	1 lb	= 0·453 kg
	1 lb/in³	= 27 650 kg/m³
	1 lb/ft³	= 16·02 kg/m³
(e)	1 lb/bhp-hr	= 0·169 kg/MJ
	1 tonf	= 9964 N
	1 tonf/in²	= 15·44 MPa
	1 tonf/ft	= 32 700 N/m
(f)	1 tonf in²	= 6·42 Nm²
(g)	1 ft lbf/lb deg F	= 2·99 Nm/kg deg K
(h)	1 Btu in/ft² h deg F	= 0·1442 Wm/m² deg K
	1 Btu ft/ft² h deg F	= 1·73 Wm/m² deg K
(i)	1 Btu/ft² h	= 3·16 J/m² s
(j)	1 Btu ft² h deg F	= 5·68 W/m² deg K
(k)	1 ft² h deg F/Btu	= 0·1765 m² deg K/W
	1 ft	= 0·305 m
	1 ft/s	= 0·305 m/s
	1 ft/s²	= 0·305 m/s²
	1 ft²	= 0·0929 m²
	1 ft³	= 0·0283 m³

APPENDIX 4

1 ft³/lb	= 0·0624	m³/kg
1 ft⁴	= 0·00863	m⁴
1 ft lbf	= 1·356	J
1 kWh	= 3·6	MJ
1 hp-hr	= 2·68	MJ
1 cal	= 4·187	J
1 Btu	= 1055	J
1 slug	= 14·59	kg
(m) 1 slug ft²	= 1·356	kg m²
1 gallon	= 4·546	litres
1 mile	= 1·609	km
(n) 1 Btu/lb	= 2325	J/kg
1 Btu/hr	= 0·293	W
1 Btu/ft²	= 11 350	J/m²
(p) 1 Btu/ft³	= 37·26	kJ/m³

(a) stress and pressure
(b) torque and bending moment
(c) strain energy
(d) second moment of area
(e) specific fuel consumption
(f) flexural rigidity
(g) characteristic constant
(h) thermal conductivity
(i) energy transfer
(j) surface coefficient
(k) thermal resistance
(m) moment of inertia
(n) calorific value, solids
(p) calorific value, gases

Finally, (miles per gallon)×(litres per 100 km) = 282·5.

5

A FEW INTERESTING DATES

Blaise Pascal	1623–1662 (see p. viii)
Robert Hooke	1635–1703
Jacob Bernoulli	1654–1705
Daniel Bernoulli	1700–1782 (son of Jacob)
Leonhard Euler	1707–1783
C. A. Coulomb	1736–1806
G. B. Venturi	1746–1822
J. A. C. Charles	1746–1823
M. I. Brunel	1769–1849 (father of I. K. B.)
Thomas Young	1773–1829 (of modulus fame)
S. D. Poisson	1781–1840
J. J. Guest	1785–1852
Michael Faraday	1791–1867
G. G. Coriolis	1792–1843
G. Lamé	1795–1870
H. P. G. Darcy	1803–1858
Eugene Bourdon	1808–1884
William Froude	1810–1879
J. P. Joule	1818–1889
W. J. M. Rankine	1820–1872 (founder of thermodynamics)
Otto Möhr	1835–1918
Clemens Herschel	1842–1930 (invented venturimeter)
C. A. Castigliano	1847–1894
Heinrich Hertz	1857–1894

Volume 2, *History of Technology*, is very informative. So are the two volumes of *Engineering Heritage*, published by the Institution of Mechanical Engineers.

6

THE 0·1 PER CENT PROOF STRESS FOR A FEW MATERIALS

Material	MPa
Mild steel	230
Monel	175
Aluminium bronze	170
Beryllium copper	160
Aluminium	140
Phosphor bronze	110
Brass	105
Copper	62

If the maximum stress in a component as calculated for the expected conditions of service is less than some certain value called the *working stress*, the component will have a margin of security against failure. The working value must be less than the proof stress because of

(1) inaccuracy of stress analysis,
(2) imperfection of material,
(3) uncertainty about operating conditions.

The margin of security may be reduced in proportion to the reliability of the material, knowledge of environment, accuracy of assumptions on which calculations are based and the degree to which failure would be unattended by loss or damage.

7

INTERNATIONAL PAPER SIZES

The range of sizes adopted by the International Organisation for Standardisation is based on a rectangle one square metre (1 m²) in area, this basic size being designated A0 and having sides in the ratio $1 : \sqrt{2}$. A series of smaller sizes, each of which has the same geometric proportion, is obtained by halving the basic size successively to give designations A1, A2, A3 and so on, the figure following the letter A denoting the number of divisions. Referring to Figure 289:

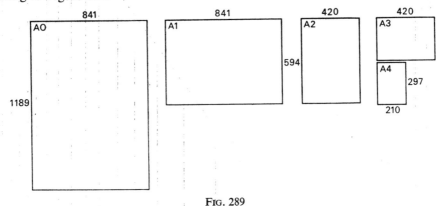

Fig. 289

Size A0 $= \dfrac{1189 \times 841}{10^6} = 1\cdot 0$ m² (where $1189 = \sqrt{2} \times 841$).

Size A4 $= \dfrac{1\cdot 0}{2^4} = 0\cdot 0625$ m².

Designation	Size in mm	Usage	Designation	Size in mm	Usage
4A0	2378×1682 ⎫	Large drawings	A4	297×210	Reports
2A0	1682×1189 ⎭		A5	210×148	Correspondence
A0	1189×841 ⎫		A6	148×105	Postcards
A1	841×594 ⎮	Normal drawings	A7	105×74	Labels
A2	594×420 ⎮				
A3	420×297 ⎭				

The two largest sizes are multiples of the basic size.

8

VALUES OF $e^{\mu\theta}$

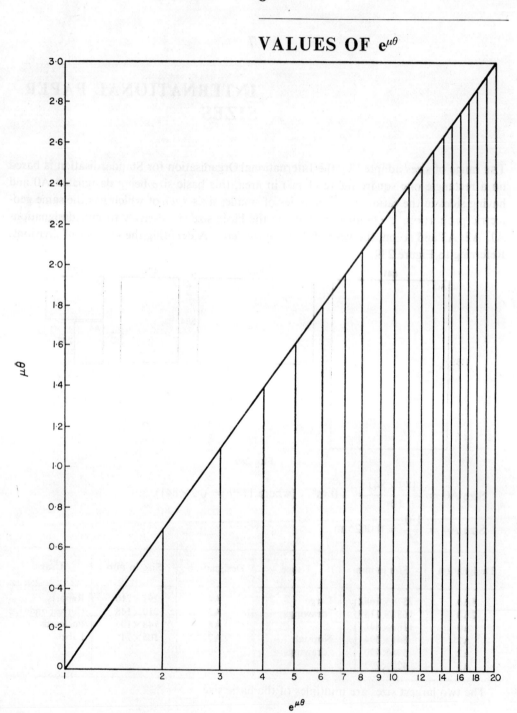

9

SOME NOTES ON PROPORTION

By proportion is meant either the relation of one principal dimension of a component to another, or the dimensional relation of part of an assembly to other parts. In engineering, proportion is judged almost inevitably in terms of function although the basing of judgement on familiar shapes can, occasionally, result in unjustified criticism of a new form.

The use of series, geometric, arithmetic, etc., can lead to many proportional relationships. One of these is based on the so-called *golden ratio*, known to the ancient Greeks. This is a mathematical ratio and is best explained by referring to Figure 290 which depicts a rectangle *ABCD*, the sides of which are in the ratio $1 : 1+x$ where $1/(1+x) = x/1$. Manipulation of this gives $x^2+x-1 = 0$ whence $x = 0.618$ so that the sides of the rectangle are in the ratio $1 : 1.618$.

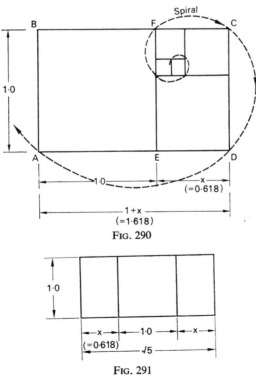

Fig. 290

Fig. 291

APPENDIX 9

[Note that: $1/1\cdot618 = 0\cdot618 = 1\cdot618 - 1\cdot0$. Thus the reciprocal of the ratio is obtained by subtracting unity from it. This is possible only with this number.]

It follows that subtraction of the square *ABFE* (of unit side) from the rectangle *ABCD* will leave a rectangle *FCDE* having the same proportions. (The line *EF* is known as the *golden mean* and is useful in proportioning control panels etc.) If further squares are drawn successively, a logarithmic spiral can be constructed by joining up corners as shown, Fig. 290.

Now,
$$1+2x = 1+(2\times 0\cdot618)$$
$$= 2\cdot236$$
$$= \sqrt{5}.$$

This leads to the rectangle shown in Figure 291 which consists of a square flanked by two "golden rectangles".

342

INDEX

Acceleration
  centripetal 102
  Coriolis component 143
  diagram 138
  linear and angular 101
  of a gear train 107

Balancing
  of couples 119
  of forces 117
Bending
  assumptions 13
  cantilever 25, 27, 43, 47
  deflection coefficient 30
  differential equation of flexure 25
  equation 15
  first theorem of Castigliano 46
  moment of resistance to 14
  neutral plane and axis 11
  relation between w, F and M 15
  section modulus 22
  shear stress due to 49
  sign convention 11
  simply supported beam 28, 30, 33, 42, 44, 48
  strain due to 11
  strain energy of 41
  stress due to 12, 14
  with direct stress 21
  with torsion 75

Centre of pressure 155, 161
Combustion of
  acetylene 212
  carbon 207
  carbon monoxide 208
  hydrocarbons 210
  hydrogen 206
  hexane 213
  methyl alcohol 221
  pump petrol 216
  sulphur 208
Control
  closed loop 312
  derivative 324
  integral 327
  open loop 313
  proportional 317
Cylinder
  rotating 94
  thick 87
  thin 93

Energy
  datum 165
  hydrostatic 164
  internal 195
  kinetic 102, 165, 203
  pressure 165
  strain 3, 11, 41
Energy transfer
  at constant pressure 198
  at constant temperature 200
  at constant volume 198
  by conduction through a lagged pipe 237
  by conduction through a uniform pipe 236
  by conduction through a uniform sheet 227
  during combustion 207
  rate of 219
  steady flow equation 196
  through boundary layer 232
  work 136, 190
Equations of motion 101

Flow
  along pipes 173, 177
  laminar 169, 172, 173
  process 189
  steady 165, 167, 173, 181, 182, 196
  turbulent 172, 177
Force
  centripetal 102
  damping 317
  friction 243
  gyroscopic 104
  inertia 147
  shear 11, 243
Friction
  at screw threads 244

INDEX

Friction (*cont.*)
  between annular surfaces 247, 271
  between conical surfaces 251
  between cylindrical surfaces 252, 255, 268, 270
  boundary 243
  Coulomb 242
  on inclined plane 244

Gears
  epicyclic 111
  simple 107
  torque conversion 115
Gyroscopic effects 103

Mechanisms
  four bar chain 128, 136, 138, 140, 142, 154
  instantaneous centre 137
  quick return 133, 134
  reciprocating 130, 141
Modulus
  of bulk 60
  of elasticity 62
  of rigidity 1
Momentum 102, 103, 104

Orifices 182

Poisson's ratio 60
Principal planes and stresses 64

Relation between E, G and K 62

Shear
  complementary 57
  moment of resistance to 2
  sign convention 11
  strain and modulus 1
  stress due to bending 49
  stress due to tension 56
  stress in complex system 64
  work 191
Simple theory of failure 73
Springs
  closed-coiled 7
  helical 4
  leaf 36
Strain
  due to bending 11
  due to shear 1
  lateral 60
  volumetric 58, 98
Strain energy
  of bending 41
  of torsion 3
Stress
  complex system of 64
  due to bending 12, 14, 21, 49
  due to compression 84
  due to rotation 94
  due to shrink fit 91
  due to tension 56
  due to torsion 3
  hoop and radial 87
  in leaf spring 38
  Möhr circle of 70
Struts
  critical load 81
  Euler theory 81
  Rankine–Gordon formula 86
  slenderness ratio 84

Torsion
  equation 3
  strain energy of 4
  with bending 75

Vibration
  damping ratio 308
  degrees of freedom 277
  flexural 284
  forced and damped 299
  normal modes 286
  of rotating shafts 291
  simple harmonic 278
  torsional 280
  viscous damped 295, 307
Viscosity
  dynamic 169, 176
  index 265
  kinematic 170, 264